Also of interest

Cementitious Materials
Herbert Pöllmann (Ed.), 2017
ISBN 978-3-11-047373-5, e-ISBN (PDF) 978-3-11-047372-8,
e-ISBN (EPUB) 978-3-11-047391-9

Symmetry. Through the Eyes of Old Masters
Emil Makovicky, 2016
ISBN 978-3-11-041705-0, e-ISBN (PDF) 978-3-11-041714-2
e-ISBN (EPUB) 978-3-11-041719-7

Rietveld Refinement. Practical Pattern Analysis using Topas 6.0
Robert E. Dinnebier, Andreas Leineweber, 2018
ISBN 978-3-11-045621-9, e-ISBN (PDF) 978-3-11-046138-1
e-ISBN (EPUB) 978-3-11-046140-4

Multi-Component Crystals. Synthesis, Concepts, Function
Edward R.T. Tiekink, Julio Zukerman-Schpector (Eds.), 2017
ISBN 978-3-11-046365-1, e-ISBN (PDF) 978-3-11-046495-5
e-ISBN (EPUB) 978-3-11-046379-8

*Electrochemical Storage Materials.
From Crystallography to Manufacturing*
Dirk C. Meyer, Tilmann Leisegang (Eds.)
ISBN 978-3-11-049137-1, e-ISBN (PDF) 978-3-11-049398-6
e-ISBN (EPUB) 978-3-11-049187-6

Highlights in Mineralogical Crystallography
Thomas Armbruster, Rosa Micaela Danisi (Eds.), 2015
ISBN 978-3-11-041704-3, e-ISBN (PDF) 978-3-11-041710-4
e-ISBN (EPUB) 978-3-11-041721-0

Soraya Heuss-Aßbichler, Georg Amthauer, Melaı
Highlights in Applied Mineralogy

Highlights in Applied Mineralogy

Edited by
Soraya Heuss-Aßbichler,
Georg Amthauer,
Melanie John

DE GRUYTER

Soraya Heuss-Aßbichler
Ludwig-Maximilians-Universität München
Department of Earth and Environmental
Sciences
Theresienstr. 41
80333 Munich
Germany
soraya@min.uni-muenchen.de

Melanie John
Ludwig-Maximilians-Universität München
Department of Earth and Environmental Sciences
Theresienstr. 41
80333 Munich
Germany
melanie@john-stadler.de

Georg Amthauer
Universität Salzburg
Fachbereich Chemie und Physik
der Materialien
Jakob Haringer Straße 2a
Salzburg, 5020
Austria
georg.amthauer@sbg.ac.at

Cover image
Electron backscatter diffraction (EBSD)-orientation and band contrast map of the ventral valve of the modern brachiopod *Magellania venosa*. Different colors in the map highlight different crystal orientations within the investigated region of the sample. Well observable are the microcrystalline, seaward, primary and the inward, fibrous, shell layers with the stacks of transversely cut calcite fibres. The cover image was provided by Erika Griesshaber and Wolfgang W. Schmahl.

ISBN 978-3-11-049122-7
e-ISBN (PDF) 978-3-11-049734-2
e-ISBN (EPUB) 978-3-11-049508-9
Set-ISBN 978-3-11-049743-4

Library of Congress Cataloging-in-Publication Data
A CIP catalog record for this book has been applied for at the Library of Congress.

Bibliographic information published by the Deutsche Nationalbibliothek
The Deutsche Nationalbibliothek lists this publication in the Deutsche Nationalbibliografie; detailed bibliographic data are available on the Internet at http://dnb.dnb.de.

© 2018 Walter de Gruyter GmbH, Berlin/ Boston
Typesetting: Compuscript Ltd. Shannon, Ireland
Printing and binding: CPI books GmbH, Leck
∞ Printed on acid-free paper
Printed in Germany

www.degruyter.com

Preface

The material sciences are undergoing tremendous progress. New products are being developed with overwhelming potential. Looking more closely, one sees that nature creates many of these materials in a wide array of variations. Mineralogy covers the physical and chemical properties of materials in the geosphere and biosphere, which are closely related to their structure. Many of these properties are now utilized in various areas of our modern society. One of the most impressive examples is garnet, a mineral observed in magmatic and metamorphic rocks. Its chemistry and crystal structure are closely related to the chemistry of the host rock as well as the pressure and temperature conditions of formation. For humans, garnet serves an important material for technology that is opening emerging application fields, e.g. its potential for energy storage.

This book focuses on what we can learn from nature for technological applications and vice versa, what we might benefit from modern technical applications and experiments to gain a better understanding of the multifaceted processes in nature. This book does not cover all the different aspects of applied mineralogy but instead focuses on three main topics: (i) high-technology materials based on minerals, (ii) environmental mineralogy, and (iii) biomineralization, biomimetics, and medical mineralogy. We have assembled renowned scientists and specialists to write a chapter on their research fields, reflecting the different facets of applied mineralogy. Each contribution in the book highlights recent activities within the specific research area of the author(s) and gives a perspective for future development in technology, environment, and medicine.

This book is addressed to a more advanced audience, including graduate students, who are interested in nature and technical progress. It also intends to make younger students curious to learn more about the diverse fields of applied mineralogy. We would be delighted if we could inspire them with regard to their future scientific and professional career in universities, public research institutions, or industrial companies.

We thank all authors and reviewers – without their contribution, "Highlights in Applied Mineralogy" would not have been possible.

Munich and Salzburg in November 2017

Soraya Heuss-Aßbichler   Georg Amthauer   Melanie John

Contents

Preface —— v

List of contributing authors —— xiii

Part I: High-technology materials

1 Lithium ion–conducting oxide garnets —— 3
1.1 Introduction —— 3
1.2 Crystal structure of garnets —— 4
1.3 Natural and synthetic garnets —— 6
1.3.1 Natural garnets —— 6
1.3.2 Synthetic garnets —— 7
1.4 Solid Li-ion conductors —— 8
1.5 Introduction on Li-oxide garnets —— 8
1.5.1 Structural features of Li-oxide garnets —— 9
1.6 Systematic of Li-oxide garnets —— 11
1.6.1 $Li_3Ln_3M^{6+}_2O_{12}$ (Ln = Y, Pr, Nd, Sm–Lu; M^{6+} = Te^{6+}, W^{6+}) (Li_3 phases) —— 11
1.6.2 $Li_5La_3M^{5+}_2O_{12}$ (M^{5+} = Nb^{5+}, Ta^{5+}, Sb^{5+}, Bi^{5+}) (Li_5 phases) —— 11
1.6.3 $Li_6A^{2+}La_2M^{5+}_2O_{12}$ (A^{2+} = Mg^{2+}, Ca^{2+}, Sr^{2+}, Ba^{2+}; M^{5+} = Nb^{5+}, Ta^{5+}) (Li_6 phases) —— 12
1.6.4 $Li_7La_3M^{4+}_2O_{12}$ (M^{4+} = Zr^{4+}, Hf^{4+}, Sn^{4+}) (Li_7 phases) —— 12
1.6.5 Li-oxide garnets with space group $I\bar{4}3d$ —— 14
1.7 Outlook —— 16
1.8 Conclusions —— 16
Acknowledgments —— 16
References —— 16

2 Olivine-type battery materials —— 23
2.1 Introduction —— 23
2.2 Olivine in nature —— 23
2.3 $LiFePO_4$ olivine —— 27
2.3.1 Synthesis —— 27
2.3.2 Impurities and defects —— 27
2.3.3 Electrochemical properties —— 29
2.3.4 Carbon coating —— 31
2.3.5 Aging —— 32
2.4 Other Li-based olivines —— 34
2.4.1 Li(Mn,Fe)PO_4 olivines —— 34
2.4.2 $LiCoPO_4$ and $LiNiPO_4$ olivines —— 35

2.5	Battery of the future: olivine-based Na-ion cells — 35	
2.6	Conclusion — 37	

References — 37

3 Natural and synthetic zeolites — 41
3.1 History and definitions — 41
3.2 The zeolite structure — 44
3.3 Natural occurrence of zeolites — 47
3.3.1 Structural variability of the alumosilicate zeolites — 48
3.3.2 Compositional variability of the framework — 48
3.3.3 Compositional variability of the cations occupying the pore volume — 49
3.4 Applications — 49
3.4.1 Ion exchange — 50
3.4.2 Adsorption — 50
3.4.3 Catalysis — 51
3.5 Synthesis of zeolites — 52
3.6 Recent developments — 53
3.6.1 Synthesis of zeolite materials — 53
3.6.2 Applications — 61
3.7 Outlook — 66

References — 67

4 Microstructure analysis of chalcopyrite-type $CuInSe_2$ and kesterite-type $Cu_2ZnSnSe_4$ absorber layers in thin film solar cells — 73
4.1 Introduction — 73
4.2 Experimental — 77
4.2.1 Depth-resolved grazing incidence X-ray diffraction — 77
4.2.2 Microstructure analysis — 78
4.3 Summary — 95

References — 96

5 Surface-engineered silica via plasma polymer deposition — 99
5.1 Plasma polymerization — 99
5.2 Polymer plasma synthesis of amine-functionalized silica particles — 100
5.3 Polymer plasma synthesis of sulfonate-functionalized silica particles — 103
5.4 Plasma polymer synthesis of hydrophobic films on silica particles — 107

References — 110

6 Crystallographic symmetry analysis in NiTi shape memory alloys —— 113
- 6.1 Introduction —— 113
- 6.2 Elementary Landau theory —— 115
- 6.3 M-point transverse distortion modes —— 123
- 6.4 Γ-point strain distortion modes —— 125
- 6.5 Mode coupling and invariants in Landau free-energy polynomial —— 126
- 6.6 Model of biquadratic order parameter coupling —— 128

Acknowledgment —— 132
References —— 133

Part II: Environmental mineralogy

7 Gold, silver, and copper in the geosphere and anthroposphere: can industrial wastewater act as an anthropogenic resource? —— 137
- 7.1 Introduction —— 137
- 7.2 Gold, silver, and copper in the geosphere —— 138
- 7.2.1 Gold (Au) —— 138
- 7.2.2 Silver (Ag) —— 139
- 7.2.3 Copper (Cu) —— 139
- 7.3 Gold, silver, and copper in the anthroposphere —— 140
- 7.3.1 Synthesis of gold, silver, and copper phases from aqueous solutions —— 140
- 7.3.2 Wastewater: an anthropogenic resource for gold, silver, and copper? —— 144
- 7.4 What do/did we learn? —— 146

Acknowledgments —— 147
References —— 147

8 Applied mineralogy for recovery from the accident of Fukushima Daiichi Nuclear Power Station —— 153
- 8.1 Introduction —— 153
- 8.2 Mineralogical issues on- and off-site of the FDNPS —— 154
- 8.2.1 On-site —— 154
- 8.2.2 Off-site —— 157
- 8.3 Case studies —— 159
- 8.3.1 Selection of adsorbents and processes for water treatment on site —— 159
- 8.3.2 Solidification of the spent adsorbents for safe and economical storage and disposal —— 161

8.3.3   Identification of the host mineral for Cs retention in off-site soils —— 163
8.4   Concluding remark —— 167
Acknowledgment —— 167
References —— 167

9   Phosphates as safe containers for radionuclides —— 171
9.1   Introduction —— 171
9.2   Nuclear waste management —— 171
9.3   Why single-phase phosphate ceramics? —— 172
9.4   Crystal structures and properties —— 174
9.4.1   Monazite type —— 174
9.4.2   Apatite type —— 175
9.4.3   Xenotime type —— 176
9.4.4   Kosnarite type —— 177
9.4.5   Florencite type —— 179
9.4.6   Th-phosphate-diphosphate (β-TPD) type —— 180
9.4.7   Specific structural features of monazite —— 181
9.5   Chemical thermodynamics of single phases and solid solutions —— 185
9.6   Computer simulations as a valuable supplement to experimental results —— 187
9.7   Summary —— 189
References —— 190

10   Immobilization of high-level waste calcine (radwaste) in perovskites —— 197
10.1   Introduction —— 197
10.2   Immobilization of radioactive waste in ceramics —— 198
10.2.1   Supercalcine ceramic —— 198
10.2.2   Applications and phase assemblages of synroc ceramics —— 199
10.3   Crystal chemistry of perovskite-type structures suitable for the fixation of radwaste —— 203
10.3.1   Perovskite minerals and their technical analogues —— 203
10.3.2   Artificial ternary actinide perovskites —— 212
10.4   Discussion and summary —— 214
References —— 215

11   Titanate ceramics for high-level nuclear waste immobilization —— 223
11.1   Introduction —— 223
11.2   Multibarrier repository —— 224
11.3   Design and fabrication of titanate ceramics —— 225
11.4   Radiation resistance —— 229

11.5	Durability of titanate ceramics —— 232
11.6	Atomistic modeling —— 233
11.7	Conclusions —— 237

References —— 237

Part III: Biomineralization, biomimetics, and medical mineralogy

12	Patterns of mineral organization in carbonate biological hard materials —— 245
12.1	Introduction —— 245
12.2	Composite nature of biological hard tissues —— 245
12.3	Characteristic basic mineral units in gastropod, bivalve, and brachiopod shells —— 248
12.4	Carbonate hard tissue microstructures and textures, difference in crystal co-orientation strength —— 252
12.5	Biomaterial functionalization through carbonate crystal orientation variation —— 260
12.5.1	Calcitic tooth of the sea urchin *P. lividus* —— 261
12.5.2	Shells of *H. ovina* and *M. edulis* —— 262
12.5.3	Calcite in the tergite and mandible exocuticula of the isopod species *Porcellio scaber* and *Tylos europaeus* —— 262
12.6	Concluding summary —— 264

Acknowledgments —— 267
References —— 267

13	Sea urchin spines as role models for biological design and integrative structures —— 273
13.1	Introduction —— 273
13.2	Properties related to composition and nanostructure —— 274
13.3	Properties related to design —— 276
13.4	Biomimetic materials for energy dissipation —— 280
13.5	Conclusions and outlook —— 281

Acknowledgments —— 282
References —— 282

14	Nacre: a biomineral, a natural biomaterial, and a source of bio-inspiration —— 285
14.1	Introduction —— 285
14.2	Nacre: a biomineral —— 285
14.2.1	Composition of nacre —— 285
14.2.2	Multiscale structure of an iridescent biomineral —— 286

14.2.3	Nacre tablet formation — 287	
14.3	Nacre: a natural biomaterial — 287	
14.3.1	Nacre implant — 288	
14.3.2	Nacre powder — 289	
14.4	Nacre: a source of bio-inspiration for new materials — 289	
14.4.1	Layered structure — 290	
14.4.2	Interfacial/interlocking structure — 294	
14.4.3	Innovative methods — 294	
14.5	Conclusion — 298	

References — 298

15 Hydroxylapatite coatings: applied mineralogy research in the bioceramics field — 301
- 15.1 Introduction — 301
- 15.2 Geological HAp vs. biological HAp — 301
- 15.2.1 Structure of geological HAp — 302
- 15.2.2 Structure of biological apatite — 302
- 15.2.3 Biocompatibility of bioceramics — 303
- 15.2.4 Work in the area of bioceramics — 304
- 15.2.5 Outlook on future developments and further research requirements — 311

References — 313

16 A procedure to apply spectroscopic techniques in the investigation of silica-bearing industrial materials — 317
- 16.1 Introduction — 317
- 16.2 Activating factors in industrial CS-bearing materials: examples — 319
- 16.2.1 Size distribution — 319
- 16.2.2 Speciation of inorganic radicals — 320
- 16.2.3 Chemical and mineralogical contamination — 322
- 16.2.4 Species selection in suspended dusts — 324
- 16.3 Procedure: sampling and investigation techniques — 325
- 16.4 Conclusions — 328

Acknowledgments — 328

References — 329

About the authors — 333

Index — 339

List of contributing authors

Behnam Akhavan
The University of Sydney
School of Physics
Physics Road
NSW 2006
Australia
behnam.akhavan@sydney.edu.au

Georg Amthauer
Universität Salzburg
Fachbereich Chemie und Physik der Materialien
Jakob Haringer Straße 2a
5020 Salzburg
Austria
georg.amthauer@sbg.ac.at

Gerald Buck
Eberhard Karls Universität Tübingen
Wilhelmstraße 56
72074 Tübingen
Germany
gerald.buck@student.uni-tuebingen.de

Fabio Capacci
Unita' Funzionale PISLL Firenze
Dipartimento della Prevenzione AUSL Toscana Centro
Via della Cupola, 64
50145 Firenze
Italy
fabio.capacci@asf.toscana.it

Antonio Checa
University of Granada
Department of Stratigraphy and Paleontology
E-18071 Granada
Spain
acheca@ugr.es

Francesco Di Benedetto
University of Florence
Department of Earth Sciences
Via La Pira, 4
50121 Florence
Italy
francesco.dibenedetto@unifi.it

Anton Eisenhauer
GEOMAR, Helmholtz-Zentrum für Ozeanforschung Kiel
Wischhofstrasse 1-3
24148 Kiel
Germany
aeisenhauer@geomar.de

Reto Gieré
University of Pennsylvania
Department of Earth and Environmental Science
240 S. 33rd Street
Philadelphia, PA 19104-6316
USA
giere@sas.upenn.edu

Hermann Gies
Ruhr-University Bochum
Institute of Geology, Mineralogy and Geophysics
Universitätsstrasse 150
44780 Bochum
Germany
hermann.gies@rub.de

Erika Griesshaber
Ludwig-Maximilians-Universität München
Department of Earth and Environmental Sciences
Theresienstrasse 41
80333 Munich
Germany
e.griesshaber@lrz.uni-muenchen.de

René Gunder
Department Structure and Dynamics of Energy Materials (EM-ASD)
Helmholtz-Zentrum Berlin für Materialien und Energie
Hahn-Meitner-Platz 1
14109 Berlin
Germany
rene.gunder@helmholtz-berlin.de

List of contributing authors

Robert B. Heimann
Am Stadtpark 2A
02826 Görlitz
Germany
robert.heimann@ocean-gate.de

Antje Hirsch
RWTH Aachen University
Institute of Crystallography
Jägerstraße 17-19
52066 Aachen
Germany
hirsch@ifk.rwth-aachen.de

Markus Hoelzel
Technische Universität München
Forschungs-Neutronenquelle Heinz Maier-Leibnitz (FRM II)
Lichtenbergstrasse 1
85747 Garching
ga68jid@mytum.de

Soraya Heuss-Aßbichler
Ludwig-Maximilians-Universität München
Department of Earth and Environmental Sciences
Theresienstrasse 41
80333 Munich
Germany
soraya@min.uni-muenchen.de

Karyn Jarvis
Swinburne University of Technology
Faculty of Science, Engineering and Technology
John Street,
Hawthorn, Victoria 3122
Australia
kljarvis@swin.edu.au

Melanie John
Ludwig-Maximilians-Universität München
Department of Earth and Environmental Sciences
Theresienstrasse 41
80333 Munich
Germany
melanie@john-stadler.de

Christian M. Julien
Sorbonne Universities
UPMC-Paris-6
Paris
France
christian.julien@courriel.upmc.fr

Peter M. Kadletz
Ludwig-Maximilians-Universität München
Department of Earth and Environmental Sciences
Theresienstrasse 41
80333 Munich
Germany
kadletz@lmu.de

Katharina Klang
Eberhard Karls Universität Tübingen
Geowissenschaften/Angewandte Mineralogie
Wilhelmstraße 56
72074 Tübingen
Germany
katharina.klang@uni-tuebingen.de

Klemens Kelm
Deutsches Zentrum für Luft und Raumfahrt (DLR)
Institut für Werkstoff-Forschung
Linder Höhe
51147 Köln
Germany

Christoph Lauer
Eberhard Karls Universität Tübingen
Geowissenschaften/Angewandte Mineralogie
Wilhelmstraße 56
72074 Tübingen
Germany
christoph.lauer@uni-tuebingen.de

Leonhard Leppin
Department Structure and Dynamics of Energy Materials (EM-ASD)
Hahn-Meitner-Platz 1
14109 Berlin
Germany
leppin@posteo.de

Gregory R. Lumpkin
Australian Nuclear Science and Technology
Organisation
Locked Bag 2001
Kirrawee DC, NSW 2232
Australia
gregory.lumpkin@ansto.gov.au

Bernd Marler
Ruhr-University Bochum
Institute of Geology, Mineralogy and Geophysics
Universitätsstrasse 150
44780 Bochum
Germany
bernd.marler@rub.de

Julien Marquardt
Department Structure and Dynamics of Energy
Materials (EM-ASD)
Hahn-Meitner-Platz 1
14109 Berlin
Germany
julien.marquardt@helmholtz-berlin.de

Alain Mauger
UPMC
Institut de minéralogie, de physique des
matériaux et de cosmochimie
Place Jussieu
75005 Paris
France
alain.mauger@impmc.upmc.fr

Peter Majewski
University of South Australia
Future Industries Institute
Mawson Lakes Campus
Mawson Lakes SA 5095, Australia
peter.majewski@unisa.edu.au

Stefan Neumeier
Forschungszentrum Jülich GmbH
Institut für Energie- und Klimaforschung
Wilhelm-Johnen-Straße
52425 Jülich
Germany
s.neumeier@fz-juelich.de

Klaus G. Nickel
Eberhard Karls Universität Tübingen
Wilhelmstraße 56
72074 Tübingen
Germany
klaus.nickel@uni-tuebingen.de

Luca A. Pardi
IPCF U.O.S. di Pisa
Area della Ricerca CNR
Via G. Moruzzi, 1
56124 Pisa
Italy
pardi@ipcf.cnr.it

Lars Peters
RWTH Aachen University
Institute of Crystallography
Jägerstraße 17-19
52066 Aachen
Germany
peters@ifk.rwth-aachen.de

Herbert Pöllmann
Martin-Luther-Universität Halle-Wittenberg
Institut für Geowissenschaften und
Geographie
Von-Seckendorff-Platz 3
06120 Halle (Saale)
Germany
herbert.poellmann@geo.uni-halle.de

Daniel Rettenwander
Massachusetts Institute of Technology
Department of Materials Science and
Engineering
Cambridge, Massachusetts 02139
USA
drett@mit.edu

Georg Roth
RWTH Aachen University
Institute of Crystallography
Jägerstraße 17-19
52066 Aachen
Germany
roth@ifk.rwth-aachen.de

Maurizio Romanelli
University of Florence
Department of Earth Sciences
Via La Pira, 4
50121 Florence
romank@unifi.it

Marthe Rousseau
U1059 INSERM - SAINBIOSE
(SAnté INgéniérie BIOlogie St-Etienne)
Campus Santé Innovation
10, rue de la Marandière
42270 Saint-Priest-en-Jarez
France
rousseam@gmx.net

Tsutomu Sato
Hokkaido University
Faculty of Engineering
Kita 13 Nishi 8, Kita-Ku
Sapporo 060-8628
Japan
tomsato@eng.hokudai.ac.jp

Korbinian Schiebel
Ludwig-Maximilians-Universität München
Department of Earth and Environmental
Sciences
Theresienstrasse 4
80333 Munich
Germany
Korbinian.Schiebel@lrz.uni-muenchen.de

Hartmut Schlenz
Steinmann-Institut für Geologie
Mineralogie und Paläontologie
Rheinische Friedrich-Wilhelms-Universität Bonn
Nussallee 8
53115 Bonn
Germany
h.schlenz@fz-juelich.de

Wolfgang W. Schmahl
Ludwig-Maximilians-Universität München
Department of Earth and Environmental
Sciences
Theresienstrasse 41
80333 Munich
Germany
Wolfgang.W.Schmahl@lrz.uni-muenchen.de

Katherine L. Smith
Australian Embassy and Permanent Mission to
the United Nations in Vienna
Mattiellistrasse 2-4
1040 Vienna
Austria
Kath.Smith@dfat.gov.au

Susan Schorr
Department Structure and Dynamics of Energy
Materials
Hahn-Meitner-Platz 1
14109 Berlin
Germany
susan.schorr@helmholtz-berlin.de

Stefan Stöber
Institut für Geowissenschaften und Geographie
Martin-Luther-Universität Halle-Wittenberg
Von-Seckendorff-Platz 3
06120 Halle (Saale)
Germany
stefan.stoeber@geo.uni-halle.de

Reinhard Wagner
Universität Salzburg
Fachbereich Chemie und Physik der Materialien
Jakob Haringer Straße 2a
5020 Salzburg
Austria
reinhard.wagner@sbg.ac.at

Xiaofei Yin
Ludwig-Maximilians-Universität München
Department of Earth and Environmental
Sciences
Theresienstrasse 4
80333 Munich
Germany

Andreas Ziegler
University of Ulm
Central Facility for Electron Microscopy
Albert-Einstein-Allee 11
89069 Ulm
Germany
andreas.ziegler@uni-ulm.de

Part I: **High-technology materials**

Reinhard Wagner, Daniel Rettenwander and Georg Amthauer
1 Lithium ion–conducting oxide garnets

1.1 Introduction

Garnets comprise a well-known mineral supergroup for geoscience-related mineralogists, petrologists, and geologists. As garnets are very common and form attractive crystals (see Fig. 1.1), they are also desired by mineral collectors and for use in jewelry. In nature, garnets occur in magmatic and metamorphic rocks. The chemical composition of natural garnets is related to the chemistry of the host rock. In addition, their crystal chemistry and structure reflect the pressure and temperature conditions during their formation. For example, pyrope-rich garnets indicate that their host rock formed at the high pressures and temperatures most probably of the upper earth mantle. Furthermore, garnet is also an important material for technical applications, such as the well-known YAG laser. Recently, a new group of garnet-type compounds, namely Li-oxide garnets, has been identified as a promising material in the field of energy storage, as they can be used as solid electrolyte in Li-based battery concepts.

All these phases have one common feature, which is their crystal structure. The so-called garnet structure is presented in detail in a separate section of this chapter. The general formula of garnets is frequently given as $X_3Y_2Z_3\varphi_{12}$, where X, Y, and Z represent cations with the coordination numbers 8, 6, and 4, respectively, while φ represents the anions, which can be either O^{2-}, OH^-, or F^-, respectively [1]. Most natural garnets are oxides. As the garnet structure is notably flexible from a crystal chemical point of view, it allows a wide variety of elements to be incorporated into this structure type. According to Grew et al., 53 elements have been reported for garnets in the Inorganic Crystal Structure Database; five additional elements were reported in synthetic garnet-structured compounds [1]. The different positions can be occupied by various cations, e.g. X by

Fig. 1.1: Natural garnet samples from the mineralogical collections of the University of Salzburg. (Left) Almandine, Ötztal, Austria (width 30 cm). (Middle) Grossular *var.* hessonite, Mussa Alp, Italy (width 8 cm). (Right) Andradite, Serifos, Greece (width 8 cm).

Tab. 1.1: Important natural (upper) and synthetic (lower) garnets.

Species name	Formula	X	Y	Z	φ
Pyrope	$Mg_3Al_2Si_3O_{12}$	Mg_3	Al_2	Si_3	O_{12}
Almandine	$Fe_3Al_2Si_3O_{12}$	Fe_3	Al_2	Si_3	O_{12}
Spessartine	$Mn_3Al_2Si_3O_{12}$	Mn_3	Al_2	Si_3	O_{12}
Uvarovite	$Ca_3Cr_2Si_3O_{12}$	Ca_3	Cr_2	Si_3	O_{12}
Grossular	$Ca_3Al_2Si_3O_{12}$	Ca_3	Al_2	Si_3	O_{12}
Andradite	$Ca_3Fe_2Si_3O_{12}$	Ca_3	Fe_2	Si_3	O_{12}
Katoite	$Ca_3Al_2(OH)_{12}$	Ca_3	Al_2	□	$(OH)_{12}$
Cryolithionite	$Na_3Al_2Li_3F_{12}$	Na_3	Al_2	Li_3	F_{12}
Yttrium-iron-garnet (YIG)	$Y_3Fe_5O_{12}$	Y_3	Fe_2	Fe_3	O_{12}
Yttrium-aluminum-garnet (YAG)	$Y_3Al_5O_{12}$	Y_3	Al_2	Al_3	O_{12}
Gadolinium-gallium-garnet (GGG)	$Gd_3Ga_5O_{12}$	Gd_3	Ga_2	Ga_3	O_{12}
Lithium-lanthanum-zirconate (LLZO)	$La_3Zr_2Li_7O_{12}$	La_3	Zr_2	Li_7	O_{12}

Mg^{2+}, Fe^{2+}, Mn^{2+}, Ca^{2+}, Y^{3+}, Gd^{3+}, La^{3+}; Y by Al^{3+}, Fe^{3+}, Cr^{3+}, Ga^{3+}, Ti^{4+}, Sn^{4+}, Zr^{4+}, Sb^{5+}, Ta^{5+}, Nb^{5+}, Te^{6+}; and Z by Li^+, Al^{3+}, Fe^{3+}, Ga^{3+} Si^{4+}, Ti^{4+} [2]. Many of these elements occur in different oxidation states at the different cation positions of the garnet structure. Therefore, numerous different compounds of natural and synthetic origin have been reported. Some of the most important garnets are listed in Tab. 1.1.

1.2 Crystal structure of garnets

The so-called garnet structure is the fundamental feature of this group of minerals and related inorganic compounds. To understand the properties of garnet-type materials, a detailed knowledge of the structure is essential. The crystal structure of garnets was first solved by the German mineralogist and crystallographer Georg Menzer in 1925 and was the first solved crystal structure of a silicate mineral [3–5]. Since then, a number of research have been performed on the garnet structure [1, 6–12].

The garnet structure belongs to the cubic system and shows space group $Ia\bar{3}d$ (no. 230), i.e. the space group with the highest symmetry. A polyhedral model of the garnet structure is shown in Fig. 1.2. Despite a wide variation of their chemical composition, all minerals of the garnet supergroup can be described by the general crystal chemical formula $\{X_3\}[Y_2](Z_3)\varphi_{12}$ [1], where X represents an eightfold-coordinated cation located at special position 24c, whereas the coordination polyhedron can be described as triangular dodecahedron with site symmetry 222. Simply put, this polyhedron can be seen as distorted cube. "Y" refers to a sixfold-coordinated cation located at the 16a site with the coordinating anions forming an octahedron, site symmetry $\bar{3}$. "Z" is a fourfold coordinated cation, located at the 24d site, with site symmetry $\bar{4}$. The anions, represented by φ in the general formula, are located at a general position 96h. In most cases, the φ-position is occupied by O^{2-}. With these four symmetrically unique

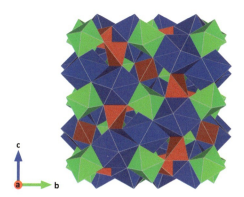

Fig. 2.1: Visualization of the garnet structure as polyhedra model, created with VESTA 3 [13]. Blue dodecahedra represent XO_8 polyhedra with X as central cation, located at Wyckoff position 24c; green octahedra depict YO_6 octahedra with Y as central cation, located at Wyckoff position 16a. Red tetrahedra correspond to ZO_4 tetrahedra with Z as central cation located at Wyckoff position 24d. From [14].

atomic positions, the garnet structure can be considered as comparatively simple structure. Using a polyhedra model, the garnet structure can be described as three-dimensional framework formed by alternating corner-sharing ZO_4 tetrahedra and YO_6 octahedra. The polyhedral framework resulting from this combination forms triangular dodecahedral cavities hosting the comparatively large X cation, which can also be referred to as XO_8 dodecahedron. Each anion is coordinated by two X cations, one Y cation, and one Z cation. As a consequence, the garnet structure shows a high percentage of shared edges (see Tab. 1.2). The high density and refractive index of minerals belonging to the garnet group can be related to this structural feature [6, 7].

Tab. 1.2: Shared polyhedral edges in the garnet structure.

Polyhedron	Shared edges
Tetrahedron (24d)	2 with triangular dodecahedra
Octahedron (16a)	6 with triangular dodecahedra
Triangular dodecahedron (24c)	2 with tetrahedra
	4 with octahedra
	4 with other triangular dodecahedra

After Novak and Gibbs [6].

The unit-cell parameter of garnets is strongly dependent on the radii of the ions [9]. For example, the unit-cell parameter of silicate garnets with Si^{4+} on the Z site and O^{2-} as anion can be predicted by using the formula

$$a_0(\text{Å}) = 1.61[r(X)] + 1.89[r(Y)] + 9.04$$

in which $r(X)$ is the ionic radius of the X cation located on the eightfold-coordinated site and $r(Y)$ is the ionic radius of the Y cation located on the octahedral site [6]. A more general formula is

$$a_0(\text{Å}) = 1.750[r(X)] + 1.653[r(Y)] + 1.904[r(Z)] + 6.225[r(\varphi)]$$

which additionally includes the atom at the tetrahedral Z site and oxide and fluoride anions at the φ site [9]. If one site hosts different kinds of ions, e.g. in solid solutions, the weighted average of the respective ionic radii has to be used.

Furthermore, also the positional parameters x, y, and z of the anion φ are dependent on the ionic radii of the cations and can be estimated by using appropriate equations [6, 9]. Novak and Gibbs performed a detailed investigation on the effects of the chemical composition of silicate garnets, i.e. garnets with Si^{4+} at the tetrahedral Z position, on the polyhedral interaction, bond lengths, and angles. They showed that the X–O and Y–O bond lengths, the unit-cell parameter a_0, as well as the O–X–O, O–Y–O, and O–Si–O angles of silicate garnets are linearly dependent on the radii of the X and Y cations. By using this information, they were able to predict a "structural stability field" for silicate garnets as a function of the ionic radii of the X and Y cations [6].

Besides compounds exhibiting the conventional cubic garnet structure, several phases with pseudocubic crystal structures, showing tetragonal space group $I4_1/acd$ (no. 142), are also listed within the garnet supergroup. The formation of the tetragonal structure is caused by a reduction in symmetry, as some of the crystallographic sites split into more symmetrically unique positions [1]. Compounds with this tetragonal structure are known from naturally occurring minerals as well as from synthetically produced materials [15, 16].

1.3 Natural and synthetic garnets

The following section introduces a selection of the most important garnets, whereas natural as well as synthetic compounds will be presented briefly.

1.3.1 Natural garnets

Natural garnets have been investigated in detail, as these minerals play an important role in geoscience. To classify the different kinds of natural garnets systematically, the International Mineralogical Association (IMA) therefore introduced the term "garnet supergroup" [1]. This term does not only include silicate garnets, which belong to the common garnet group, but also minerals from other chemical classes, such as arsenates, vanadates, oxides, and fluorides, which are isostructural with classic silicate garnets. Thus, the garnet supergroup includes all minerals that show the garnet-type structure and currently consists of 32 approved species and 5 additional species needing further studies for their approval [1].

Silicate garnets play an important role in geosciences. Natural silicate garnets can be divided into two major solid solution series. For members of the pyralspite series, i.e. pyrope ($Mg_3Al_2Si_3O_{12}$), almandine ($Fe^{2+}_3Al_2Si_3O_{12}$), and spessartine ($Mn_3Al_2Si_3O_{12}$), the octahedral site is occupied with Al^{3+}, while the dodecahedral site hosts divalent cations such as Mg^{2+}, Fe^{2+}, and Mn^{2+}. The other solid solution series is called ugrandite series, which is an acronym for uvarovite ($Ca_3Cr^{3+}_2Si_3O_{12}$), grossular ($Ca_3Al_2Si_3O_{12}$), and andradite ($Ca_3Fe^{3+}_2Si_3O_{12}$). For ugrandites, the dodecahedral position is occupied with Ca^{2+}, while the octahedral site can host various trivalent cations. Samples of various members of the silicate garnet group are shown in Fig. 1.1. The SiO_4^{4-} tetrahedron can be replaced by 4 OH^- groups; these compounds are also referred to as hydrogarnets. Katoite, $Ca_3Al_2(OH)_{12}$, is the most common hydrogarnet. As a curiosity, the fluoride garnet cryolithionite, $Na_3Al_2Li_3F_{12}$, has to be mentioned.

Natural silicate garnets are used as abrasive materials, e.g. as abrasive blasting material for water jet cutting and as abrasive powder [17]. Because of to their brilliance, hardness, and rich color variety, garnets are very popular gemstones.

1.3.2 Synthetic garnets

The garnet structure has also been reported for numerous synthetic compounds. Some of these synthetic garnets are of technological importance [11, 12]. Yttrium iron garnet (YIG), $Y_3Fe^{3+}_2Fe^{3+}_3O_{12}$, is an example for a magnetic garnet. Fe^{3+} is located on octahedral and tetrahedral sites, exhibiting differently orientated spins and therefore leading to a ferrimagnetic behavior. The high Verdet constant of YIG leads to a strong Faraday effect. Thus, YIG is used for various magneto-optical, optical, acoustic, and microwave applications such as microwave filters.

Yttrium-aluminum-garnet (YAG), $Y_3Al_2Al_3O_{12}$, shows a high refractive index and is thus used as a gemstone. By doping with certain elements, a large variety of colors can be achieved for this material. YAG doped with rare earth element, in particular Nd-doped YAG, is an important material for solid-state lasers [18].

Further synthetic garnets, e.g. gadolinium-gallium-garnet (GGG, $Gd_3Ga_2Ga_3O_{12}$), are of technological importance for special applications. Finally, there is also a wide variety of other synthetic garnet-structured compounds that do not have a technological relevance at the moment, but many show interesting effects.

One of the most recent developments in synthetic garnets is the research on Li-ion–conducting oxide garnets.

1.4 Solid Li-ion conductors

Much effort is undertaken to find and improve solid Li-ion conductors in order to enable the realization of new battery concepts. Batteries require an ionically conductive electrolyte, which has to be an electrical insulator. The electrolyte enables the transport of ions from the negative electrode (anode) to the positive electrode (cathode) and vice versa. The electrolyte also physically separates the anode and cathode in a battery and thus prevents a short circuit. Currently, Li-ion batteries are the most common battery type. These batteries usually contain organic, polymer-based, and gel-like electrolytes, which have numerous disadvantages, such as flammability, toxicity, as well as danger of leakage and short circuits. In addition, the chemical and electrochemical stability of these organic electrolytes is not satisfactory [19–21].

Using solid Li-ion conductors as electrolyte material, most of these problems could be solved. As mentioned above, the key requirement for electrolyte materials is high Li-ion conductivity, while, at the same time, the electronic conductivity should be negligible. Furthermore, the electrolyte materials should show a high electrochemical stability window and should be stable against Li metal and Li alloy anodes as well as against high-voltage cathode materials. In addition, thermal and mechanical stability as well as a low charge-transfer resistance at the electrode-electrolyte interface are required. Finally, inexpensive and environmentally friendly thin-film preparation methods are needed in order to ensure large-scale manufacturing [19, 20].

Numerous compounds have been identified as inorganic solid Li-ion conductors. Based on their crystal structure, they can be categorized into several groups. The most important solid Li-ion conductors are NASICON-type materials such as LATP, $Li_{1+x}Al_xTi_{2-x}(PO_4)_3$, LISICON-type materials, $Li_{3+x}(P_{1-x}M_x)O_4$ (M = Si^{4+}, Ge^{4+}), thio-LISICON-type materials, e.g. $Li_{10}GeP_2S_{12}$ (LGPS), perovskite-type materials such as LLT, $Li_{3x}La_{(2/3)-x}\square_{(1/3)-2x}TiO_3$ (0<x<0.16), Li-argyrodites, e.g. Li_6PS_5X (X = Cl, Br, I), LIPON compounds, and Li-oxide garnets [22]. The latter group of materials is of particular interest, as it shows very promising properties.

1.5 Introduction on Li-oxide garnets

Garnet-structured Li-ion conductors were first identified by Thangadurai et al. in 2003 [23]. In the meantime, numerous garnet-type Li-ion conductors with different compositions have been discovered. Most of these compounds are based on the cubic garnet structure with space group $Ia\bar{3}d$, but some members of this group actually show structures with lower symmetry, which has been discovered by Wagner et al. [24, 25] and Rettenwander et al. in 2016 [26]. These modifications can be derived from the common $Ia\bar{3}d$ garnet structure [16, 24–26]. Different from common garnets, most garnet-type Li-ion conductors contain more than three Li-ions per formula unit at the tetrahedral site. Thus, these compounds are also known as Li-stuffed oxide garnets. Garnet-type Li-ion conductors exhibit ionic conductivities of up to 10^{-3} S/cm at room temperature [25–28]. Furthermore, they also show superior chemical and electrochemical stability, as they are stable against metallic Li and exhibit a wide electrochemical stability window of up to 6 V [19–21, 29].

In contrast to the general crystal-chemical garnet formula $\{X_3\}^{[8]}[Y_2]^{[6]}(Z_3)^{[4]}\varphi_{12}$, which is commonly used in mineralogy, the chemical formulae of Li-stuffed oxide garnets are frequently written differently. For Li-oxide garnets, Li$^+$, which is located at the tetrahedrally coordinated 24d site as well as at additional 48g and 96h sites of space group $Ia\bar{3}d$, is written as the first element. The cation located at the dodecahedrally coordinated site, i.e. 24c, is given as the second element, and the cation located at the octahedrally coordinated 16a site is given as the last cation, resulting in a general formula $(Li_{3+x})^{[4]}\{X_3\}^{[8]}[Y_2]^{[6]}O_{12}$.

1.5.1 Structural features of Li-oxide garnets

As mentioned above, Li-stuffed oxide garnets contain more than three cations, namely Li$^+$, at the tetrahedral site to accomplish charge balance. The Li content can be adjusted by changing the valence of the X and Y cations. These additional Li$^+$-ions, as shown in Fig. 1.3, are located at two "interstitial" sites, which are not occupied in the conventional garnet structure [30]. One of these interstitial sites is the sixfold-coordinated 48g site of the space group $Ia\bar{3}d$, with point symmetry 2. The coordinating polyhedron of the 48g site is also described as distorted octahedron. The second interstitial site is an additional

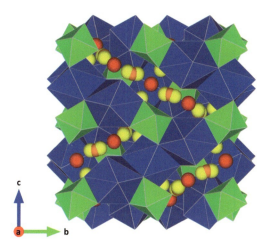

Fig. 1.3: Crystal structure of cubic Li$_7$La$_3$Zr$_2$O$_{12}$ with space group $Ia\bar{3}d$, created with VESTA 3 [13]. This figure illustrates the characteristic structural features of Li-stuffed oxide garnets. Blue LaO$_8$ dodecahedra with La^{3+} located at Wyckoff position 24c; and green ZrO$_6$ octahedra with Zr^{4+} located at the 16a site. Li$^+$ is located on 3 different positions which are partially occupied: Li$^+$ occupies the 24d position, as is well known from the conventional garnet structure, shown in red; as well as two additional sites, 48g, shown in orange; 96h, represented by yellow spheres. Modified from [14].

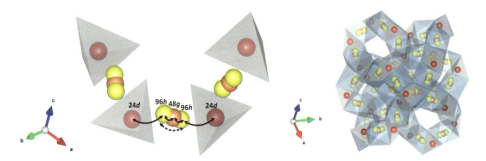

Fig. 1.4: Illustration of Li-ion diffusion in Li-stuffed oxide garnets, visualized using the VESTA 3 program [13]. The Li$^+$-ion diffusion mechanism, involving the three different Li$^+$ positions, is shown on the left side. The image on the right illustrates the three-dimensional network of different Li$^+$ sites within the unit cell, enabling fast Li$^+$ diffusion. From [14], modified.

distorted fourfold-coordinated 96h site of the space group $Ia\bar{3}d$, with point symmetry 1. The 96h site can be interpreted as the split position of the 48g site, resulting from a slight displacement of Li$^+$ at the 48g site. The coordinating polyhedron of the 96h site can be described as strongly distorted LiO$_4$ tetrahedron; sometimes, it is also described as strongly distorted octahedron, with two O^{2-} ions showing a significantly longer Li–O bond length. As the 48g and 96h positions are located very closely, some structural models omit the 48g position. Due to electrostatic interactions between Li$^+$ cations, the occupation of the sixfold-coordinated Li$^+$ site is not possible if both neighboring tetrahedral 24d sites are occupied by Li$^+$. This leads to a theoretical upper limit of 7.5 Li$^+$ per formula unit [31]. Even for Li-oxide garnets with high Li contents close to 7 Li$^+$ per formula unit, the Li sites are only partially occupied. The additional interstitial Li$^+$ sites 48g and 96h are considered as essential factors for enabling fast Li-ion conductivity [30, 32]. From a structural point of view, each LiO$_6$ polyhedron, hosting Li$^+$ on the 48g site, connects two LiO$_4$ tetrahedra with Li$^+$ located at 24d sites by shared faces. Each of these LiO$_4$ tetrahedra with Li$^+$ at the 24d site as central component is connected to four LiO$_6$ polyhedra (with Li$^+$ at the 48g site) by shared faces. This results in a three-dimensional polyhedron network consisting of alternating LiO$_4$ (24d)–LiO$_6$ (48g) polyhedra, which are only partially occupied by Li$^+$ and thus enabling fast Li-ion conduction [20]. This diffusion pathway is shown in Fig. 1.4. The Li$^+$-ion diffusion pathway can therefore be described as 24d–96h–48g–96h–24d, affecting all three Li positions if the 96h–48g–96h split-atom site is taken into consideration [33]. For Li-oxide garnets with space group $Ia\bar{3}d$, the movement of Li$^+$-ions from the tetrahedral 24d site to the distorted four-fold coordinated 96h site is seen as the limiting factor within the diffusion pathway, as this step requires the highest activation energy of the diffusion pathway described above. Direct movements between two octahedral 48g voids, bypassing the bottleneck on the 24d site, are considered as improbable due to the high activation energy needed for this step [33–35].

For Li-stuffed oxide garnets with very high Li contents of seven Li$^+$ per formula unit, a tetragonal garnet structure of space group $I4_1/acd$ was reported [16, 36, 37].

As already mentioned, this space group symmetry has also been reported for some natural garnets [1, 15]. In the $I4_1/acd$ modification of $Li_7La_3M_2O_{12}$ (M = Zr^{4+}, Hf^{4+}, Sn^{4+}), Li^+ is located at three different sites: Li1 is located at the tetrahedrally coordinated $8a$ position, Li2 occupies the distorted sixfold-coordinated $16f$ site, and Li3 is located at distorted sixfold-coordinated $32g$ sites [16]. These positions are similar to the Li^+ positions reported for cubic Li-stuffed oxide garnets; however, in the tetragonal modification, the Li distribution is "ordered". This is a major difference compared to cubic Li-stuffed oxide garnets with space group $Ia\bar{3}d$. The latter modification shows only a partial occupation of Li sites, resulting in a statistic distribution of Li^+-ions, leading to a higher symmetry.

1.6 Systematic of Li-oxide garnets

Since the discovery of fast Li-ion–conducting Li-oxide garnets by Thangadurai et al. in 2003, numerous Li-oxide garnet materials have been developed [23]. The following overview on these materials, modified from [14], is based on the reviews by Thangadurai et al. [19, 20] and Zeier et al. [21].

1.6.1 $Li_3Ln_3M^{6+}{}_2O_{12}$ (Ln = Y, Pr, Nd, Sm–Lu; M^{6+} = Te^{6+}, W^{6+}) (Li_3 phases)

Rare earth containing Li-oxide garnets with the general formula $Li_3Ln^{3+}{}_3M^{6+}{}_2O_{12}$ (M^{6+} = Te^{6+}, W^{6+}) were first described by Kasper in 1969 [38]. The applicability of Li-oxide garnets with Te^{6+} at the octahedral $16a$ site and trivalent rare earth element cations at the dodecahedral $24c$ site as Li-ion conductors has been studied by O'Callaghan et al. [39] and Cussen et al. [40]. In addition, similar phases with W^{6+} instead of Te^{6+} at the octahedral $16a$ site can be attributed to the group of Li_3 phases [41]. As opposed to Li-oxide garnets containing more than 3 Li^+ per formula unit, Li^+-ions of Li_3 phases are exclusively located at the tetrahedral $24d$ site of the garnet structure, which is similar to natural garnets. Li-oxide garnets with 3 Li^+ per formula unit show comparatively poor Li-ion conductivities as well as high activation energies, leading to the conclusion that Li^+-ions located at tetrahedral $24d$ sites are less mobile. Li-oxide garnets containing more than 3 Li^+ per formula unit, also referred to as Li-stuffed oxide garnets, are therefore seen as superior alternative.

1.6.2 $Li_5La_3M^{5+}{}_2O_{12}$ (M^{5+} = Nb^{5+}, Ta^{5+}, Sb^{5+}, Bi^{5+}) (Li_5 phases)

The existence of a $Li_5La_3M^{5+}{}_2O_{12}$ (M^{5+} = Ta^{5+}, Nb^{5+}) phase in the system La_2O_3–M_2O_5–Li_2O was initially suggested by Abbattista et al. in 1987 [42], followed by further studies [43, 44]. In 2003, Thangadurai et al. identified $Li_5La_3M_2O_{12}$ (M = Ta^{5+}, Nb^{5+}) as fast Li-ion conductor [23].

This was the discovery of the fast ionic conductivity of Li-stuffed oxide garnets, giving rise to numerous studies on these materials [23]. Besides Nb^{5+} and Ta^{5+}, other pentavalent cations such as Sb^{5+} and Bi^{5+} have been identified as suitable constituents for the formation of Li_5 phases [41, 45]. The crystal structure has been a matter of intensive discussion, as besides the garnet-type structure with space group $Ia\bar{3}d$ initially suggested by Mazza and colleagues in 1988, the non-centrosymmetric $I2_13$ space group has also been considered for this type of materials [43, 44]. Finally, in 2006, Cussen solved the crystal structure for $Li_5La_3M_2O_{12}$ (M = Ta^{5+}, Nb^{5+}) using neutron diffraction and confirmed the space group $Ia\bar{3}d$ [46]. In contrast to Li_3 phases, the high Li^+ content of these Li_5 phases requires the occupation of additional Li^+ sites that are not required in the conventional garnet structure. These additional Li sites located at the 48g and 96h positions are only partially occupied and they are considered as essential for enabling fast Li-ion conduction in garnet-type materials. Cussen and Yip compared d^0 and d^{10} phases and concluded that the replacement of a d^0 cation (e.g. W^{6+}, Ta^{5+}, Nb^{5+}) with a d^{10} cation (e.g. Te^{6+}, Sb^{5+}) of the same charge causes an increase of the unit-cell parameter, which is related to polarization effects [20, 41].

1.6.3 $Li_6A^{2+}La_2M^{5+}{}_2O_{12}$ (A^{2+} = Mg^{2+}, Ca^{2+}, Sr^{2+}, Ba^{2+}; M^{5+} = Nb^{5+}, Ta^{5+}) (Li_6 phases)

The Li^+ content of Li_5 phases can be increased by a partial replacement of La^{3+} with divalent cations such as Mg^{2+}, Ca^{2+}, Sr^{2+}, and Ba^{2+}, which leads to a further increase of the Li content of the garnets [45, 47–50]. This results in a change of the Li^+ distribution over the three 24d, 48g and 96h sites. The occupation of the tetrahedral 24d site decreases with increasing Li^+ contents, while the occupation of distorted octahedrally coordinated 48g and distorted fourfold-coordinated 96h sites increases. The partial substitution of La^{3+} by divalent ions also affects the unit-cell parameters, as these cations have different ionic radii. This topic was studied in detail by Zeier et al. [51].

1.6.4 $Li_7La_3M^{4+}{}_2O_{12}$ (M^{4+} = Zr^{4+}, Hf^{4+}, Sn^{4+}) (Li_7 phases)

Li-oxide garnets with exceptionally high Li contents of 7 Li^+ per formula unit were first described by Murugan et al. in 2007, as they reported cubic garnet-type $Li_7La_3Zr_2O_{12}$ (LLZO) with space group $Ia\bar{3}d$ [52]. This study showed that the garnet structure can accommodate 7 Li^+-ions per formula unit. Furthermore, also the ionic conductivity of 7.74×10^{-4} S/cm at 25°C was superior compared to most other solid Li-ion conductors. In contrast to this study, Awaka et al. reported a tetragonal polymorph of LLZO and $Li_7La_3Hf_2O_{12}$ showing space group $I4_1/acd$ [37]. The same structure was also reported for $Li_7La_3Sn_2O_{12}$ [16, 36]. The ionic conductivity of the tetragonal $I4_1/acd$ modification of LLZO is approximately two orders of magnitude lower compared to the cubic $Ia\bar{3}d$ modification. For $Li_7La_3Sn_2O_{12}$, a phase transition from the tetragonal $I4_1/acd$ modification to the garnet-like cubic $Ia\bar{3}d$ modification was observed above 750°C, indicating that the cubic modification might be interpreted as high-temperature modification [36]. A similar phase transition from $I4_1/acd$ to $Ia\bar{3}d$ has been reported for LLZO at temperatures around 650°C [53].

The discussion about these two structural modifications has finally been solved by Logéat et al., who investigated the solid solution series of $Li_{7-x}La_3Zr_{2-x}Ta_xO_{12}$ with $x = 0–2$ [54]. At room temperature, Li-stuffed oxide garnets with 7 Li^+ per formula unit, such as pure LLZO, thermodynamically prefer the tetragonal $I4_1/acd$ structure with an "ordered" distribution of the Li^+ cations. Li occupies all sixfold-coordinated $16f$ and $32c$ sites of the tetragonal structure as well as one third of the tetrahedrally coordinated $8a$ sites. The ordering of the Li^+-ions leads to comparatively low ionic conductivities. The introduction of supervalent cations causes the formation of Li^+ vacancies and leads to a disordered, statistical distribution of Li^+-ions among sixfold-coordinated and tetrahedrally coordinated sites. These Li-oxide garnets with a statistically disordered Li distribution therefore exhibit a higher symmetry, namely the cubic $Ia\bar{3}d$ garnet structure [54].

Indeed, it has been shown that a stabilization of the cubic $Ia\bar{3}d$ modification of "pure" LLZO at room temperature might be caused by the unintended incorporation of Al^{3+} into the samples from the alumina crucible used during sintering. This leads to compounds with the chemical composition $Li_{7-3x}Al_xLa_3Zr_2O_{12}$ [55, 56]. Subsequently, further supervalent cations have been identified to stabilize the cubic $Ia\bar{3}d$ modification of LLZO at room temperature. These stabilizing cations can be introduced at different positions into the LLZO structure. It has been reported that Al^{3+}, Fe^{3+}, Ga^{3+}, and Zn^{2+} can be incorporated on Li^+ positions to stabilize the cubic LLZO modification at room temperature [55–65]. A substitution of La^{3+} with Ce^{4+} on the eightfold-coordinated $24c$ position also causes a stabilization of the garnet-structured LLZO modification [66]. Furthermore, a partial replacement of Zr^{4+} on the octahedrally coordinated $16a$ site by pentavalent and hexavalent cations such as Ta^{5+}, Nb^{5+}, Sb^{5+}, Bi^{5+}, Te^{6+}, W^{6+}, and Mo^{6+} leads to the stabilization of the cubic $Ia\bar{3}d$ structure for LLZO-type materials [29, 33, 54, 63, 67–76]. Several of these substituents that stabilize the cubic modification have been discovered by Rettenwander et al. [29, 60]. All these substitutions with supervalent cations lead to a reduction of the Li content, which causes a disorder in the Li^+ distribution and thus stabilizing the cubic garnet-structured LLZO modification at room temperature. This emphasizes the importance of Li^+ vacancies for the stabilization of the cubic LLZO modification. The "critical" Li content required for the stabilization seems to be around 6.5–6.6 Li^+ per formula unit [77]. Another stabilization mechanism is related to the interaction of LLZO with CO_2 and H_2O; this modification is sometimes also referred to as low-temperature cubic phase, exhibiting comparatively poor Li-ion conductivity [53, 78–80].

LLZO materials show superior properties compared to other Li-stuffed oxide garnets. The Li-ion conductivity of cubic LLZO samples reaches up to 10^{-3} S/cm at room temperature [27, 28]. The ionic conductivity of LLZO garnets is affected by the chemical composition as well as by the structural parameters, such as Li content, occupancy disorder, and unit-cell parameter, but the microstructure and density of the samples also have a significant influence on the Li-ion conductivity [21, 81–87].

Besides the high ionic conductivity, LLZO also shows other properties that are considered essential for application as solid-state electrolyte. The chemical stability of electrolyte materials is one of these parameters. Compared to other solid-state electrolytes, LLZO is a chemically very stable material. In particular, its stability against Li metal and high-voltage metal oxide Li cathodes is seen as significant advantage [19, 20, 52, 88–90]. This allows the preparation of LLZO-based solid-state batteries using Li metal anodes, which enable much higher energy densities compared to batteries using other anode materials such as graphite. However, the stability of LLZO in air is still under discussion, as reactions with H_2O and CO_2 were reported to cause the formation of LiOH and Li_2CO_3. Also, protonation of the LLZO garnet phase due to the exchange of Li^+ for H^+ is mentioned in literature [75, 80, 83, 91–99].

The electrochemical inertness is also seen as a critical factor for electrolyte materials. Cyclic voltammetry experiments on cells containing LLZO-type electrolytes indicate a very wide electrochemical stability window of up to 6 V [19, 29]. Ohta et al. (2011) reported a potential window of up to 9 V vs. Li^+/Li [70]. Computational predictions actually report a smaller electrochemical stability window [100, 101]. However, these results also show satisfying values for the potential application of LLZO as electrolyte in solid-state batteries.

LLZO materials exhibit an excellent mechanical and thermal stability, enabling the application of LLZO electrolytes over a wide range of temperatures [90, 102, 103]. This is of special interest for certain battery systems such as molten Li metal halide batteries, which have to be operated at high temperatures [90, 104].

In summary, it can be stated that LLZO-type materials are particularly suited to be used as solid-state electrolytes. Therefore, a lot of effort is currently undertaken to develop new battery concepts using LLZO as solid-state Li-ion conductor.

1.6.5 Li-oxide garnets with space group $I\bar{4}3d$

A new category of Li-oxide garnets with acentric cubic space group $I\bar{4}3d$ has been identified by single crystal X-ray diffraction by Wagner and colleagues [24–26]. The introduction of certain trivalent substituents, namely Ga^{3+} and Fe^{3+}, leads to the formation of a different structural modification, showing the acentric cubic space group $I\bar{4}3d$. The stabilization of the cubic LLZO modification by the substitution $3\,Li^+ \rightarrow M^{3+} + 2\,\square_{Li}$ (M^{3+} = Ga^{3+}, Fe^{3+}) has already been reported earlier [60–62, 105]. However, as the X-ray powder diffraction patterns of the conventional garnet-type LLZO modification with space group $Ia\bar{3}d$ and the ones of the acentric cubic $I\bar{4}3d$ modification are almost identical, these former studies assumed the garnet-type $Ia\bar{3}d$ structure.

The reduction in symmetry, compared to the $Ia\bar{3}d$ structure, seems to be caused by the site preference of the substituents. This leads to the splitting of the tetrahedrally coordinated 24d position, which is the common tetrahedral Z site of the $Ia\bar{3}d$ structure. This tetrahedral site splits into two alternating tetrahedral sites, 12a and 12b. The coordinating tetrahedron of the 12a site is slightly smaller and more distorted compared to the one of the 12b site. The trivalent substituents distinctly prefer the 12a site [24, 25]. A graphical representation of the $I\bar{4}3d$ modification is given in Fig. 1.5.

These garnet-type compounds with space group $I\bar{4}3d$ show exciting properties concerning their Li-ion mobility. ^{7}Li NMR spectroscopy of Ga^{3+}-substituted LLZO with space group $I\bar{4}3d$ showed features that might be attributed to an additional Li-ion diffusion process that is absent in Al^{3+}-substituted LLZO with space group $Ia\bar{3}d$ [24]. Another study by Rettenwander et al. compared the solid solution $Li_{6.40}Al_{0.20-x}Ga_xLa_3Zr_2O_{12}$ and observed the phase transformation from the $Ia\bar{3}d$ modification to the $I\bar{4}3d$ modification for $x \geq 0.15$, which coincides with a reduction of the activation energy required for the Li^+ diffusion process [26]. This study also reported high ionic conductivities of 1.2×10^{-3} S/cm at room temperature for $Li_{6.40}Ga_{0.20}La_3Zr_2O_{12}$ with space group $I\bar{4}3d$. For Fe^{3+}-substituted LLZO, which also shows the $I\bar{4}3d$ modification, ionic conductivities of up to 1.38×10^{-3} S/cm at 23.5°C have been reported by Wagner et al. [25]. This is the highest ionic conductivity that has been reported for Li-oxide garnet materials. The results of these studies on garnet-type Li-oxide materials with space group $I\bar{4}3d$ indicate that these garnet-related compounds actually show superior Li-ion dynamics compared to the conventional garnet-structured compounds with space group $Ia\bar{3}d$. These findings highlight the significance of structure-property relationships.

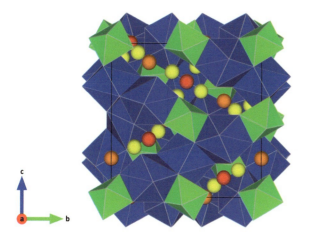

Fig. 1.5: Crystal structure of Ga-substituted LLZO with space group $I\bar{4}3d$, created with VESTA 3 [13]. Blue dodecahedra represent 8-fold coordinated La^{3+} (located at the Wyckoff position 24d); green octahedra 6-fold coordinated Zr^{4+} (Wyckoff position 16c). The red spheres represent tetrahedrally coordinated Li^+ at Wyckoff position 12a (Li1), orange spheres correspond to tetrahedrally coordinated Li^+ located at the 12b site (Li2); yellow spheres represent to distorted 6-fold coordinated Li^+ at Wyckoff position 48e (Li3). From [14], modified.

1.7 Outlook

Systematic research has led to the discovery of Li-oxide garnets that are suited to be used as solid Li electrolytes, in particular for thin film batteries. In order to enable the large-scaled production of such batteries, suitable preparation methods have to be developed. Therefore, different approaches such as magnetron sputtering, pulsed laser deposition, aerosol deposition, sol-gel based dip coating, spin coating and other methods are currently investigated [89, 106–110]. However, the deposition of such thin films seems to be delicate, as the formation of Li-deficient phases such as $La_2Zr_2O_7$ and other phases shows that the deposition of the correct Li content is challenging [111]. Therefore, further investigations are needed in order to improve the deposition techniques [106, 112–117].

1.8 Conclusions

Li-oxide garnets are among the most promising materials to be used as solid Li electrolytes. These materials show the garnet structure with space group $Ia\bar{3}d$. For Li-stuffed oxide garnets, containing more than three Li-ions per formula unit, additional Li sites are introduced compared to the conventional garnet structure. This three-dimensional network of partially occupied Li positions enables fast Li-ion diffusion. In particular, cubic $Li_7La_3Zr_2O_{12}$ (LLZO) shows excellent electrochemical properties. The introduction of certain substituents, such as Ga^{3+} and Fe^{3+}, into the LLZO structure leads to a slightly different cubic modification with the non-centrosymmetric space group $I\bar{4}3d$. These compounds show superior ionic conductivities. Detailed knowledge of structure property relationships is essential for the understanding and improvement of this class of materials.

Acknowledgments

The authors thank Thomas Armbruster (University of Bern) for his suggestions, which improved the manuscript. Furthermore, the authors thank the Austrian Science Fund (FWF), Vienna, for financial support within the project P25702 "Li-oxide garnets".

References

[1] Grew ES, Locock AJ, Mills SJ, Galuskina IO, Galuskin EV, Halenius U. Nomenclature of the garnet supergroup. Am Mineral 2013;98:785–810.
[2] Tolksdorf W, Hansen P, Klages C-P, Mateika D, Welz F. Flüssigphasen-Epitaxie magnetooptischer Granatschichten. In: Philips – Unsere Forschung in Deutschland. Vol. IV. Hamburg: Philips; 1988.
[3] Menzer G. Die Kristallstruktur von Granat. Centralbl Min 1925;344–5.

[4] Menzer G. Die Kristallstruktur von Granat. Zeitschr Kristallogr 1926;68:157–8.
[5] Menzer G. Die Kristallstruktur der Granate. Zeitschr Kristallogr 1928;69:300–96.
[6] Novak GA, Gibbs GV. The crystal chemistry of the silicate garnets. Am Mineral 1971;56: 791–825.
[7] Gibbs GV, Smith JV. Refinement of crystal structure of synthetic pyrope. Am Mineral 1965;50:2023–39.
[8] Hawthorne FC. Some systematics of the garnet structure. J Solid State Chem 1981;37:157–64.
[9] Langley RH, Sturgeon GD. Lattice parameters and ionic radii of the oxide and fluoride garnets. J Solid State Chem 1979;30:79–82.
[10] Geller S. Crystal chemistry of the garnets. Zeitschr Kristallogr 1967;125:1–47.
[11] Enke K, Fleischhauer J, Gunber W, Hansen P, Nomura S, Tolksdorf W, Winkler G, Wolfmeier U. Magnetic and other properties of oxides and related compounds. Berlin/Heidelberg: Springer: 1978. Vol. 12.
[12] Winkler G. Magnetic garnets. Vol. 5. Braunschweig/Wiesbaden: Vieweg; 1981.
[13] Momma K, Izumi F. VESTA 3 for three-dimensional visualization of crystal, volumetric and morphology data. J Appl Crystallogr 2011;44:1272–6.
[14] Wagner R. Structural and crystal chemical investigations on lithium oxide garnets with general composition $Li_7La_3Zr_2O_{12}$. Unpublished PhD thesis, Universität Salzburg, 2016.
[15] Armbruster T, Kohler T, Libowitzky E, Friedrich A, Miletich R, Kunz M, Medenbach O, Gutzmer J. Structure, compressibility, hydrogen bonding, and dehydration of the tetragonal Mn^{3+} hydrogarnet, henritermierite. Am Mineral 2001;86:147–58.
[16] Awaka J, Kijima N, Hayakawa H, Akimoto J. Synthesis and structure analysis of tetragonal $Li_7La_3Zr_2O_{12}$ with the garnet-related type structure. J Solid State Chem 2009;182:2046–52.
[17] Olson DW. Garnet, industrial. In: US Geological Survey minerals yearbook 2001. Reston (VA): USGS; 2001:30.1–30.4.
[18] Lupei V, Lupei A. Nd:YAG at its 50th anniversary: still to learn. J Luminescence 2016;169:426–39.
[19] Thangadurai V, Pinzaru D, Narayanan S, Baral AK. Fast solid-state Li ion conducting garnet-type structure metal oxides for energy storage. J Phys Chem Lett 2015;6:292–9.
[20] Thangadurai V, Narayanan S, Pinzaru D. Garnet-type solid-state fast Li ion conductors for Li batteries: critical review. Chem Soc Rev 2014;43:4714–27.
[21] Zeier WG. Structural limitations for optimizing garnet-type solid electrolytes: a perspective. Dalton T 2014;43:16133–8.
[22] Bachman JC, Muy S, Grimaud A, Chang HH, Pour N, Lux SF, Paschos O, Maglia F, Lupart S, Lamp P, Giordano L, Shao-Horn Y. Inorganic solid-state electrolytes for lithium batteries: mechanisms and properties governing ion conduction. Chem Rev 2016;116:140–62.
[23] Thangadurai V, Kaack H, Weppner WJF. Novel fast lithium ion conduction in garnet-type $Li_5La_3M_2O_{12}$ (M = Nb, Ta). J Am Ceram Soc 2003;86:437–40.
[24] Wagner R, Redhammer GJ, Rettenwander D, Senyshyn A, Schmidt W, Wilkening M, Amthauer G. Crystal structure of garnet-related Li-ion conductor $Li_{7-3x}Ga_xLa_3Zr_2O_{12}$: fast Li-ion conduction caused by a different cubic modification? Chem Mater 2016;28:1861–71.
[25] Wagner R, Redhammer GJ, Rettenwander D, Tippelt G, Welzl A, Taibl S, Fleig J, Franz A, Lottermoser W, Amthauer G. Fast Li-ion-conducting garnet-related $Li_{7-3x}Fe_xLa_3Zr_2O_{12}$ with uncommon $I\bar{4}3d$ structure. Chem Mater 2016;28:5943–51.
[26] Rettenwander D, Redhammer G, Preishuber-Pflügl F, Cheng L, Miara L, Wagner R, Welzl A, Suard E, Doeff MM, Wilkening M, Fleig J, Amthauer G. Structural and electrochemical consequences of Al and Ga cosubstitution in $Li_7La_3Zr_2O_{12}$ solid electrolytes. Chem Mater 2016;28:2384–92.
[27] Bernuy-Lopez C, Manalastas W, Lopez del Amo JM, Aguadero A, Aguesse F, Kilner JA. Atmosphere controlled processing of Ga-substituted garnets for high Li-ion conductivity ceramics. Chem Mater 2014;26:3610–7.

[28] Li YT, Han JT, Wang CA, Xie H, Goodenough JB. Optimizing Li$^+$ conductivity in a garnet framework. J Mater Chem 2012;22:15357–61.

[29] Rettenwander D, Welzl A, Cheng L, Fleig J, Musso M, Suard E, Doeff MM, Redhammer GJ, Amthauer G. Synthesis, crystal chemistry, and electrochemical properties of Li$_{7-2x}$La$_3$Zr$_{2-x}$Mo$_x$O$_{12}$ (x = 0.1–0.4): stabilization of the cubic garnet polymorph via substitution of Zr^{4+} by Mo^{6+}. Inorg Chem 2015;54:10440–9.

[30] Awaka J, Takashima A, Kataoka K, Kijima N, Idemoto Y, Akimoto J. Crystal structure of fast lithium-ion-conducting cubic Li$_7$La$_3$Zr$_2$O$_{12}$. Chem Lett 2011;40:60–2.

[31] Xie H, Alonso JA, Li YT, Fernandez-Diaz MT, Goodenough JB. Lithium distribution in aluminum-free cubic Li$_7$La$_3$Zr$_2$O$_{12}$. Chem Mater 2011;23:3587–9.

[32] Rettenwander D, Wagner R, Langer J, Maier ME, Wilkening M, Amthauer G. Crystal chemistry of "Li$_7$La$_3$Zr$_2$O$_{12}$" garnet doped with Al, Ga, and Fe: a short review on local structures as revealed by NMR and Mößbauer spectroscopy studies. Eur J Mineral 2016;28:619–29.

[33] Bottke P, Rettenwander D, Schmidt W, Amthauer G, Wilkening M. Ion dynamics in solid electrolytes: NMR reveals the elementary steps of Li$^+$ hopping in the garnet Li$_{6.5}$La$_3$Zr$_{1.75}$Mo$_{0.25}$O$_{12}$. Chem Mater 2015;27:6571–82.

[34] Wang YX, Klenk M, Page K, Lai W. Local structure and dynamics of lithium garnet ionic conductors: a model material Li$_5$La$_3$Ta$_2$O$_{12}$. Chem Mater 2014;26:5613–24.

[35] Miara LJ, Ong SP, Mo YF, Richards WD, Park Y, Lee JM, Lee HS, Ceder G. Effect of Rb and Ta doping on the ionic conductivity and stability of the garnet Li$_{7+2x-y}$(La$_{3-x}$Rb$_x$)(Zr$_{2-y}$Ta$_y$)O$_{12}$ ($0 \le x \le 0.375$, $0 \le y \le 1$) superionic conductor: a first principles investigation. Chem Mater 2013;25:3048–55.

[36] Percival J, Kendrick E, Smith RI, Slater PR. Cation ordering in Li containing garnets: synthesis and structural characterisation of the tetragonal system, Li$_7$La$_3$Sn$_2$O$_{12}$. Dalton Trans 2009;26:5177–81.

[37] Awaka J, Kijima N, Kataoka K, Hayakawa H, Ohshima K-i, Akimoto J. Neutron powder diffraction study of tetragonal Li$_7$La$_3$Hf$_2$O$_{12}$ with the garnet-related type structure. J Solid State Chem 2010;183:180–5.

[38] Kasper HM. Series of rare earth garnets Ln$^{3+}_3$M$_2$Li$^+_3$O$_{12}$ (M = Te, W). Inorg Chem 1969;8:1000–2.

[39] O'Callaghan MP, Lynham DR, Cussen EJ, Chen GZ. Structure and ionic-transport properties of lithium-containing garnets Li$_3$Ln$_3$Te$_2$O$_{12}$(Ln = Y, Pr, Nd, Sm-Lu). Chem Mater 2006;18:4681–9.

[40] Cussen EJ, Yip TWS, O'Neill G, O'Callaghan MP. A comparison of the transport properties of lithium-stuffed garnets and the conventional phases Li$_3$Ln$_3$Te$_2$O$_{12}$. J Solid State Chem 2011;184:470–5.

[41] Cussen EJ, Yip TWS. A neutron diffraction study of the d^0 and d^{10} lithium garnets Li$_3$Nd$_3$W$_2$O$_{12}$ and Li$_5$La$_3$Sb$_2$O$_{12}$. J Solid State Chem 2007;180:1832–9.

[42] Abbattista F, Vallino M, Mazza D. Remarks on the binary systems Li$_2$O-Me$_2$O$_5$ (Me = Nb, Ta). Mater Res Bull 1987;22:1019–27.

[43] Mazza D. Remarks on a ternary phase in the La$_2$O$_3$-Me$_2$O$_5$-Li$_2$O system (Me = Nb, Ta). Mater Lett 1988;7:205–7.

[44] Hyooma H, Hayashi K. Crystal structures of La$_3$Li$_5$M$_2$O$_{12}$ (M = Nb, Ta). Mater Res Bull 1988;23:1399–407.

[45] Murugan R, Weppner W, Schmid-Beurmann P, Thangadurai V. Structure and lithium ion conductivity of bismuth containing lithium garnets Li$_5$La$_3$Bi$_2$O$_{12}$ and Li$_6$SrLa$_2$Bi$_2$O$_{12}$. Mater Sci Eng B Solid 2007;143:14–20.

[46] Cussen EJ. The structure of lithium garnets: cation disorder and clustering in a new family of fast Li$^+$ conductors. Chem Commun 2006;4:412–413.

[47] Murugan R, Thangadurai V, Weppner W. Lattice parameter and sintering temperature dependence of bulk and grain-boundary conduction of garnet-like solid Li-electrolytes. J Electrochem Soc 2008;155:A90–101.

[48] Thangadurai V, Weppner W. Li$_6$ALa$_2$Ta$_2$O$_{12}$ (A = Sr, Ba): novel garnet-like oxides for fast lithium ion conduction. Adv Funct Mater 2005;15:107–12.

[49] Thangadurai V, Weppner W. $Li_6ALa_2Nb_2O_{12}$ (A = Ca, Sr, Ba): a new class of fast lithium ion conductors with garnet-like structure. J Am Ceram Soc 2005;88:411–8.
[50] Percival J, Apperley D, Slater PR. Synthesis and structural characterisation of the Li ion conducting garnet-related systems, $Li_6ALa_2Nb_2O_{12}$ (A = Ca, Sr). Solid State Ionics 2008;179:1693–6.
[51] Zeier WG, Zhou SL, Lopez-Bermudez B, Page K, Melot BC. Dependence of the Li-ion conductivity and activation energies on the crystal structure and ionic radii in $Li_6MLa_2Ta_2O_{12}$. Acs Appl Mater Inter 2014;6:10900–7.
[52] Murugan R, Thangadurai V, Weppner W. Fast lithium ion conduction in garnet-type $Li_7La_3Zr_2O_{12}$. Angew Chem Int Edit 2007;46:7778–81.
[53] Matsui M, Takahashi K, Sakamoto K, Hirano A, Takeda Y, Yamamoto O, Imanishi N. Phase stability of a garnet-type lithium ion conductor $Li_7La_3Zr_2O_{12}$. Dalton T 2014;43:1019–24.
[54] Logéat A, Köhler T, Eisele U, Stiaszny B, Harzer A, Tovar M, Senyshyn A, Ehrenberg H, Kozinsky B. From order to disorder: the structure of lithium-conducting garnets $Li_{7-x}La_3Ta_xZr_{2-x}O_{12}$ (x = 0–2). Solid State Ionics 2012;206:33–8.
[55] Buschmann H, Dolle J, Berendts S, Kuhn A, Bottke P, Wilkening M, Heitjans P, Senyshyn A, Ehrenberg H, Lotnyk A, Duppel V, Kienle L, Janek J. Structure and dynamics of the fast lithium ion conductor "$Li_7La_3Zr_2O_{12}$". Phys Chem Phys 2011;13:19378–92.
[56] Geiger CA, Alekseev E, Lazic B, Fisch M, Armbruster T, Langner R, Fechtelkord M, Kim N, Pettke T, Weppner W. Crystal chemistry and stability of "$Li_7La_3Zr_2O_{12}$" garnet: a fast lithium-ion conductor. Inorg Chem 2011;50:1089–97.
[57] Rettenwander D, Blaha P, Laskowski R, Schwarz K, Bottke P, Wilkening M, Geiger CA, Amthauer G. DFT study of the role of Al^{3+} in the fast ion-conductor $Li_{7-3x}Al_x^{3+}La_3Zr_2O_{12}$ garnet. Chem Mater 2014;26:2617–23.
[58] Jin Y, McGinn P. Al-doped $Li_7La_3Zr_2O_{12}$ synthesized by a polymerized complex method. J Power Sources 2011;196:8683–7.
[59] Rangasamy E, Wolfenstine J, Sakamoto J. The role of Al and Li concentration on the formation of cubic garnet solid electrolyte of nominal composition $Li_7La_3Zr_2O_{12}$. Solid State Ionics 2012;206:28–32.
[60] Rettenwander D, Geiger CA, Amthauer G. Synthesis and crystal chemistry of the fast Li-ion conductor $Li_7La_3Zr_2O_{12}$ doped with Fe. Inorg Chem 2013;52:8005–9.
[61] Rettenwander D, Geiger CA, Tribus M, Tropper P, Wagner R, Tippelt G, Lottermoser W, Amthauer G. The solubility and site preference of Fe^{3+} in $Li_{7-3x}Fe_xLa_3Zr_2O_{12}$ garnets. J Solid State Chem 2015;230:266–71.
[62] Rettenwander D, Geiger CA, Tribus M, Tropper P, Amthauer G. A synthesis and crystal chemical study of the fast ion conductor $Li_{7-3x}Ga_xLa_3Zr_2O_{12}$ with x = 0.08 to 0.84. Inorg Chem 2014;53:6264–9.
[63] Allen JL, Wolfenstine J, Rangasamy E, Sakamoto J. Effect of substitution (Ta, Al, Ga) on the conductivity of $Li_7La_3Zr_2O_{12}$. J Power Sources 2012;206:315–9.
[64] Wolfenstine J, Ratchford J, Rangasamy E, Sakamoto J, Allen JL. Synthesis and high Li-ion conductivity of Ga-stabilized cubic $Li_7La_3Zr_2O_{12}$. Mater Chem Phys 2012;134:571–5.
[65] Chen Y, Rangasamy E, Lang CD, An K. Origin of high Li^+ conduction in doped $Li_7La_3Zr_2O_{12}$ garnets. Chem Mater 2015;27:5491–94.
[66] Rangasamy E, Wolfenstine J, Allen J, Sakamoto J. The effect of 24c-site (A) cation substitution on the tetragonal-cubic phase transition in $Li_{7-x}La_{3-x}A_xZr_2O_{12}$ garnet-based ceramic electrolyte. J Power Sources 2013;230:261–6.
[67] Mukhopadhyay S, Thompson T, Sakamoto J, Huq A, Wolfenstine J, Allen JL, Bernstein N, Stewart DA, Johannes MD. Structure and stoichiometry in supervalent doped $Li_7La_3Zr_2O_{12}$. Chem Mater 2015;27:3658–65.

[68] Li YT, Wang CA, Xie H, Cheng JG, Goodenough JB. High lithium ion conduction in garnet-type $Li_6La_3ZrTaO_{12}$. Electrochem Commun 2011;13:1289–92.

[69] Buschmann H, Berendts S, Mogwitz B, Janek J. Lithium metal electrode kinetics and ionic conductivity of the solid lithium ion conductors "$Li_7La_3Zr_2O_{12}$" and $Li_{7-x}La_3Zr_{2-x}Ta_xO_{12}$ with garnet-type structure. J Power Sources 2012;206:236–44.

[70] Ohta S, Kobayashi T, Asaoka T. High lithium ionic conductivity in the garnet-type oxide $Li_{7-x}La_3(Zr_{2-x}, Nb_x)O_{12}$ ($X = 0$–2). J Power Sources 2011;196:3342–5.

[71] Ramakumar S, Satyanarayana L, Manorama SV, Murugan R. Structure and Li^+ dynamics of Sb-doped $Li_7La_3Zr_2O_{12}$ fast lithium ion conductors. Phys Chem Phys 2013;15:11327–38.

[72] Deviannapoorani C, Dhivya L, Ramakumar S, Murugan R. Lithium ion transport properties of high conductive tellurium substituted $Li_7La_3Zr_2O_{12}$ cubic lithium garnets. J Power Sources 2013;240:18–25.

[73] Dhivya L, Janani N, Palanivel B, Murugan R. Li^+ transport properties of W substituted $Li_7La_3Zr_2O_{12}$ cubic lithium garnets. Aip Adv 2013;3:082115.

[74] Xia Y, Ma L, Lu H, Wang X-P, Gao Y-X, Liu W, Zhuang Z, Guo L-J, Fang Q-F. Preparation and enhancement of ionic conductivity in Al-added garnet-like $Li_{6.8}La_3Zr_{1.8}Bi_{0.2}O_{12}$ lithium ionic electrolyte. Front Mater Sci 2015;9:366–72.

[75] Wagner R, Rettenwander D, Redhammer GJ, Tippelt G, Sabathi G, Musso ME, Stanje B, Wilkening M, Suard E, Amthauer G. Synthesis, crystal structure, and stability of cubic $Li_{7-x}La_3Zr_{2-x}Bi_xO_{12}$. Inorg Chem 2016;55:12211–9.

[76] Gao YX, Wang XP, Lu H, Zhang LC, Ma L, Fang QF. Mechanism of lithium ion diffusion in the hexad substituted $Li_7La_3Zr_2O_{12}$ solid electrolytes. Solid State Ionics 2016;291:1–7.

[77] Thompson T, Wolfenstine J, Allen JL, Johannes M, Huq A, David IN, Sakamoto J. Tetragonal vs. cubic phase stability in Al – free Ta doped $Li_7La_3Zr_2O_{12}$ (LLZO). J Mater Chem A 2014;2:13431–6.

[78] Matsui M, Sakamoto K, Takahashi K, Hirano A, Takeda Y, Yamamoto O, Imanishi N. Phase transformation of the garnet structured lithium ion conductor: $Li_7La_3Zr_2O_{12}$. Solid State Ionics 2014;262:155–9.

[79] Toda S, Ishiguro K, Shimonishi Y, Hirano A, Takeda Y, Yamamoto O, Imanishi N. Low temperature cubic garnet-type CO_2-doped $Li_7La_3Zr_2O_{12}$. Solid State Ionics 2013;233:102–6.

[80] Larraz G, Orera A, Sanjuan ML. Cubic phases of garnet-type $Li_7La_3Zr_2O_{12}$: the role of hydration. J Mater Chem A 2013;1:11419–28.

[81] Cheng L, Park JS, Hou HM, Zorba V, Chen GY, Richardson T, Cabana J, Russo R, Doeff M. Effect of microstructure and surface impurity segregation on the electrical and electrochemical properties of dense Al-substituted $Li_7La_3Zr_2O_{12}$. J Mater Chem A 2014;2:172–81.

[82] Cheng L, Chen W, Kunz M, Persson K, Tamura N, Chen G, Doeff M. Effect of surface microstructure on electrochemical performance of garnet solid electrolytes. ACS Appl Mater Interfaces 2015;7:2073–81.

[83] Cheng L, Wu CH, Jarry A, Chen W, Ye YF, Zhu JF, Kostecki R, Persson K, Guo JH, Salmeron M, Chen GY, Doeff M. Interrelationships among grain size, surface composition, air stability, and interfacial resistance of al-substituted $Li_7La_3Zr_2O_{12}$ solid electrolytes. Acs Appl Mater Inter 2015;7:17649–55.

[84] David IN, Thompson T, Wolfenstine J, Allen JL, Sakamoto J. Microstructure and Li-ion conductivity of hot-pressed cubic $Li_7La_3Zr_2O_{12}$. J Am Ceram Soc 2015;98:1209–14.

[85] Janani N, Deviannapoorani C, Dhivya L, Murugan R. Influence of sintering additives on densification and Li^+ conductivity of Al doped $Li_7La_3Zr_2O_{12}$ lithium garnet. RSC Adv 2014;4:51228–38.

[86] Suzuki Y, Kami K, Watanabe K, Watanabe A, Saito N, Ohnishi T, Takada K, Sudo R, Imanishi N. Transparent cubic garnet-type solid electrolyte of Al_2O_3-doped $Li_7La_3Zr_2O_{12}$. Solid State Ionics 2015;278:172–6.

[87] Wachter-Welzl A, Wagner R, Rettenwander D, Taibl S, Amthauer G, Fleig J. Microelectrodes for local conductivity and degradation measurements on Al stabilized $Li_7La_3Zr_2O_{12}$ garnets. J Electroceram 2016;https://doi.org/10.1007/s10832-017-0102-1

[88] Tan JJ, Tiwari A. Synthesis of cubic phase $Li_7La_3Zr_2O_{12}$ electrolyte for solid-state lithium-ion batteries. Electrochem Solid St 2012;15:A37–9.

[89] Tan JJ, Tiwari A. Fabrication and characterization of $Li_7La_3Zr_2O_{12}$ thin films for lithium ion battery. ECS Solid State Lett 2012;1:Q57–60.

[90] Wolfenstine J, Allen JL, Read J, Sakamoto J. Chemical stability of cubic $Li_7La_3Zr_2O_{12}$ with molten lithium at elevated temperature. J Mater Sci 2013;48:5846–51.

[91] Liu C, Rui K, Shen C, Badding ME, Zhang GX, Wen ZY. Reversible ion exchange and structural stability of garnet-type Nb-doped $Li_7La_3Zr_2O_{12}$ in water for applications in lithium batteries. J Power Sources 2015;282:286–93.

[92] Orera A, Larraz G, Rodriguez-Velamazan JA, Campo J, Sanjuan ML. Influence of Li^+ and H^+ distribution on the crystal structure of $Li_{7-x}H_xLa_3Zr_2O_{12}$ ($0 \le x \le 5$) garnets. Inorg Chem 2016;55:1324–32.

[93] Jin Y, McGinn PJ. $Li_7La_3Zr_2O_{12}$ electrolyte stability in air and fabrication of a $Li/Li_7La_3Zr_2O_{12}/Cu_{0.1}V_2O_5$ solid-state battery. J Power Sources 2013;239:326–31.

[94] Larraz G, Orera A, Sanz J, Sobrados I, Diez-Gómez V, Sanjuán ML. NMR study of Li distribution in $Li_{7-x}H_xLa_3Zr_2O_{12}$ garnets. J Mater Chem A 2015;3:5683–91.

[95] Wang YX, Lai W. Phase transition in lithium garnet oxide ionic conductors $Li_7La_3Zr_2O_{12}$: the role of Ta substitution and H_2O/CO_2 exposure. J Power Sources 2015;275:612–20.

[96] Xia W, Xu B, Duan H, Guo Y, Kang H, Li H, Liu H. Ionic conductivity and air stability of al-doped $Li_7La_3Zr_2O_{12}$ sintered in alumina and Pt crucibles. ACS Appl Mater Interfaces 2016;8:5335–42.

[97] Galven C, Dittmer J, Suard E, Le Berre F, Crosnier-Lopez M-P. Instability of lithium garnets against moisture. Structural characterization and dynamics of $Li_{7-x}H_xLa_3Sn_2O_{12}$ and $Li_{5-x}H_xLa_3Nb_2O_{12}$. Chem Mater 2012;24:3335–45.

[98] Galven C, Fourquet JL, Crosnier-Lopez MP, Le Berre F. Instability of the lithium garnet $Li_7La_3Sn_2O_{12}$: Li^+/H^+ exchange and structural study. Chem Mater 2011;23:1892–900.

[99] Gam F, Galven C, Bulou A, Le Berre F, Crosnier-Lopez MP. Reinvestigation of the total Li^+/H^+ ion exchange on the garnet-type $Li_5La_3Nb_2O_{12}$. Inorg Chem 2014;53:931–4.

[100] Miara LJ, Richards WD, Wang YE, Ceder G. First-principles studies on cation dopants and electrolyte|cathode interphases for lithium garnets. Chem Mater 2015;27:4040–7.

[101] Richards WD, Miara LJ, Wang Y, Kim JC, Ceder G. Interface stability in solid-state batteries. Chem Mater 2016;28:266–73.

[102] Cussen EJ. Structure and ionic conductivity in lithium garnets. J Mater Chem 2010;20:5167–73.

[103] Afyon S, Krumeich F, Rupp JLM. A shortcut to garnet-type fast Li-ion conductors for all-solid state batteries. J Mater Chem A 2015;3:18636–48.

[104] Wolfenstine J, Jo H, Cho YH, David IN, Askeland P, Case ED, Kim H, Choe H, Sakamoto J. A preliminary investigation of fracture toughness of $Li_7La_3Zr_2O_{12}$ and its comparison to other solid Li-ion conductors. Mater Lett 2013;96:117–20.

[105] Howard MA, Clemens O, Kendrick E, Knight KS, Apperley DC, Anderson PA, Slater PR. Effect of Ga incorporation on the structure and Li ion conductivity of $La_3Zr_2Li_7O_{12}$. Dalton T 2012;41:12048–053.

[106] Lobe S, Dellen C, Finsterbusch M, Gehrke HG, Sebold D, Tsai CL, Uhlenbruck S, Guillon O. Radio frequency magnetron sputtering of $Li_7La_3Zr_2O_{12}$ thin films for solid-state batteries. J Power Sources 2016;307:684–9.

[107] Tadanaga K, Egawa H, Hayashi A, Tatsumisago M, Mosa J, Aparicio M, Duran A. Preparation of lithium ion conductive Al-doped $Li_7La_3Zr_2O_{12}$ thin films by a sol-gel process. J Power Sources 2015;273:844–7.

[108] Takano R, Tadanaga K, Hayashi A, Tatsumisago M. Low temperature synthesis of Al-doped $Li_7La_3Zr_2O_{12}$ solid electrolyte by a sol-gel process. Solid State Ionics 2014;255:104–7.

[109] Ahn CW, Choi JJ, Ryu J, Hahn BD, Kim JW, Yoon WH, Choi JH, Park DS. Microstructure and ionic conductivity in $Li_7La_3Zr_2O_{12}$ film prepared by aerosol deposition method. J Electrochem Soc 2015;162:A60–3.

[110] Djenadic R, Botros M, Benel C, Clemens O, Indris S, Choudhary A, Bergfeldt T, Hahn H. Nebulized spray pyrolysis of Al-doped $Li_7La_3Zr_2O_{12}$ solid electrolyte for battery applications. Solid State Ionics 2014;263:49–56.

[111] Rawlence M, Garbayo I, Buecheler S, Rupp JL. On the chemical stability of post-lithiated garnet Al-stabilized $Li_7La_3Zr_2O_{12}$ solid state electrolyte thin films. Nanoscale 2016;8:14746–53.

[112] El-Shinawi H, Paterson GW, MacLaren DA, Cussen EJ, Corr SA. Low-temperature densification of Al-doped $Li_7La_3Zr_2O_{12}$: a reliable and controllable synthesis of fast-ion conducting garnets. J Mater Chem A 2017;5:319–329.

[113] Gordon ZD, Yang T, Gomes Morgado GB, Chan CK. Preparation of nano- and microstructured garnet $Li_7La_3Zr_2O_{12}$ solid electrolytes for li-ion batteries via cellulose templating. ACS Sustain Chem Eng 2016;4:6391–6398.

[114] Yang T, Gordon ZD, Li Y, Chan CK. Nanostructured garnet-type solid electrolytes for lithium batteries: electrospinning synthesis of $Li_7La_3Zr_2O_{12}$ nanowires and particle size-dependent phase transformation. J Phys Chem C 2015;119:14947–53.

[115] Loho C, Djenadic R, Bruns M, Clemens O, Hahn H. Garnet-type $Li_7La_3Zr_2O_{12}$ solid electrolyte thin films grown by CO_2-laser assisted CVD for all-solid-state batteries. J Electrochem Soc 2016;164:A6131–9.

[116] Yoshima K, Harada Y, Takami N. Thin hybrid electrolyte based on garnet-type lithium-ion conductor $Li_7La_3Zr_2O_{12}$ for 12 V-class bipolar batteries. J Power Sources 2016;302:283–90.

[117] Botros M, Djenadic R, Clemens O, Möller M, Hahn H. Field assisted sintering of fine-grained $Li_{7-3x}La_3Zr_2Al_xO_{12}$ solid electrolyte and the influence of the microstructure on the electrochemical performance. J Power Sources 2016;309:108–15.

Christian M. Julien and Alain Mauger

2 Olivine-type battery materials

2.1 Introduction

Among the minerals, olivine LiFePO$_4$ has been the subject of intense research and is now commonly used as a cathode material of lithium (Li)-ion batteries, which have been developed for electrical vehicles and plug-in hybrid vehicles. Such batteries are also needed to buffer intermittence problems before integration on the grid of the electricity produced by solar cells or wind turbines. The purpose of this chapter is to report the outstanding story of this mineral, which has been first discovered and studied by mineralogists almost 200 years ago. Chemists and physicists only realized its potential 20 years ago and still struggled for more than 10 years to reach the present state of the art where iron-phosphate batteries are commercialized all over the world. The extension of the research to other olivine phosphates for their potential use as cathodes for Li-ion batteries is also discussed. Section 2.5 is devoted to the research for the future at longer scale, with the promising application of iron phosphate olivine as cathode element of the future sodium (Na)-ion batteries.

2.2 Olivine in nature

Olivine is a common mineral of volcanic rocks, chemical formula (Mg, Fe)$_2$SiO$_4$ (chrysolite, *golden stone* in Greek), that belongs to the silicates subgroup of nesosilicates and got its name from its usual olivine-green color. It is the median term in the fayalite-forsterite series. The forsterite Mg$_2$SiO$_4$ is the mineral that does not contain iron, while fayalite Fe$_2$SiO$_4$ is magnesium-free. Note that fayalite and forsterite create a solid solution series. Other olivine minerals are monticellite CaMgSiO$_4$, kirschsteinite CaFeSiO$_4$, and tephroite Mn$_2$SiO$_4$. Olivine crystallizes in the orthorhombic crystal system (*Pmna* S.G.) (Fig. 2.1). A strong relief can be observed under a polarizing microscope, as well as a high birefringence with bright shades from second- to third-order polarizing microscope analysis (hardness 6.5–7). Olivine is the dominant mineral in peridotites, rocks making up the upper mantle that crystallizes first when magma cools. This mineral is often found in basaltic lava. Being grown at high temperatures and lacking water, olivines are very sensitive to atmospheric agents, hydrothermal alteration, and low-degree metamorphism involving hydration, oxidation, silicification, or carbonation.

In 1834, a German professor of mineralogy at the Ludwig-Maximilians Universität München, Johann Nepomuk von Fuchs, discovered a new mineral that he calls *triphylin* [1]. This discovery, however, went largely unnoticed, since this professor, also a conservator of the Mineralogical State Collection Munich, became notorious

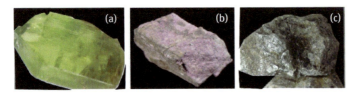

Fig. 2.1: Pictures of olivine-type crystals: (a) chrysolite $(Mg,Fe)_2SiO_4$, (b) purpurite Fe^3PO_4 containing Mn^3 impurities, and (c) triphylite $LiFe^2PO_4$.

mainly for his work on an alkaline silicate used in stereochromy (a method of wall painting in which water glass is used either as a painting medium or as a final fixative coat). His name is even commemorated by the fuchsite, a chromium-rich variety of the mineral muscovite, belonging to the mica group of phyllosilicate minerals, with the following formula: $K(Al,Cr)_2(AlSi_3O_{10})(OH)_2$. Triphylin took its name from the Greek τοίς, *tria*, for threefold, plus φυλή *phylon*, family, because it contains three cations (Fe, Li, and Mn). Indeed, in nature, Fe and Mn are always mixed in an alloy composition $LiFe_{1-x}Mn_xPO_4$ that exists in any composition x. Later on, the name was translated in English as triphylite. It is a scarce orthophosphate primary mineral found in phosphatic pegmatites and pegmatite dikes. Nowadays, chemists keep the name triphylite for $LiFePO_4$ (despite the fact that there are only two cations, rendering the name unjustified), while lithiophilite (from the greek φίλος, *philos*, for friend) refers to the other end member $LiMnPO_4$, a name that is well justified since it identifies the material as a friend of Li. Li itself was identified only soon before, in 1817, by Johan August Arfwedson from the analysis of the mineral petalite $(LiAlSi_4O_{10})$ and the name is derived from the Greek λιθος *lithos*, meaning stone. Since these end members are artificial ceramics, mineralogists refer to triphylite and lithiophilite as Fe- and Mn-rich $LiFe_{1-x}Mn_xPO_4$, respectively. The mineral with intermediate composition is known as sicklerite. Lithiophylite is a resinous reddish to yellowish brown mineral, while triphilite is a gray-blue primary phosphate mineral that is rather rare.

The delithiated form Fe^3PO_4 of triphylite, named heterosite, which crystallizes in the orthorhombic system (*Pmna* S.G.) (hardness 4–4.5) is a secondary mineral in the oxidized zone of complex granite pegmatites, replacing primary phosphate minerals. Purpurite, a Mn-rich $Mn^3_{1-x}Fe^3_xPO_4$ mineral, is identified by its dark violet-brown to bright purple color. The existence of this delithiated $Fe_{1-x}Mn_xPO_4$ solid solution gives evidence that Li can be extracted from the $LiFe_{1-x}Mn_xPO_4$ series. The natural degradation of the olivine-type Fe-Mn phosphates has been studied by mineralogists. Ferri- and mangano-sicklerite $Li_{1-n}(Fe_{1-n}Mn_n)PO_4$ are secondary Li phospho-olivine minerals, a weathered product of the lithiophilite-triphylite series occurring in Li-rich pegmatite [2].

Since their discovery, almost no other publication or study concerning these materials can be found in the literature until 1967, when Santoro and Newnham from the Electrical Engineering Department of MIT determined their structural and magnetic properties [3]. This is, to our knowledge, the first time that physicists or chemists expressed interest in LiFePO$_4$ and LiMnPO$_4$. Still, this is some sort of anachronism, since it was also the last time until 1997, when John Goodenough stunned the electrochemical community by claiming that LiFePO$_4$ was a good candidate as a new cathode material for Li-ion batteries, as it realizes a capacity of ~170 mAh/g^{-1} at moderate current densities [4]. The community, at that time, was skeptical because LiFePO$_4$ is an insulator, which seems incompatible with the use as a cathode. Also, triphylite alters easily into other phosphate minerals, and geologists show a lot of respect for it, for making other phosphate minerals possible. However, this feature and the fact that pure and stoichiometric LiFePO$_4$ does not exist in nature means that good-quality and well-crystallized are difficult to make, which is bad news for chemists and physicists. Moreover, the material ages. One aging mechanism of Li(Fe$_{1-x}$Mn$_x$)PO$_4$ under soft conditions involves Li$^+$ ions extraction from the lattice with a progressive oxidation of Fe and/or Mn, leading to the transformation of triphylite into ferri-sicklerite and then heterosite according the relation [5, 6]

$$\text{Li}(\text{Fe}^2_{1-x}\text{Mn}^2_x)\text{PO}_4 \rightarrow \text{Li}_x(\text{Fe}^3_{1-x}\text{Mn}^2_x)\text{PO}_4 \rightarrow (\text{Fe}^3_{1-x}\text{Mn}^3_x)\text{PO}_4. \qquad (2.1)$$

A mineral well known to form in nature as an alteration product of the triphylite is the strengite FePO$_4 \cdot$2H$_2$O. The role of water can result as a second possible iron oxidation by the incorporation of H$_2$O and OH$^-$ groups in the triphylite framework leading to a tavorite-like LiFe^{3}PO$_4$(OH) compound [7]. The strengite occurs if the oxidation conditions are high enough to oxidize the iron of triphylite from a ferrous (+2) state to the ferric (+3) state, which is actually the iron state in strengite. Such aging phenomena are not surprising since they only result from the fact that the iron "likes" to be in the trivalent state. These properties that we know from the mineralogists are the indication that the material needs to be protected against humidity.

Another difficulty comes from the crystal structure illustrated in Fig. 2.2. Corner-shared FeO$_6$ octahedra are linked together in the bc-plane; the LiO$_6$ octahedra form edge-sharing chains along the b-axis. Thus described, the material looks lamellar. This, however, is not the case, because these atomic bc-planes are linked together by PO$_4$ units that bridge neighboring layers of FeO$_6$ octahedra by sharing a common edge with one FeO$_6$ octahedron and two edges with LiO$_6$. Owing to these strong bridges that give LiFePO$_4$ remarkable thermal stability, the material is a truly three-dimensional lattice.

Between two adjacent bridges, there is room for Li$^+$ ions that are thus distributed along one-dimensional channels. These Li$^+$ ions can thus be removed easily, and during this delithiation process, the Fe^{2+} is oxidized in Fe^{3+} in order to keep the electrical neutrality locally. This is, in essence, the reversible electrochemical reaction when LiFePO$_4$ is used as a cathode compound for Li-ion batteries. Meanwhile, this structure also implies difficulty. Since the Li$^+$ ions must be removed by emptying the Li channels, any impurity or defect involving a Li site will block not only this Li, but

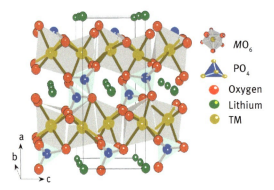

Fig. 2.2: Crystal structure of LiMPO$_4$ olivine (M = transition metal). Corner-shared MO$_6$ octahedra are linked together in the bc-plane; LiO$_6$ octahedra form edge-sharing chains along the b-axis. The tetrahedral PO$_4$ groups bridge neighboring layers of MO$_6$ octahedra by sharing a common edge with one MO$_6$ octahedra and two edges with LiO$_6$ octahedra.

Tab. 2.1: Lattice parameters of olivine LiMPO$_4$ (M = transition metal) materials.

Crystal	a (Å)	b (Å)	c (Å)	Unit cell volume (Å^3)
LiFePO$_4$	10.332(4)	6.010(5)	4.692(2)	291.3(9)
LiMnPO$_4$	10.431(0)	6.094(7)	4.736(6)	301.1(2)
LiCoPO$_4$	10.200(1)	5.919(9)	4.690(0)	286.9(8)
LiNiPO$_4$	10.027(5)	5.853(7)	4.676(3)	274.4(8)

also the whole channel to which it belongs. As a consequence, the electrochemical performance of this cathode material is drastically dependent on its purity.

Despite these difficulties, the pioneering work of John Goodenough in 1997 was the motivation for chemists to work on the synthesis and investigation of the electrochemical properties of LiFePO$_4$ and the whole phospho-olivine LiMPO$_4$ (M indicates transition metal ions, including not only Fe and Mn, but also Co, Ni). Lattice parameters of olivine LiMPO$_4$ materials are listed in Tab. 2.1. The first problem that prevented any application as a cathode for Li-ion batteries was the much too small electrical conductivity. It took 4 years before the problem was solved by serendipity, when plastic introduced by mistake with the precursors during the synthesis liberated carbon during the annealing process, and this conductive carbon coated the LiFePO$_4$ particles, drastically improving the electrical conductivity of the powder [8]. Of course, the carbon coat does not modify the conductivity inside the LiFePO$_4$ particles. However, once an electron has reached the surface, it can be driven to the collector through the framework of the carbon coat, provided that the carbon is percolating through the structure, which is the case if the particles are in contact and if the carbon coat covers uniformly the particles. Therefore, a good electrical conductivity can be obtained provided that the LiFePO$_4$ powder has two properties: (i) the size of the particles, which is the typical length of the electron path inside any LiFePO$_4$ particle to reach the surface, must be as small as possible (inn practice, this requires particles of size in the order of 100 nm); (ii) the carbon coat must be conducting and must cover the surface of the particles uniformly. Hence, enormous work has been done to synthesize LiFePO$_4$ powders that fulfill these two requirements.

2.3 LiFePO$_4$ olivine

2.3.1 Synthesis

Pure and well-crystallized LiFePO$_4$ does not exist in nature, and we have to synthesize this material. Many synthesis routes have been used to prepare LiFePO$_4$, which have been reviewed in [9]. Here we have selected two processes that are used to synthesize this material at the industrial scale for battery applications. The first process is the solid-state synthesis method. FePO$_4$·2(H$_2$O) and Li$_2$CO$_3$ are used as the precursors. After ball-milling in 2-propanol overnight, the blend is dried and mixed with 5 wt.% of a polymeric carbon additive. The choice of the polymeric additive is not critical; it can be polyethylene-block-poly(ethylene glycol) 50% ethylene oxide like that of Ravet et al. [10] or lactose [11]. After annealing at 700°C–750°C under flowing argon atmosphere, LiFePO$_4$ particles are obtained, with a 3-nm-thick carbon coat. This powder pressed at 3750 kg/cm at room temperature presents an electronic conductivity higher than 10^{-8} S/cm [10]. The reason why the choice of polymeric additive is broad originates from the strong affinity between Fe and C [12], and it was very fortunate since it allowed for the first carbon coating that was not done on purpose initially.

The hydrothermal route is also successful to prepare well-crystallized LiFePO$_4$ with controlled size of the particles within a reaction time of 5–12 h [13, 14], with the advantage that the synthesis temperature can be as small as 230°C [15]. Then, the choice of complexing agent [16] and optimization of the stirring speed used in the hydrothermal chamber [17] have made possible the synthesis of LiFePO$_4$ of very high quality. Nevertheless, annealing under argon atmosphere at 700°C is still needed to obtain a coating of the particles with conductive carbon, for reasons that are explained in a following section devoted to the optimization of the carbon deposit.

2.3.2 Impurities and defects

As mentioned earlier, the problem of low conductivity of the LiFePO$_4$ was solved in 2001. Nevertheless, at that time, the electrochemical properties of the LiFePO$_4$ powder were irreproducible and varied a lot from one sample to another, which still prevented any commercial development. This was due to the lack of control of the purity of the sample. As we have mentioned earlier, good electrochemical properties require the preparation of LiFePO$_4$ free of any impurity or defect that might block the Li channels. Actually, impurities were damaging at concentrations that are so small that they cannot be detected by conventional tools such as X-ray diffraction for instance, so that the chemists thought they were dealing with a pure and stoichiometric material, which was not the case. That is in essence why the solution of this problem took another few years, owing to the analysis of magnetic properties. Indeed, Fe^{2+} is a magnetic ion, so that any impurity in its vicinity may modify its magnetic properties. The basic idea is then to measure the perturbation of the magnetic properties induced by an impurity or a defect as a probe to identify it and to determine its concentration. Once this has been done, the synthesis parameters can be modified accordingly to get rid of the impurity step by step. When it is pure, LiFePO$_4$ is an antiferromagnet with a Néel temperature of 52 K [3]. If the annealing temperature during the synthesis is higher than 800°C, then a ferromagnetic component to the magnetization is detected, even at room temperature, which is the signature of the presence of a γ-Fe$_2$O$_3$ impurity. Then from the value of the remanent magnetization, we can deduce the amount of γ-Fe$_2$O$_3$ in the material, and by a fit of the magnetization curves in a model of superparamagnetim, we could even determine that the impurity is under the form of nano-sized γ-Fe$_2$O$_3$ particles [18, 19]. This has been confirmed later on by the observation of such γ-Fe$_2$O$_3$ particles on images obtained by high-resolution transmission electron microscopy. In another case, a remanent magnetization is observed only below 245 K, which is the signature of the Fe$_2$P impurity, as Fe$_2$P orders ferromagnetically at this temperature [18, 19]. Actually, this impurity almost stopped the research on the application of this material as a cathode for Li-ion batteries. Some groups had found that there was a dissolution of iron in the electrolyte used for the batteries and concluded that LiFePO$_4$ had no future as it could not endure the corrosion by the electrolyte. Fortunately, we determined that the presence of Fe in the electrolyte was detected only when Fe$_2$P impurity was present, and not when LiFePO$_4$ was pure [20]. This gave evidence that LiFePO$_4$ can be used as a cathode element, but only when it is prepared free of this impurity, i.e. when it is prepared with an annealing temperature lower than 800°C; otherwise, the corrosion of the nano-particles of Fe$_2$P liberates iron inside the electrolyte, which results in the aging of the battery. In the opposite case, i.e. if the annealing temperature is too low, or in absence of reducing atmosphere, the presence of a Fe^{3+} impurity is observed, namely the NASICON-like phase Li$_3$Fe$_2$(PO$_4$)$_3$ that can be identified by its first-order magnetic transition near 25 K [21]. The amplitude of the jump of magnetization at this temperature gives access to the amount of this impurity present in the sample. Then it was straightforward to determine the amount of these different impurities as a function of the different synthesis parameters, and finally determine their optimized values leading to pure LiFePO$_4$.

Some defects are the result of deviation from stoichiometry [22]. An excess in Li among the precursors leads to the formation of Li$_3$PO$_4$. Since it is not magnetic, the presence of this impurity and its amount in the material can be determined by the decrease in the magnetic susceptibility with respect to that of pure $LiFePO_4$. The presence of this impurity, however, is not dramatic, since it just acts as an inert mass in the electrochemical process. Meanwhile, a deficiency of Li among the precursors during the synthesis results in the formation of an antisite defect that can be written as $Fe^{\bullet}_{Li}+V'_{Li}$ in Kröger-Vink notation. First, antisite defects are understood as an exchange of crystallographic positions. Hence, Li takes place on Fe site and vice versa, while defect-free $LiFePO_4$ has an $M_1M_2PO_4$ formula antisite defect $LiFePO_4$ that would look like $(Li_{1-x}Fe_x)(Fe_{1-x}Li_x)PO_4$. An excess of Fe in $LiFePO_4$ would look like $(Li_{1-2/3x}Fe_x)FePO_4$, depending on the valence state of the Fe on the Li-position, assuming the Fe on the M_2 place has oxidation state +2 [22–24]. Since iron keeps the divalent state in the antisite defect, it was not possible to detect this defect with magnetic measurements, but it has been identified and its concentration determined by other techniques such as Rietveld refinement of the X-ray diffraction pattern and inductive coupled plasma spectroscopy [22]. This defect turns out to be very damaging to the electrochemical performance of $LiFePO_4$, because each Fe^{3+} ion on a Li site blocks the Li-channel, i.e. prevents all the other Li⁺ ions of the corresponding channel from participating to the electrochemical process. This will be illustrated in the next section.

2.3.3 Electrochemical properties

Let us first recall some bases of the electrochemistry of $LiFePO_4$. The redox potential of Fe^{2+}/Fe^{3+} with respect to Li^0/Li^+ is 3.45 V, which is the voltage of a cell with $LiFePO_4$ and Li as electrodes. The electrolyte is usually a mixture of ethylene carbonate and diethylene carbonate (EC/DEC 1:1) and the Li salt is $LiPF_6$. The reasons for the choice of this electrolyte and salt is beyond the scope of the present report and can be found in a book devoted to Li-ion batteries [25]. We report herein the electrochemical properties of such cells. Among different techniques, one simple probe of the performance of the cell is the measurement of the potential of the cell versus its capacity. After it has been built, the cell is discharged. It means that the first action to do is to charge the cell, i.e. extract the Li⁺ ions from $LiFePO_4$ to send them to the Li electrode through the electrolyte. Meanwhile, the same amount of electrons is moved through the external electrical circuit connecting the two electrodes through the charger to ensure charge neutrality in the crystal. If all the Li-ions can be extracted from $LiFePO_4$, we can obtain the theoretical specific capacity of the cell, 170 mAh/g, calculated simply as the number of moles of electrons in 1 g of $LiFePO_4$ times the charge of the electron. In practice, the capacity will be at most equal to this value and will depend on the speed at which the cell will be charged or discharged. When the speed is increased, we do not give enough time to all the Li-ions to move from one electrode to the other, which results in the reduction of the experimental value of the capacity. Therefore, when reporting the measurement of the capacity, we must always specify the speed used to charge or discharge the cell. This is specified by the C-rate, which means that the charge or discharge is performed when the electrical current fixed during the experiment is such that the theoretical capacity should be achieved in a time of 1/C hours. For instance, a 2C rate means that the current is fixed at such a value that full charge or discharge should be obtained after half an hour. In practice, however, the measurement will reveal that it is not possible to maintain this current so long, so that the actual capacity measured at this C-rate will be smaller. The performance of all the cells reported here have been prepared after carbon coating; otherwise, the electrical conductivity is so small that it is impossible to extract the Li within a reasonable time. Also, the performance of the cells depends on the size of the $LiFePO_4$ particles, as mentioned earlier in this report. The results reported hereunder have been obtained on powders of typical size 100 nm.

The voltage-capacity curve obtained with pure $LiFePO_4$ measured at C/10 rate is reported in Fig. 2.3 together with those obtained with other $LiMPO_4$ olivine materials (M = Mn, Co, Ni) for comparison. We recover the plateau at 3.45 V, characteristic of the redox potential of Fe^{2+}/Fe^{3+} with respect to Li^0/Li^+, and a capacity close to the theoretical one. However, in presence of antisite defects, the capacity is severely decreased, despite the very small C-rate of C/20, as shown in Fig. 2.4. The degradation of the performance of the cell resulting from the presence of Fe_2P impurity is evidenced, despite the fact that only 0.5% of the iron belonged to this impurity. This is due to the fact that the presence of Fe_2P decreases the ionic conductivity so that both the capacity and cycling rate are degraded with respect to the pure carbon-coated $LiFePO_4$.

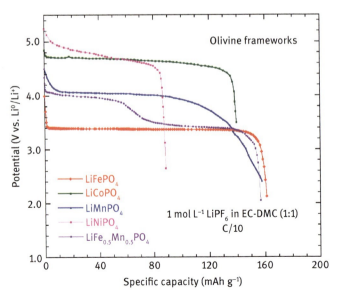

Fig. 2.3: Typical discharge profile of Li//$LiMPO_4$ cells. Measurements were carried at C/10 rate at room temperature with electrolyte 1 mol/L $LiPF_6$ in EC:DMC (1:1).

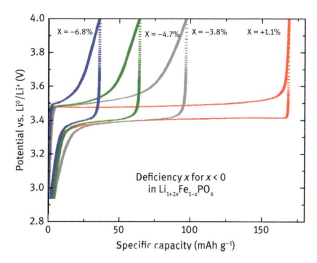

Fig. 2.4: Electrochemical performance of Li-deficient ($Li_{1+2x}Fe_{1-x}PO_4$ with $x < 0$) iron phosphate [22].

2.3.4 Carbon coating

Owing to the dramatic impact of the impurities that are generated when the annealing temperature exceeds 750°C, it is mandatory to keep this temperature at 750°C or lower. Unfortunately, the conductive form of carbon is graphite, which is synthesized above 1100°C. The question of how conductive is the carbon coat deposited on the surface of $LiFePO_4$ then arises. The tool to characterize this coat is Raman spectroscopy for two reasons. First, carbon can be identified by two broad Raman bands structure: one at 1583 cm^{-1} corresponds to the G line associated with the optically allowed E_{2g} zone-center mode of crystalline graphite and the other one at 1345 cm^{-1} corresponds to the D line associated with disorder-allowed zone-edge modes of graphite. Second, the penetration depth of the laser beam in the particles is about 30 nm, which means that the analysis of the reflected beam in the Raman experiment gives access to the full carbon coat plus only a small amount of $LiFePO_4$. This feature makes the Raman experiment the ideal tool to probe the carbon layer [26]. Most of the time, the D/G ratio is considered to predict the electrical conductivity, since the G band is attributed to graphite, which is a good electrical conductor. The type of carbon, e.g. MWNT, can only be determined by looking at high wavenumbers (>2300 cm^{-1}) in the Raman spectrum. As a result, we have determined from these spectra that the carbon is similar to coke, which is a conductive form of carbon. We have also shown that the electrical conductivity of this carbon layer deposited on $LiFePO_4$ at a temperature of 700°C is comparable to that of a carbon film deposited on silicon by pyrolysis at 850°C, measured in Ref. [27], which explains the efficiency of the carbon coating of $LiFePO_4$ in improving electrical conductivity. This is fortunate, on the one hand, the carbon coating at 850 °C was impossible due to the formation of the impurities; on the other hand, if the carbon deposited at 700 °C was as insulating as the carbon deposited on a silicon wafer by pyrolysis at this temperature, the carbon coat would have had no effect. Again, this outstanding result is presumably attributable to the particular affinity between carbon and iron that is known to play an important role in biology. The carbon coating, however, has another beneficial effect: it prevents the formation of Fe^{3+}-based impurities. The reason was subject to controversy. One hypothesis was that it was a carbothermal effect. Indeed, this process of reduction of iron by carbon is used to purify iron in blast furnaces since the 19th century. However, this reduction process occurs above 1000 °C; thus, it is very unlikely in the present case when the temperature does not exceed 750 °C. This hypothesis has been definitely ruled out by an experiment showing that the choice of pure carbon as the carbon additive with the precursors in the synthesis process failed to give pure and well-crystallized $LiFePO_4$; only the choice of an organic precursor does the job [21]. The reason is that the organic additive contains hydrogen, and the hydrogenous gas that is liberated creates the reducing atmosphere needed to synthesize $LiFePO_4$ and avoid the formation of Fe^{3+}-based impurities [21]. Meanwhile the carbon part of the organic precursor forms the carbon coat. Finally, the annealing process at 700 °C–750 °C has another beneficial effect: it re-crystallized the surface layer of the $LiFePO_4$ that, without the carbon coating, is disordered if not amorphous [11]. It is remarkable that this understanding of the synthesis process has been obtained a posteriori and dates from only a few years. Somehow, Santoro and Newham [3] were very lucky to obtain pure and well-crystallized samples in 1967. At that time, they prepared the samples using a solid-state reaction with phosphate and iron precursors that contained hydrogen and Li carbonate that contains carbon. Presumably, the hydrogen liberated in the annealing provided the reducing atmosphere, whereas the carbon in the Li precursor eventually coated the $LiFePO_4$ particles. Also, the annealing temperature that they have chosen was 800 °C, which was probably slightly overestimated, so that they barely avoided the formation of magnetic impurities. It took years to again obtain pure $LiFePO_4$ on a regular basis, since the role of the choice of the precursors and of the synthesis parameters was not understood and also because the measurements of the magnetic properties that turned out to be the best tool to detect the impurities were not used by the electrochemists.

2.3.5 Aging

Circa 2010, the problems mentioned above were all solved, and the industrial production of $LiFePO_4$ developed fast. However, some of the companies that bought $LiFePO_4$ complained to their suppliers that the product was not good, whereas other companies did not complain. The reason was due to the fact that not enough attention was paid to the work of mineralogists who had mentioned the sensitivity of these olivine materials to water exposure [5]. This degradation when kept in a humid atmosphere was evidenced in two steps. First, we have detected Fe^{3+} in the surface layer of the $LiFePO_4$ particles, which gave evidence of a delithiation of this layer over a thickness of few nanometers [12]. In a second step, we have analyzed the evolution of the capacity as a function of the time spent in dry (5% humidity) atmosphere and in ambient atmosphere (55% humidity) at different temperatures ranging from 20°C to 60°C. The results [28] are shown in Fig. 2.5. The exposure to humidity results in a decrease of the capacity, because the Li in $LiFePO_4$ that has reacted with H_2O no longer participates in the electrochemical process. The degradation increases with temperature because the kinetics of the reaction increases with temperature.

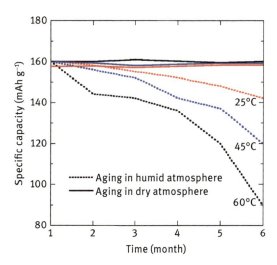

Fig. 2.5: Capacity of the C-$LiFePO_4$/$LiPF_6$-EC-DEC/Li cells as a function of time spent in dry atmosphere and in ambient atmosphere (55% relative humidity) at three different temperatures [28].

The different opinions on the product coming from the same companies actually came from the different storage conditions. Therefore, the recommendation to keep the product in a dry chamber has been added to the technical specifications of the suppliers of $LiFePO_4$. Since then, the material is developing fast and is included in the composition of many batteries used today in Li-ion batteries. In particular, owing to the very good thermal stability of $LiFePO_4$, it is now possible to make batteries with a $LiFePO_4$ cathode and titanate $Li_4Ti_5O_{12}$ anode that can be charged or discharged very fast. Such a battery with 90-nm-thick carbon-coated particles displays a charge capacity of 650 mAh at low C-rate and retains more than 80% of rated capacity at 60C charge rate (1 min) over the 3500 cycles that have been tested [29]. This result makes this battery the most powerful in the market and with the longest life. Its energy density, however, is smaller, due to loss of 1.4 V in the voltage of the cell. Note that the power is too large to be available to individuals, since the companies in charge of the

2.4 Other Li-based olivines

2.4.1 Li(Mn,Fe)PO$_4$ olivines

Figure 2.3 shows that LiFePO$_4$ has the lowest redox potential vs. Li^0/Li$^+$ among the olivine phosphate family. This is why many efforts have been made to replace Fe by Mn in increasing the potential to 4.1 V, since higher potential means more energy density. Unfortunately, LiMnPO$_4$ (LMP) is even more insulating than LiFePO$_4$, so that electrochemical activity requires a reduction of the size of the particles to few tens of nanometers, which is expensive. Among the best results obtained so far, carbon-coated LMP particles of 30 nm delivered a specific capacity of 110 mAh/g at 1C rate [30, 31]. Nano-rods synthesized by modified poll and resin coating processes to enhance their conductivity delivered a capacity of 120 mAh/g at 1C rate [32]. Another difficulty comes from the fact that, in the delithiation process, Mn^{2+} shifts to the Mn^{3+} valence configuration. Mn^{3+} is a Jahn-Teller ion, meaning that this change of valence is accompanied by a strong local trigonal distortion of the lattice. The resulting strain and stress fields, although easier to accommodate in nano-particles, remains a source of aging upon cycling. We can then conclude that LiMnPO$_4$ is not competitive with LiFePO$_4$ so far. A way that has been explored to overcome these problems is to use the fact that, from triphylite to lithiophilite, the solid-solution LiFe$_{1-x}$Mn$_x$PO$_4$ exists for any x, as mentioned in the first section. We have investigated this series, determined their structural and magnetic properties [33, 34], and related them to the electrochemical properties [34, 35]. The magnetic properties gave evidence that the strain field increases importantly for $x > 0.6$, and ultimately prevents the full delithiation as soon as $x = 0.08$. These results were supported by the analysis of the electrochemical properties so that we have recommended that the optimum concentration of the solid solution is $x = 0.6$ and must be kept lower than 0.8. Under such conditions, Mn contributes efficiently to the electrochemical process at 4.1 V, which results in an improvement of the energy density with respect to LiFePO$_4$, but only at low C-rates. The last difficulty with Mn is that, in contrast with Fe, it does not interact with carbon, so that the carbon coating is more difficult. In another approach, we have synthesized composites made of LiMnPO$_4$ as the core of the particles, coated with a layer of LiFePO$_4$ that was itself coated with carbon [36]. This approach was indeed successful in the sense that the electrochemical properties of this composite were improved with respect to the solid solution with the same proportion of Mn. Nevertheless, the synthesis of such a composite is not cheap and not suited for industrial-scale mass production.

2.4.2 LiCoPO$_4$ and LiNiPO$_4$ olivines

With Co and Ni, the potential of the cell increases (4.8–4.9 and 5.1–5.3 V vs. Li^0/Li$^+$, respectively), but it is detrimental to safety. For instance, Li$_x$CoPO$_4$ (x = 0.6) and CoPO$_4$ appearing during the delithiation of LiCoPO$_4$ are unstable upon heating and decompose readily in the range of 100°C–200°C, while the decomposition of Li-poor phases leads to gas evolution [37]. With LiNiPO$_4$ electrode, we even reach the voltage where the standard electrolytes become unstable. This, however, is not necessarily crippling, since sulfo-based electrolytes extend the oxidizing stability to 5.8 V. The different processes that lead to the synthesis of pure LiNiPO$_4$ can be found in Ref. [38]. Nevertheless, the study of the electrochemical properties of LiNiPO$_4$ is very limited. On a general basis, the laws of thermodynamics tell us that the higher potential, the higher the partial oxygen pressure at equilibrium in transition metal oxides [39], which means increased risk of loss of oxygen. This is a major cause of instability, in particular when the anode is graphitic carbon, since the reaction of carbon and oxygen to produce CO and CO$_2$ is exothermal, resulting in battery fires. That is why we do not recommend batteries with too high concentration of nickel in the cathode. This is a recommendation that industries have difficulties heeding, as they are all trying to make batteries with energy density as high as possible to win the market.

2.5 Battery of the future: olivine-based Na-ion cells

In the column just below hydrogen and Li in the table of elements, one finds Na, which is very cheap and has a huge quantity. Therefore, many efforts in research are now focused on Na-ion batteries. NaMPO$_4$ (M = Fe, Mn) compounds synthesized under high-temperature conditions crystallize in the maricite structure, in which case they are electrochemically inactive [40]. Meanwhile, olivine NaFePO$_4$ can be obtained at low-temperature synthesis, by chemical or electrochemical delithiation followed by sodiation of olivine [41–43]. As an alternative to the organic-based electrochemical ion-exchange process which is disadvantaged by sluggish dynamics and co-intercalation of Li$^+$, Tang et al. [44] investigated an aqueous-based, electrochemical-driven ion-exchange process to transform olivine LiFePO$_4$ into highly pure olivine NaFePO$_4$, which shows superior electrochemical performance. Other techniques employed to synthesize NaFePO$_4$ include solid-state sintering at 350°C and chemical oxidation-reduction using NaI and FePO$_4$ [45]. Fang et al. [46] prepared olivine NaFePO$_4$/C microsphere cathode by a facile aqueous electrochemical displacement method from LiFePO$_4$/C precursor. The NaFePO$_4$/C cathode shows a high discharge capacity of 111 mAh/g, excellent cycling stability with 90% capacity retention over 240 cycles at 0.1C rate, and high rate capacity (46 mAh/g at 2C rate).

In situ X-ray diffraction measurements disclose that the Na//NaFePO$_4$ cell has different charge and discharge mechanisms. Contrary to LiFePO$_4$, the Na extraction process from NaFePO$_4$ occurs with an intermediate Na$_{0.7}$FePO$_4$, which has been described by a Na$^+$/vacancy ordering [41, 47, 48]. As shown in Fig. 2.6, a single plateau is observed on discharge profiles [41, 49, 50]. Owing to the larger volumetric mismatch between NaFePO$_4$ and FePO$_4$ (~17.6% difference in unit volume), the asymmetry Na-ion intercalation/de-intercalation mechanism is observed [47].

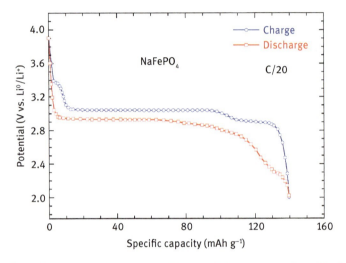

Fig. 2.6: Charge-discharge curves of Na//NaFePO$_4$ electrochemical cells. Measurements were carried at C/20 rate at room temperature with electrolyte 1 mol/L NaClO$_4$ in EC/PC (1:1).

Olivine NaFePO$_4$ has the theoretical specific capacity of 154 mAh/g, which makes it an attractive positive electrode material for Na-ion batteries. The uncoated NaFePO$_4$ electrode operates with a working potential range of 2.2–4.0 V vs. Na^0/Na$^+$ and delivers a discharge capacity of ca. 108 mAh/g at a current density of 10 mA g^{-1}. Coating with polythiophene (a conductive polymer) improved the electrochemical performance, as a discharge capacity of 142 mAh/g and a stable cycle life over 100 cycles, with a capacity retention of 94% were observed [51]. A specific capacity of 147 mAh/g for NaFePO$_4$ was reported during the first cycle for the cell operated at 60°C and C/24 rate, but the cycle ability was limited to 4–5 cycles [40]. Oh et al. [49] prepared a NaFePO$_4$ electrode by electrochemical Li-Na exchange of LiFePO$_4$ and showed that the material keeps its original structure over 50 charge-discharge processes at C/20 rate and room temperature with a capacity of 125 mAh/g. Amorphous-phase NaFePO$_4$ exhibits a high discharge capacity of ~150 mAh/g but operates at a low working potential (~2.4 V) with an s-shape discharge profile [52]. Wongittharom et al. [50] demonstrated the good electrochemical performance of olivine NaFePO$_4$ using an ionic liquid (IL) electrolyte (NaTFSI-(BMP)TFSI). At 50°C, an optimal capacity of 125 mAh/g (at 0.05C rate) is found for NaFePO$_4$ in a 0.5-mol/L NaTFSI-incorporated IL electrolyte. Despite significant similarities, NaFePO$_4$ differs from LiFePO$_4$ in the detail of its surface structures. An electrochemically converted NaFePO$_4$ olivine cathode and a nano-structured Sn-C anode were combined in an efficient Na-ion battery, characterized by a capacity of 150 mAh/g, voltage of 3 V, and consequently, a practical energy density estimated to be of the order of 150 Wh/kg [53]. These results give evidence of a rapid progress in the performance of NaFePO$_4$ as a cathode of a new generation of Na-ion battery.

2.6 Conclusion

In 1834, the phosphate olivine family was discovered by the mineralogists who recognized primary minerals in it. Chemists and physicists, however, only recognized its promising application as a cathode element of Li-ion batteries in 1997, and it took another decade before it became a reality. Probably, a better communication between these two scientific communities would have accelerated the process. Nevertheless, the problems that needed to be solved, starting from the mineral to the synthesis of pure nano-sized samples without any defect or impurity at the industrial scale were complex and illustrate that the delay between fundamental research and applications can be long. Today, one element of the family, LiFePO$_4$, is commonly used in Li-ion batteries all over the world, and yet the family has not said its last world, since Na phosphate olivine is a promising candidate as a cathode element for next-generation Na-ion batteries in the future.

References

[1] Fuchs JN. Ueber ein neues Mineral (Triphylin). J Prakt Chem 3 (1834) 98–102.
[2] Zhang R. New mineral from lithium-rich pegmatite mine. Kexue Tongbao 1981; 26:334–7.
[3] Santoro RP, Newham RE. Antiferromagnetism in LiFePO$_4$. Acta Cryst 1967; 22:344–7.
[4] Padhi AK, Nanjundaswamy KS, Goodenough JB. Phospho-olivines as positive-electrode materials for rechargeable lithium batteries. J Electrochem Soc 1997;144:1188–94.
[5] Hatert F, Schmid-Beurmann P. European Conference on Mineralogy, E. Schweizerbart Sci. Publ., Stuttgart; 2005.
[6] Palomares V, Goñi A, Iturrondobeitia A, Lezama L, de Meatza I, Bengoechea M, Rojo T. Structural, magnetic and electrochemical study of a new active phase obtained by oxidation of a LiFePO$_4$/C composite. J Mater Chem 2012;22:4735–43.
[7] Moore PB. Pegmatite Minerals of P(V) and B(III). Mineralogical Association of Canada, Quebec;1985.
[8] Ravet N, Chouinard Y, Magnan JF, Besner S, Gauthier M, Armand M. Electroactivity of natural and synthetic triphylite. J Power Sources 2001;97–98:503–7.
[9] Zaghib K, Mauger A, Julien CM. Olivine-based cathode materials. In: Zhang Z, Zhang SS, editors. Rechargeable batteries. Chapter 3. Green Energy and Technology, Springer International Publishing Switzerland, 2015:25–65.
[10] Ravet N, Besner S, Simoneau M, Vallée A, Armand M, Magnan JF. Electrode materials with high surface conductivity. US Patent 6,962,666. Nov. 8, 2005.

[11] Trudeau ML, Laul, Veillette R, Serventi AM, Mauger A, Julien CM, Zaghib K. In-situ HRTEM synthesis observation of nanostructured LiFePO$_4$. J Power Sources 2011;196:7383–94.
[12] Zaghib K, Mauger A, Gendron F, Julien CM. Surface effects on the physical and electrochemical properties of thin LiFePO$_4$ particles. Chem Mater 2008;20:462–9.
[13] Azib T, Ammar S, Nowak S, Lau-Truing S, Groult H, Zaghib K, Mauger A, Julien CM. Crystallinity of nano C-LiFePO$_4$ prepared by the polyol process. J Power Sources 2012;217:220–8.
[14] Saravanan K, Reddy MV, Balaya P, Gong H, Chowdari BVR, Vittal JJ. Storage performance of LiFePO$_4$ nanoplates. J Mater Chem 2009;19:605–10.
[15] Chen J, Vacchio MJ, Wang S, Chernova N, Zavalij PY, Whittingham MS. The hydrothermal synthesis and characterization of olivines and related compounds for electrochemical applications. Solid State Ionics 2008;178:1676–96.
[16] Brochu F, Guerfi A, Trottier J, Kopeć M, Mauger A, Groult H, Julien CM, Zaghib K. Structure and electrochemistry of scaling nano C-LiFePO$_4$ synthesized by hydrothermal route: complexing agent effect. J Power Sources 2012;214:1–6.
[17] Vediappan K, Guerfi A, Gariépy V, Demopoulos GP, Hovington P, Trottier J, Mauger A, Julien CM, Zaghib K. Stirring effect in hydrothermal synthesis of nano C-LiFePO$_4$. J Power Sources 2014;266:99–106.
[18] Ait-Salah A, Mauger A, Zaghib K, Goodenough JB, Ravet N, Gauthier M, Gendron F, Julien CM. Reduction Fe^{3+} of impurities in LiFePO$_4$ from pyrolysis of organic precursor used for carbon deposition. J Electrochem Soc 2006;153:A1692–701.
[19] Ait Salah A, Mauger A, Julien CM, Gendron F. Nano-sized impurity phases in relation to the mode of preparation of LiFePO$_4$. Mater Sci Eng B 2006;129:232–44.
[20] Zaghib K, Mauger A, Goodenough JB, Gendron F, Julien CM. In: Garche J, Dyer CK, Moseley P, Ogumi Z, Rand DAJ, Scrosati B, editors, Encyclopedia of electrochemical power sources. Vol. 5. Elsevier, Amsterdam; 2009:264–96.
[21] Ravet N, Gauthier M, Zaghib K, Goodenough JB, Mauger A, Gendron F, Julien CM. Mechanism of the Fe^{3+} reduction at low temperature for LiFePO$_4$ synthesis from a polymeric additive. Chem Mater 2007;19:2595–602.
[22] Axmann P, Stinner C, Wohlfahrt-Mehrens M, Mauger A, Gendron F, Julien CM. Nonstoichiometric LiFePO$_4$: defects and related properties. Chem. Mater. 2009;21:1636–44.
[23] Jensen KMO, Christensen M, Gunnlaugsson HP, Lock N, Bojesen ED, Proffen T, Iversen BB. Defects in hydrothermally synthesized LiFePO$_4$ and LiFe$_{1-x}$Mn$_x$PO$_4$ cathode materials. Chem Mater 2013;25:2282–90.
[24] Schoiber J, Tippelt G, Redhammer GJ, Yada C, Dolotko O, Berger RJF, Husing N. Defect and surface area control in hydrothermally synthesized LiMn$_{0.8}$Fe$_{0.2}$PO$_4$ using a phosphate based structure directing agent. Cryst Growth Des 2015;15:4213–8.
[25] Julien CM, Mauger A, Vijh A, Zaghib K. Lithium batteries science and technology, Springer, Cham; 2016.
[26] Julien CM, Zaghib K, Mauger A, Massot M, Ait-Salah A, Selmane M, Gendron F. Characterization of the carbon coating onto LiFePO$_4$ particles. J Appl Phys 2006;100:063511.
[27] Kostecki R, Schnyder B, Alliata D, Song X, Kinoshita K, Kotz R. Surface studies of carbon films from pyrolyzed photoresist. Thin Solid Films 2001;396:36–43.
[28] Zaghib K, Dontigny M, Charest P, Labrecque JF, Guerfi A, Kopec M, Mauger A, Gendron F, Julien CM. Aging of LiFePO$_4$ upon exposure to H$_2$O. J Power Sources 2008;185:698–710.
[29] Zaghib K, Dontigny M, Guerfi A, Trottier J, Hamel-Paquet J, Gariepy V, Galoutov K, Hovington P, Mauger A, Groult H, Julien CM. An improved high-power battery with increased thermal operating range: C-LiFePO$_4$//C-Li$_4$Ti$_5$O$_{12}$. J Power Sources 2012;216:192–200.
[30] Wang D, Buqa H, Crouzet M, Deghenghi G, Drezen T, Exnar I, Kwon NH, Miners J, Poletto LM. Gratzel M. High-performance, nanostructured LiMnPO$_4$ synthesized via a polyil method. J Power Sources 2009;189:624–8.

[31] Martha SK, Markovsky B, Grinblat J, Gofer Y, Haik O, Zinigrad E, Aurbach D, Drezen T, Wang D, Deghengh G, Exnar I. LiMnPO$_4$ as an advanced cathode materialm for rechargeable lithium batteries. J Electrochem Soc 2009;156:A541–52.

[32] Kumar PR, Venkateswarlu M, Misra M, Mohanty AK, Satyanarayana N. Carbon coated LiMnPO$_4$ nanorods for lithium batteries. J Electrochem Soc 2011;158:A227–30.

[33] Kopec M, Yamada A, Kobayashi G, Nishimura S, Kanno R, Mauger A, Gendron F, Julien CM. Structural and magnetic properties of Li$_x$(Mn$_y$Fe$_{1-y}$)PO$_4$ electrode material for Li-ion batteries. J Power Sources 2009;189:1154–63.

[34] Zaghib K, Mauger A, Gendron F, Massot M, Julien CM. Insertion properties of LiFe$_{0.5}$Mn$_{0.5}$PO$_4$ electrode materials for Li-ion batteries. Ionics 2008;14:371–6.

[35] Trottier J, Mathieu MC, Guerfi A, Zaghib K, Mauger A, Julien CM. LiMn$_y$Fe$_{1-y}$PO$_4$ (0.5≤ y ≤0.8) cathode materials grown by hydrothermal route: electrochemical performance. ECS Trans 2013;50–24:109–14.

[36] Zaghib K, Trudeau M, Guerfi A, Trottier J, Mauger A, Veillette R, Julien CM. New advanced cathode material: LiMnPO$_4$ encapsulated with LiFePO$_4$. J Power Sources 2012;204:177–81.

[37] Bramnik NN, Nikolowski K, Trots DM, Ehrenberg H. Thermal stability of LiCoPO$_4$ cathodes. Electrochem Solid State Lett 2008;11:A89–93.

[38] Rommel SM, Schall N, Brünig C, Weihrich R. Challenges in the synthesis of high voltage electrode materials for lithium-ion batteries: a review on LiNiPO$_4$. Monatsh Chem 2014;145:385–404.

[39] Huggins RA. Do you really want an unsafe battery? J Electrochem Soc 2013;160:A3001–5.

[40] Zaghib K, Trottier J, Hovington P, Brochu F, Guerfi A, Mauger A, Julien CM. Characterization of Na-based phosphate as electrode materials for electrochemical cells. J Power Sources 2011;196:9612–7.

[41] Moreau P, Guyomard D, Gaubicher J, Boucher F. Structure and stability of sodium intercalated phases in olivine FePO$_4$. Chem Mater 2010;22:4126–8.

[42] Zhu Y, Xu Y, Liu Y, Luo C, Wang C. Comparison of electrochemical performances of olivine NaFePO$_4$ in sodium-ion batteries and olivine LiFePO$_4$ in lithium-ion batteries. Nanoscale 2013;5:780–7.

[43] Casas-Cabanas M, Roddatis VV, Saurel D, Kubiak P, Carretero-González J, Palomares V, Serras P, Rojo T. Crystal chemistry of Na insertion/deinsertion in FePO$_4$-NaFePO$_4$. J Mater Chem 2012;22:17421–3.

[44] Tang W, Song X, Du Y, Peng C, Lin M, Xi S, Tian B, Zheng J, Wu Y, Pan F, Loh KP. High-performance NaFePO$_4$ formed by aqueous ion-exchange and its mechanism for advanced sodium ion batteries. J Mater Chem A 2016;4:4882–92.

[45] Lu J, Chung SC, Nishimura S, Yamada A. Phase diagram of olivine Na$_x$FePO$_4$ (0<x<1). Chem Mater 2013;25:4557–65.

[46] Fang Y, Liu Q, Xiao L, Ai X, Yang H, Cao Y. High-performance olivine NaFePO$_4$ microsphere cathode synthesized by aqueous electrochemical displacement method for sodium ion batteries. ACS Appl Mater Interfaces 2015;7:17977–84.

[47] Galceran M, Saurel D, Acebedo B, Roddatis VV, Martin E, Rojo T, Casas-Cabanas M. The mechanism of NaFePO$_4$ (de)sodiation determined by in situ X-ray diffraction. Phys Chem Chem Phys 2014;16:8837–42.

[48] Whiteside A, Fisher CAJ, Parker SC, Islam MS. Particles shapes and surface structures ofnolivine NaFePO$_4$ in comparison to LiFePO$_4$. Phys Chem Chem Phys 2014;16:21788–94.

[49] Oh SM, Myung ST, Hassoun J, Scrosati B, Sun YK. Reversible NaFePO4 electrode for sodium secondary batteries. Electrochem Commun 2012;22:149–152.

[50] Wongittharom N, Lee T, Wang C, Wang Y, Chang J. Electrochemical performance of Na/NaFePO$_4$ sodium-ion batteries with ionic liquid electrolytes. J Mater Chem A 2014;2:5655–61.

[51] Ali G, Lee JH, Susanto D, Choi SW, Cho BW, Nam KW, Chung KY. Polythiophene-wrapped olivine $NaFePO_4$ as a cathode for Na-ion batteries. ACS Appl. Mater. Interfaces 2016;8:15422.
[52] Li C, Miao X, Chu W, Wu P, Tong DG. Hollow amorphous $NaFePO_4$ nanospheres as a high-capacity and high-rate cathode for sodium-ion batteries. J Mater Chem A 2015;3:8265–71.
[53] Hasa I, Hassoun J, Sun YK, Scrosati B. Sodium-ion battery based on an electrochemically converted $NaFePO_4$ cathode and nanostructured tin-carbon anode. ChemPhysChem 2014;15:2152–5.

Bernd Marler and Hermann Gies
3 Natural and synthetic zeolites

3.1 History and definitions

Fig. 3.1: Needle-like crystals of zeolite natrolite grown on a basaltic rock.

Zeolites (Fig. 3.1) are a group of minerals first recognized in 1756. The name "zeolite" was termed by the Swedish mineralogist Axel Fredrik Cronstedt [1] to describe stilbite, the first zeolite mineral classified. "Zeolite" is a combination of the two Greek words "zeo", meaning "to boil", and "lithos", meaning "a stone" (Fig. 3.2). This describes vividly one of the important properties of zeolites: the ability to reversibly release and take up water depending on the temperature and moisture level of the external environment. During the following years, more and more natural zeolites were discovered: chabazite (1792) [2], mordenite (1864) [3], clinoptilolite (1932) [4], and tounkite-like mineral (2004) [5], to name a few. Up to now, 67 different zeolite minerals are known [6].

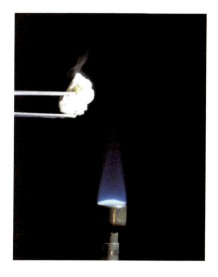

Fig. 3.2: Heated natural stilbite showing the release of water vapor.

The classical description of a zeolite based on natural zeolites is a crystalline alumosilicate possessing a silicate framework with micropores which are occupied by inorganic cations and water molecules (Fig. 3.3). The general formula of a zeolite mineral is $M^I_x M^{II}_y [Al_{(x+2y)} Si_{n-(x+2y)} O_{2n}] \cdot wH_2O$, with M^I = Li, Na, K; M^{II} = Mg, Ca, Sr, Ba, and Si ≥ Al (in analogy to [7]).

The composition of the tetrahedral framework is presented in square brackets. Each Al framework atom leads to one negative charge (distributed about the oxygen atoms bonded to the Al atom) which has to be compensated by an extra-framework cation, M^+ or M^{2+}, which are attached to the charged oxygen atoms. Water molecules are also included in the pore system of the zeolite bonded to the extra-framework cations, to the framework, or to other water molecules.

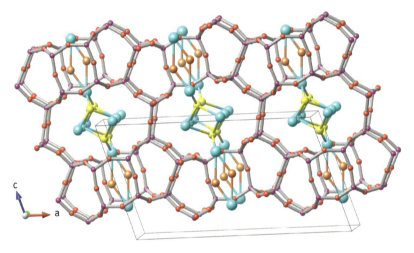

Fig. 3.3: Structure of zeolite dachiardite showing the silicate framework (gray bonds, Si/Al: purple, O: red) und the constituents of the pore volume (Ca^{2+}: yellow, K^+: orange, water molecules: light blue).

Originally, all our knowledge on zeolites results from natural zeolite minerals. In the middle of the 19th century, the water adsorption ability was recognized [8]; some decades later, the ion exchange properties [9] and the ability of activated (= water-free) zeolites to absorb small molecules was discovered [10]. The shape-selective nature of the uptake of small molecules was realized in 1925.

Among the three most important applications of zeolites, ion exchange, (shape-selective) adsorption and catalysis, only the ability of zeolites to act as efficient catalysts was studied first on synthetic materials (Mobil Oil started to use zeolite X as a cracking catalyst in 1962 [11]).

Detailed information on natural zeolites can be obtained from the websites of the Commission on Natural Zeolites of the International Zeolite Association (IZA) [6] and the International Natural Zeolite Association (INZA) [12] or from books presenting reviews on natural zeolites [7, 13].

The world production of natural zeolites in 2015 was estimated to be about 3 million tons. In addition, many different synthetic zeolite materials are produced industrially in a range of about 2 million tons per year [14]; together, these zeolites serve as catalysts, ion exchangers or sorbents for a large variety of purposes in industry, agriculture and household.

The extensive research on zeolite synthesis for at least 79 years (the first confirmed synthesis of a zeolite, mordenite, was published in 1948 [15]) led to hundreds of different synthetic products. Although natural zeolites contain predominantly Si and Al (and to a much lesser extend B and Be) as framework cations, synthetic zeolite-type materials possess a large variety of framework cations: Si, Al, P, Ge, B, Zn, Ga, Fe, Li, Co, Mg, Mn and even Ti, Cu, Cd, V, Cr. Nearly all elements that can be tetrahedrally coordinated by oxygen atoms have successfully been inserted into a zeolite framework. With respect to the non-framework constituents, the main difference between zeolite minerals and synthetic products is the fact that synthetic zeolites quite often contain organic molecules (e.g. amines) or organic cations as non-framework constituents, exclusively or in addition to inorganic cations.

A more general definition is therefore: "A zeolite is crystalline solid possessing a tetrahedral framework with micropores (2–20 Å in diameter) which may be occupied by (partly) exchangeable inorganic or organic cations, water, organic molecules and even anions."

An example of a complex synthetic zeolite is the so-called CoAPO-50 [16, page 270], a microporous metal-aluminophosphate (MeAPO) containing cobalt, with the formula $(C_6H_{15}N)_3[Co_3Al_5P_8O_{32}] * 6.7\ H_2O$ (Fig. 3.4).

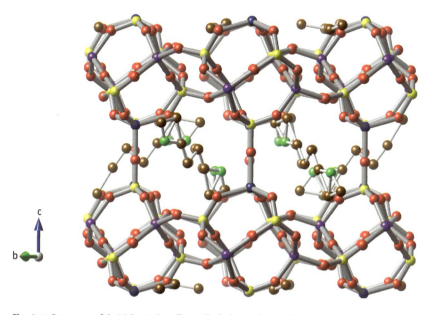

Fig. 3.4: Structure of CoAPO-50. P: yellow, Al: dark purple, Co: blue, O: red, N: green, C: brown.

Synthetic zeolites are typically named by an acronym consisting of three (seldom two) capital letters designating the laboratory which has produced the zeolite, followed by a number indicating a specific zeolite material. For example, "RUB-58" was made at the **R**uhr **U**niversity **B**ochum and 58 indicated that it is the 58th distinct material produced in that laboratory. Three bold capital letters (e.g. **MFI**) designate a particular zeolite framework type with a specific topology (see Ref. 17).

3.2 The zeolite structure

The framework of a zeolite may be described as a three-dimensional 4-connected net consisting of corner sharing $[TO_4]$ tetrahedra (T = Si, Al, B, P,...) which are linked via common oxygen bridges. Usually, two tetrahedral nodes always share one oxygen atom, but some interrupted frameworks where some of the oxygen atoms are bonded to only one T-atom are also known. There is an infinite number of different possible tetrahedral framework topologies [18]. In April 2017, 232 different framework structure types are acknowledged by the zeolite structure commission of the IZA. As a reference for the zeolite framework types (designated by a three letter code), see Baerlocher et al. [17].

If examining all the different framework types, it becomes obvious that certain structural fragments are common to several distinct zeolite frameworks. These structural fragments are either "secondary building units" (SBU) or "composite building units" (CBU).

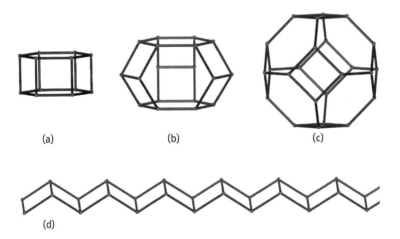

Fig. 3.5: Selected examples of CBUs: double 6-ring, cancrinite cage, sodalite cage, double zigzag chain.

SBUs consisting of up to 16 interconnected TO_4 tetrahedra are chosen such that an entire framework is made up of this particular type of SBU only. For example, the SBU

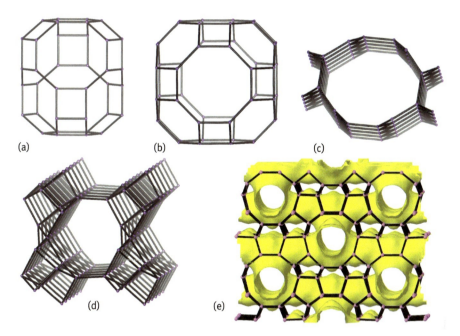

Fig. 3.6: Zeolite pores of different geometry: (a) chabazite cage, (b) alpha cage of the **LTA**-type zeolite, (c) cannel-like pore of zeolite ZSM-12, (d) cannel-like pore of zeolite RUB-41, (e) intersecting channels of the **MEL** framework.

designated as "4-2" is a unit consisting of two 4-rings being interconnected along one common edge. From this SBU, the framework types **OFF** (offretite), **MAZ** (mazzite), **GME** (gmelinite), and others can be constructed.

CBUs, e.g. double 6-ring, cancrinite cage, sodalite cage, or double zigzag chain (Fig. 3.5), also may appear in several different framework structures. These units are very useful in identifying structural relationships between framework types.

For an overview, please refer to the *Atlas of the Zeolite Framework Types* [17]; detailed lists are published in the *Compendium of Zeolite Framework Types. Building Schemes and Type Characteristics* by van Koningsveld [19].

The most interesting and characteristic feature of a zeolite is its (micro-)porosity. The pores might be cage-like (with a finite extension of ca. 5 to 20 Å) or channel-like (of infinite length) (see Fig. 3.6).

With respect to possible applications, particular structural features are of interest: the pore openings (i.e. the free diameter of a channel or of a cage-window), the total pore volume, and the dimensionality of the pore system (zero- to three-dimensional).

Zeolites possessing 8-rings as the largest pore openings are regarded to be small-pore zeolites possessing free pore diameters of ca. 4.0 Å, whereas a 10-ring pore zeolite represents a medium-sized pore with free diameter of ca. 5.5 Å. Zeolite frameworks having 12-ring pores (ca. 7.4 Å) are designated as large-pore zeolites, and a few synthetic zeolites are known to possess extra-large pores (mainly 14-, 16-, 18-rings).

Some zeolites like afghanite, $Ca_{9.8}Na_{22}[Al_{24}Si_{24}O_{96}/Cl_2/(SO_4)_{5.3}/CO_3] \cdot 4\,H_2O$, contain cage-like voids with 6-rings as largest "windows". They may be regarded as clathrate compounds with a zero-dimensional pore system since occluded molecules cannot pass through these windows without being broken down into fragments.

The pore volume can be occupied by various ions, atoms or molecules. Inorganic cations like Na^+, K^+, Ca^{2+} etc. can be

(a) coordinated by framework oxygen only, as for example Na^+ in a double 6-ring,
(b) coordinated by two framework oxygen at approximately opposite sides with additional water molecules,
(c) bound to framework oxygen atoms on one side with water molecules completing the coordination sphere at the other side,
(d) surrounded by water molecules only.

Occluded organic cations or molecules often fill the complete cage-like void or a complete section of a channel-like pore and form van der Waals contacts or hydrogen bonds with the oxygen of the framework (Fig. 3.7).

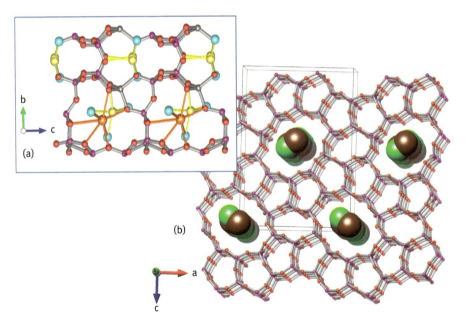

Fig. 3.7: Typical siting of inorganic and organic cations in zeolite structures: (a) hydrated Na^+ (yellow) and K^+ (orange) attached to the oxygen of the framework of zeolite RUB-17 (Si: purple, O: red, Zn: gray, water: light blue), (b) space filling molecule in the 10-ring channel of zeolite ZSM-48.

Water molecules can be part of the coordination sphere of a cation, can be bound to the walls of a hydrophilic framework, or can just fill free space in the pore system of the zeolite with contacts to other water molecules only.

3.3 Natural occurrence of zeolites

Natural zeolites are widespread in a variety of geological environments. They generally form under very mild conditions. Sedimentary rocks are the host systems of most zeolite occurrences. The environments enclose saline, alkaline lakes, soils and land surfaces, deep see sediments and low-temperature open and closed tephra systems. Zeolites are typically formed under alkaline conditions (pH: 7–10), volcanic glass being the major source of the zeolites.

The most abundant zeolites are analcime, clinoptilolite, heulandite, laumontite, and phillipsite. Chabazite, erionite, mordenite, natrolite, and wairakite are somewhat less frequently observed in nature [20].

Zeolites are also formed during burial diagenesis and low-grade metamorphosis. Again, zeolitization occurs predominantly in volcanoclastic material at low temperature (ca. 50–200°C) and low pressure (depths of 1–5 km). Among the mineral facies classification of metamorphic rocks, the zeolite facies covers the metamorphism at the lowest temperatures and pressures and represents the transition between diagenesis and the prehnite-pumpellyite facies. Often, a sequence of typical zeolite minerals with increasing crystallization temperature is observed: clinoptilolite and mordenite => analcime and heulandite => laumontite and wairakite.

Interesting with respect to envisaged zeolite syntheses is the zeolite formation by hydrothermal alterations. In this case, the crystallization of minerals or the alteration of existing mineral precursors occurs in the presence of hot solutions whose temperature originates from a close-by magmatic body. The composition of the solution is mainly influenced by the host rocks in which the zeolites form [21].

Hydrothermal zeolites appear predominantly in volcanic arcs, ocean ridges or hot spots. With respect to the conditions under which these zeolites formed, it is instructive to take a look on typical temperatures and pH values of the waters of active geothermal areas. The temperatures vary in a range of ca. 50–275°C and the pH values are about 8–9.5. These aqueous solutions are rich in SiO_2, Na^+, K^+, Ca^{2+} CO_2, SO_4^{2-} but vary considerably concerning the Cl^- content.

Nature presents us a large variety of different zeolites. The web site of the commission of IZA currently lists 67 different zeolite minerals [6]. The differences include in particular the framework structure, the chemical composition of the framework and the type of the extra-framework cations.

3.3.1 Structural variability of the alumosilicate zeolites

Zeolites in nature possess quite different framework structures. The rare mineral faujasite (first described in 1842 [22]) has a framework consisting of different types of cage-like void (Fig. 3.8). Faujasite has a very high porosity, the volume accessible by water molecules amounts to 27.4% [17b]. It is also a large pore zeolite with 12-ring pore openings which allow molecules with a kinetic diameter of up to 7.4 Å to pass through. In contrast, the structure of phillipsite is characterized by a three-dimensional pore system of channel like pores. This narrow pore zeolite possesses 8-ring pore openings with a free diameter of only about 3.6 x 4.2 Å and an accessible volume of only 9.9% [17b].

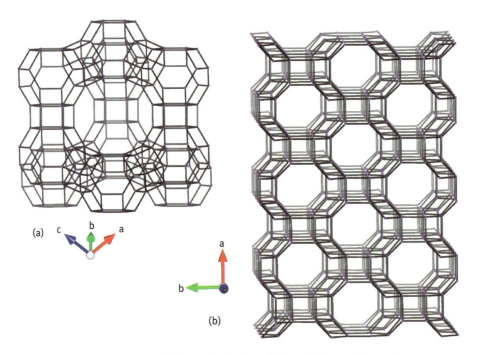

Fig. 3.8: Framework structures of (a) faujasite (**FAU**) consisting of D6Rs, sodalite- and super-cages and (b) phillipsite (**PHI**) possessing channel-like pores. Oxygen atoms are omitted for clarity.

3.3.2 Compositional variability of the framework

Although most natural zeolites possess alumosilicate frameworks, some zeolite minerals have significantly deviating framework compositions: One example is melanophlogite, a very rare mineral described from localities such as Racalmuto, Italy. Melanophlogite has a microporous structure with cage-like voids which are occupied by small organic molecules and sometimes by sulfur compounds [23, 24]. The framework consists of pure SiO_2 leading to a unit cell composition of $[Si_{46}O_{92}] \cdot x\ (CH_4, N_2, CO_2)$. Nature shows us that zeolites with a neutral, all-silica framework are feasible if only neutral extra framework species are occluded in the pore volume.

Another interesting mineral is gaultite, $Na_4[Zn_2Si_7O_{18}] \cdot 5\ H_2O$, from the type locality Poudrette Quarry, Mont Saint-Hilaire, Quebec, Canada [25]. In this case, trivalent Al^{3+} is replaced by Zn^{2+} in the silicate framework proving that M^{2+} cations can be part of stable zeolite frameworks.

Natural zeolite-like materials are not even limited to silicates: Weinebeneite is a secondary mineral in a spodumene pegmatite in the Carinthian area of Koralpe, Austria [26]. It is a microporous beryllophosphate with formula $Ca_4[Be_{12}P_8O_{32}(OH)_8] \cdot 16\ H_2O$. Here, Al as well as Si are completely replaced by Be and P indicating that it might be possible to produce zeolite-like materials with exotic types of T-atoms which, nevertheless, possess similar structural features and a similar stability as alumosilicate zeolites.

3.3.3 Compositional variability of the cations occupying the pore volume

Moreover, nature provides us with a series of different minerals which have identical alumosilicate frameworks with respect to structure and composition, but differ concerning the charge compensating cations. Natrolite is the sodium end-member $Na_6[Al_6Si_9O_{30}] \cdot 6\ H_2O$, mesolite $Na_2Ca_2[Al_6Si_9O_{30}] \cdot 8\ H_2O$ is an intermediate form, and scolezite $Ca_3[Al_6Si_9O_{30}] \cdot 9\ H_2O$ represents the calcium end-member. In a similar way, the zeolite heulandite has been found as sodium, potassium, calcium, strontium or barium dominated mineral. In all cases a very complex mixture of cations is present, e.g. the heulandite-Ca sample from Strathclyde, Scotland, has the empirical formula: $Ca_{3.57}Sr_{0.05}Ba_{0.06}Mg_{0.01}Na_{1.26}K_{0.43}[Al_{9.37}Si_{26.7}O_{72}] \cdot 26\ H_2O$ (webmineral.com). These observations prove that cations in zeolites can largely replace each other. It also suggests that the composition of natural zeolites may change considerably after crystallization if the zeolites later are exposed to a changing environment.

3.4 Applications

Zeolites are used in many areas: in industry, agriculture, transportation, or for domestic use. Three important properties lead to the three main applications: ion exchange properties, adsorption properties, and the ability to act as a catalyst or catalyst support. Natural zeolites and synthetic zeolites have their distinct domains of application; they hardly compete with each other. Low-cost natural zeolites which always contain other minerals as impurities have found applications in wastewater treatment to take up mine and industrial wastes, sewage treatment, as soil conditioner and animal feed supplement, or in the construction industry to produce light-weight concrete or in the form of altered (zeolitized) tuff as low-density construction stone. The most commonly mined natural zeolites are clinoptilolite, chabazite, and mordenite. Natural zeolites have also been used for decontamination in the case of nuclear power plant disasters: Tons of zeolites were used in the clean-up after the Chernobyl and the Fukushima nuclear disaster to take up radioactive material [27].

Synthetic zeolites are produced as well-defined pure materials. They find application for purposes where very specific and uniform properties are required.

3.4.1 Ion exchange

Cations of zeolites with open frameworks are often readily exchanged for other cations in aqueous solution. The main application of zeolites as ion exchangers (in particular, the synthetic low-silica zeolite A) is given for water softening (Na ⇔ Ca, Mg) in commercial detergents.

An advantage of zeolites over resins as ion exchangers is the selectivity with respect to the type of cation exchanged. For example, in the case of zeolite A, $Na_{12}[Al_{12}Si_{12}O_{48}] \cdot 27\,H_2O$, D. W. Breck determined the selectivity for two series of cations, M^+ and M^{2+}, as follows: Ag > Tl > Na > K > NH_4 > Li > Cs and Zn > Sr > Ba > Ca > Co > Ni > Cd > Hg > Mg, respectively [28].

3.4.2 Adsorption

The shape-selective properties of zeolites lead to the ability to readily adsorb certain molecules from a gas or a liquid while other molecules are rejected or adsorbed reluctantly. The specific size and shape of the pores may hinder branched hydrocarbons to enter the pores while linear hydrocarbons may easily pass. This property can be adjusted by modifying the structure of the zeolite (the framework as well as the size and number of cations). Also, the chemical nature of the zeolite can be adapted to the separation process by choosing specific cations or adjusting the polarity of the framework. This leads to adsorption sites of the desired strength in the structure.

Among a wide range of molecular sieving processes two important applications are mentioned here.

There is considerable interest in obtaining pure *p*-xylene due to its industrial relevance. Xylenes are often produced in a first step as a mixture of isomers. *p*-xylene (kinetic diameter: 5.8 Å) can be separated from the *m*- and *o*-xylene isomers (kinetic diameter: 6.8 Å) by the use of shape-selective zeolite X, zeolite Y (both of framework type **FAU**), or **MFI**-type zeolites [29].

Oxygen needs to be produced in large quantities, e.g. for applications in medicine. Oxygen concentrators use specific zeolite materials to adsorb nitrogen while oxygen (and argon) can pass straight through. When nitrogen and oxygen molecules come close to exposed cations of the zeolite, charge induced dipoles form. Since nitrogen is more polarizable than oxygen, the zeolite preferentially adsorbs nitrogen. To preserve exposed cations (having an unsaturated coordination sphere), the zeolite and the air stream have to be free of water. Concentrators operate on the principle of pressure swing adsorption (PSA) and produce an oxygen purity of ca. 95%. Among zeolite materials, LiAgX shows a particularly good performance in removing nitrogen [30].

3.4.3 Catalysis

Zeolites are very useful catalysts for a large number of chemical reactions, especially in the petroleum and chemical industry. Zeolites have three properties which make them particularly suitable as catalysts:

1. Zeolites possess very large internal surfaces which allow for a large number of molecules to undergo catalytically activated reactions at the inner surface at the same time. For example, the faujasite-type zeolites RE-Y (rare earth-Y) and US-Y (ultra-stable-Y) having one of the lowest framework densities among all silicate zeolites and possessing an internal surface of ca. 800 m^2/g are used for shape-selective catalytic cracking of high-mass hydrocarbon of crude oil in the petroleum refining industry [31].

2. The composition of many zeolites can be varied in a wide range. This applies both to the cations in the pore volume (adjustable in an ion exchange process) and to the T-atoms of the framework which can be adjusted to a certain extend during the crystallization of the zeolite. Cu-exchanged high-silica chabazite zeolites are very efficient catalysts for NOx decomposition (in a reaction with urea) and are employed on diesel vehicles as effective catalyst for NOx removal from the exhaust gas [32]. Hydrogen-exchanged zeolites, e.g. H-ZSM-5 (zeolite framework type **MFI**), which possess a very high acidity are used in crude oil cracking [33]. Ti-ZSM-5 with Ti replacing a small amount of Si atoms at T-sites serve as oxidation catalysts in the production of propylene oxide from propylene [34].

3. As zeolites are crystalline materials, the pore dimensions (which are in the same order as the dimensions of simple molecules) are well defined and identical all over the material. The particular pore system has a selectivity with respect to the uptake of certain molecules which then may react inside the pore volume. The shape selectivity of zeolite catalysts often leads to a lower amount of undesirable byproducts compared to other catalysts.

Other applications include
1. Desiccants: Low-silica zeolites are extensively used in this field due to their high affinity for water. For example, the "zeolite dishwasher" is equipped with a certain amount of zeolite material. After washing, the remaining water is adsorbed by the zeolite supporting the drying of the clean dishes. As a byproduct a substantial heat of adsorption is generated, increasing the temperature inside the dishwasher.
2. Energy storage and transformation: Heat storage systems exploiting the adsorption heat of water on zeolites, microporous aluminophosphates (AlPOs), or silica-aluminophosphates (SAPOs) have a large potential to utilize solar and geothermal energy as well as industrial waste heat. Zeolites are used for thermal energy storage, e.g. in the "Cool-keg" a self-cooling beer keg [35] or as a heat pump for a home heating system in combination with a gas adsorption heater [36].
3. The use as membranes for molecular sieving: A current industrial application of a zeolite membrane is the production of dry ethanol using K^+-exchanged zeolite A [37, 38]. Overviews on recent developments were published by Rangnekar et al. [39] and Feng et al. [40].

3.5 Synthesis of zeolites

By far, most zeolites (in quantity and also in variety) are produced utilizing the hydrothermal synthesis. These syntheses are predominantly performed in the temperature range of ca. 80 to 240°C. With respect to the temperature, the solvent, the chemical composition of the educts (Na, K, Ca; Si, Al, O) and the pH value, the hydrothermal synthesis has a close relationship to the growth of natural zeolites during hydrothermal alterations of volcanic rocks.

For the synthesis, sealed vessels are necessary to avoid the loss of the volatile compounds. In a temperature range below 100°C closed polypropylene bottles are sufficient to prevent water from evaporation. At higher temperatures, sturdy vessels are required to withstand the autogenous pressure generated mainly by the water content of the reaction mixture. These vessels are usually stainless steel autoclaves which often contain an inert Teflon liner to avoid the contact between reaction mixture and the steel.

Most zeolites are formed as thermodynamically metastable phases and may transform during prolonged synthesis time into more stable phases in accordance with Ostwald's rule of successive phase transformations. One example is the crystallization sequence: amorphous gel => zeolite Na-Y (zeolite framework type **FAU**) => zeolite Na-P (**GIS**) [41].

Not only the synthesis time is important but also many other synthesis parameters have an impact on the formation of a specific zeolite phase. These parameters include the alkalinity of the reaction mixture, the Si/Al ratio, temperature, water content, the presence and type of organic compounds, the duration of heating, aging of the reaction mixture, and the agitation of the mixture during crystallization.

With respect to the synthesis conditions the products might be distinguished into
(a) "Low-silica zeolites" like zeolite A, $Na_{12}[Al_{12}Si_{12}O_{48}] \cdot 27\ H_2O$,
(b) "High-silica zeolites" like ZSM-8 $(C_{12}H_{39}N)_4Na_{0.9}[Al_{2.9}Si_{93.1}O_{192}] \cdot 4\ H_2O$
(c) "Zeolite analogues" like AlPOs, MeAPOs, germanates, nitridophosphates, and others.

Low-silica zeolites are synthesized from reaction mixtures containing high contents of Al and, additionally, an inorganic base like NaOH or KOH. Crystallization takes place at relatively low temperatures (80–150°C). Typical reaction mixtures contain a silica source (e.g. silica gel), a base to mobilize the silica source (NaOH, KOH, LiOH,...), a cation source (e.g. NaCl, $SrCl_2$, $CaCl_2$,..., which is sometimes identical with the base), an aluminum source ($NaAlO_2$, AlOOH, $Al(OH)_3$,...), and water.

High-silica zeolites are synthesized with a low content of Al (or sometimes B, Be, Ga,...), a base to mobilize the silica source, and usually with a specific organic compound, the so-called organic structure directing agent (OSDA). OSDAs are predominantly quaternary ammonium cations or amines. Typical synthesis temperatures are ca. 140 to 220°C.

Zeolite analogues are also predominantly obtained by the hydrothermal synthesis route. However, some zeolite analogous are made by other, very specific synthesis routes. For example, microporous AlPOs can be synthesized using ionic liquids as solvents [42, 43].

Recipes to synthesize specific zeolites can be found in the atlas on *Verified Syntheses of Zeolitic Materials*, published on behalf of the Synthesis Commission of the IZA [44].

There are other, less frequently used routes for zeolite synthesis which will not be presented in this review. An overview on many recent developments is given by Wilson [45].

3.6 Recent developments

3.6.1 Synthesis of zeolite materials

Due to its unique topology, each zeolite framework type has its distinct properties which might lead to novel or improved applications. Therefore, we experience a continuing search for new zeolites with so far unknown compositions, framework topologies and pore geometries. Below, some selected examples of zeolite synthesis procedures will be highlighted.

3.6.1.1 Hydrothermal synthesis using OSDAs

The use of OSDAs has demonstrated to be a very successful synthesis route which led to a large number of new zeolite framework types [46–50]. The structure directing effect can be demonstrated on the basis of the most simple organic cation, the tetramethylammonium cation (TMA$^+$), which was the very first organic cation introduced to zeolite synthesis by Barrer and Denny in 1961 [51] (Tab. 3.1).

Despite the large variety of different structures being formed in the presence of TMA$^+$, it is remarkable that most structures show a close geometric correspondence between the TMA cation (approximately sphere-like) and the various cavities occluding this cation in the structure of the silicate. TMA$^+$ has a high preference to stabilize structures with small cavities during crystallization, either as a complete cage in a zeolite framework or as a cup-like fragment of a cage in layer silicates.

In the system TMAOH/SiO$_2$/water (without further additives), silicates of different dimensionality have been synthesized depending on the temperature of synthesis: group silicates with isolated Si$_8$O$_{20}^{8-}$ anions (dimensionality: 0), layered silicates (dimensionality: 2) or framework silicates like RUB-22 (dimensionality: 3).

As said above, the zeolite synthesis is a complex process affected by various parameters. The particular structure type crystallizing with TMA$^+$ as the OSDA depends

Tab. 3.1: Collection of relevant crystalline products obtained by hydrothermal syntheses in the system TMAOH/SiO$_2$/water (+ additional compound) at different temperatures.

Synthesis temperature	Additional T-atom	Additional cation	Additional anion	Silicate anion	Product
20°C	–	–	–	Group	[52, 53]
Ca. 120°C	–	–	–	Layer	HUS-1 [54]
Ca. 140°C	–	–	–	Layer	RUB-15 [55]
160°C	–	–	–	Layer	RUB-55 [56]
160°C	–	–	–	Interrupted layer	RUB-20 [57]
160°C	–	–	–	Interrupted framework	RUB-22 [57]
160–200°C	–	–	–	Framework with random defects	Defective D3C [un-published results]
150°C	–	Na	–	Layer	Helix layered silicate [58]
180°C	–	K	–	Layer	ERS-12 [59]
100–150°C	Al	–	–	Framework	FAU [51]
100–150°C	Al	–	–	Framework	LTA [51]
130°C	Al	–	–	Framework	GIS [60]
170°C	Al (B, Fe, Ga)	–	–	Framework	RUT [61]
180°C, 200°C	Al	–	–	Framework	SOD [51, 62]
250°C	Al	–	–	Framework	PHI [51]
160–220°C	B	–	–	Framework	RUT [61, 63]
180°C	–	–	F	Framework	AST [64]
190°C	–	–	F	Framework	GIS [65]
190°C	–	–	F	Framework	MTN [65]

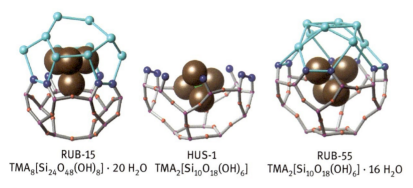

RUB-15
TMA$_8$[Si$_{24}$O$_{48}$(OH)$_8$] · 20 H$_2$O

HUS-1
TMA$_2$[Si$_{10}$O$_{18}$(OH)$_6$]

RUB-55
TMA$_2$[Si$_{10}$O$_{18}$(OH)$_6$] · 16 H$_2$O

Fig. 3.9: Cup-like voids of different silicate layers occupied by tetramethylammonium ions. Sections of the structures of RUB-15, HUS-1, and RUB-55. Hydrogen bonds and water molecules: light blue, OH-groups: blue.

on the temperature, the Si/TMA$^+$-ratio and on the type of additives (Na$^+$, Al^{3+}, F$^-$) being part of the reaction mixture.

In the case of the layered silicates HUS-1, RUB-15, and RUB-55, TMA$^+$ occupies cup-like voids of the silicate framework which represent half-cages of the sodalite structure and which include silanol (Si-OH) and siloxy (Si-O$^-$) groups (Fig. 3.9). These silicate half-cages are completed by hydrogen-bonded water molecules.

In the case of framework structures, charge compensation for TMA$^+$ is usually achieved by Me^{3+} ions (Al, Fe, B, Ga) replacing Si^{4+} at T sites. This allows to form complete silicate cages around the TMA$^+$ (Fig. 3.10).

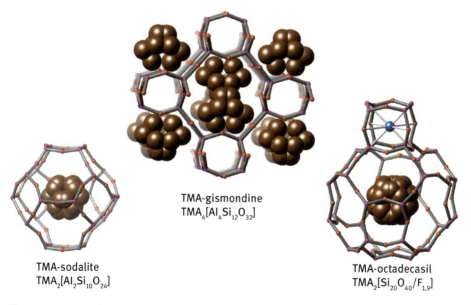

TMA-sodalite
TMA$_2$[Al$_2$Si$_{10}$O$_{24}$]

TMA-gismondine
TMA$_4$[Al$_4$Si$_{12}$O$_{32}$]

TMA-octadecasil
TMA$_2$[Si$_{20}$O$_{40}$/F$_{1.9}$]

Fig. 3.10: Sections of zeolite structures synthesized with TMA$^+$ as the OSDA: (a) sodalite (**SOD**), (b) gismondine (**GIS**), (c) octadecasil (**AST**). The TMA$^+$ cation is in all cases rotationally disordered.

RUB-22, in contrast, has a microporous framework with ordered defects (i.e. "hydroxyl nests" surrounding a vacancy □(OH-Si)$_3$($^-$O-Si)) to achieve charge compensation since no Me^{3+} was available in the synthesis system TMAOH/SiO$_2$/water (Fig. 3.11).

If the cation Na$^+$ **plus** Al^{3+} as a potential T-atom are added to the system TMAOH/SiO$_2$/water, zeolites of many different framework types are obtained such as **FAU**, **LTA**, **MAZ**, **ERI**, **SOD**, and **GIS** [28]; the geometric correspondence between TMA ion and the framework is less obvious.

It is surprising that the use of tetramethylammonium, which has extensively been used in hydrothermal synthesis for more than 55 years, led recently in the very simple system TMAOH/SiO$_2$/water to new zeolite related layered silicates, RUB-20 [57] and RUB-55 [56] and to a new zeolite, RUB-22 [57].

This indicates that it will still be possible to obtain new zeolites with novel framework types by a suitable combination of synthesis parameters even if common OSDAs are used.

While predominantly commercially available organic compounds have been used as OSDAs for zeolite synthesis, in more recent times, specifically designed OSDAs

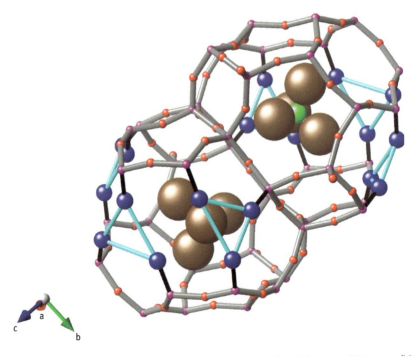

Fig. 3.11: Peanut-shaped double-cage of RUB-22. Hydrogen bonds between OH-groups (blue spheres) of the hydroxyl nests are indicated in light blue.

have been used as well [66, 67]. One example is the recent synthesis of the new silicogermanate PKU-20 [68]. The use of custom-built OSDA triethylisopropylammonium hydroxide (synthesized from isopropylamine and iodoethane) in combination with fluoride anions allowed to obtain PKU-20 from a SiO_2 and GeO_2 containing hydrogel. In this way, it was possible to obtain a new zeolite that has a novel and unique framework constructed from an alternative stacking of sti (stilbite-type) and asv (ASU-7-type) layers.

3.6.1.2 Zeolite synthesis using monoatomic inorganic structure directing agents

Low-silica zeolites are usually synthesized without OSDAs but in the presence of alkali-metal hydroxides meant to help mobilizing the silica source. It was observed that reaction mixtures containing for example K^+ favors the formation of specific zeolite structure types (e.g. **EDI, KFI, LTL, MER**) which is in contrast to Na^+ containing mixtures which led to the crystallization of a large variety of framework types: **NAT, CAN, CHA, EMT, FAU, FER, GIS, LTA, MAZ, MOR, SOD** and others [69]. Alkali-metal cations Li, Na, K, Rb, and Cs are attributed a limited structure directing role as an inorganic structure-directing agent (ISDA) which is believed to be related either to the "naked" Me^+ ion stabilizing, for example, the cancrinite cage, or to an ordering of the water structure around the cation which in turn acts as a "template" during the formation of the alumosilicate framework.

Li cations are notably interesting constituents for the synthesis of new zeolite framework types, although they have less frequently been used compared to other cations like Na^+ and K^+. Li^+ can be incorporated in the zeolite structure as a charge balancing cation in the pore volume and by that may act as a ISDA, but Li^+ can as well be part of the silicate framework as a $[LiO_4]$-tetrahedron [70, 71]. In the past, Li has been used in the system Li_2O-Al_2O_3-SiO_2-H_2O leading to zeolites which contain Li only in the pore system, as for examples in synthetic Li-A(BW) $Li_4[Al_4Si_8O_{24}] \cdot 4\ H_2O$ [72]. Also, a zeolite mineral, bikitaite, $Li_2[Al_2Si_4O_{12}] \cdot 2\ H_2O$, exists which contains Li in the channel-like voids [73].

If, however, Al^{3+} or other cations which might occupy T-sites in a silicate framework, are excluded from the reaction mixture during synthesis, Li is in some cases incorporated in the zeolite framework as a T-atom. Moreover, $[LiO_4]$-tetrahedra can form strain-free 3-rings with $[SiO_4]$-tetrahedra expanding the possibilities to form novel framework types. A recent example is RUB-12, $Li_4H_2[(Li_2Si_8O_{20})] \cdot 8\ H_2O$, which was synthesized under hydrothermal conditions in the system $LiOH/SiO_2/H_2O/OA$ at 140–160°C [74]. As organic additives (OA) different amines were used as weak organic bases. RUB-12 has a unique structure with a one-dimensional pore system of non-intersecting 10-ring channels running parallel to the c-axis. The pore system is occupied by water molecules, protons and Li cations. The silicate framework can

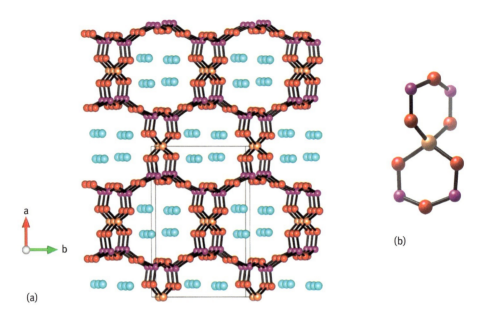

Fig. 3.12: (a) Structure of RUB-12. (b) Spiro-5-unit.

be described as being constructed exclusively from interconnected spiro-5-units with tetrahedral Li⁺ in the center of the units (Fig. 3.12).

RUB-12 is related to two other lithosilicate zeolites: RUB-23, $Cs_{10}(Li,H)_{14}[Li_8Si_{40}O_{96}] \cdot 12\,H_2O$ [75] and RUB-29, $Cs_{14}Li_{24}[Li_{18}Si_{72}O_{172}] \cdot 14\,H_2O$ [76]. All three zeolites contain spiro-5 units with Li at the central spiro position.

The replacement of one Si^{4+} by a Li^+ cation at a T-site of a silicate framework introduces three negative charges which are difficult to compensate by extra-framework cations. For comparison, if Si^{4+} is partly replaced by Al^{3+} (which is standard practice) only one negative charge per Al is introduced to the framework. This is probably one reason why only few zeolites containing Li at T-sites are known. RUB-12 has an extremely high framework charge of −0.6/T-atom, even if compared to typical zeolitic ion exchangers like zeolite A, $Na_{12}[Al_{12}Si_{12}O_{48}] \cdot 27\,H_2O$, with a framework charge of only −0.5/T-atom.

3.6.1.3 Zeolite synthesis using highly condensed ISDAs: the seed-assisted OSDA-free synthesis

Adding seed crystals to reaction mixtures has been common practice in zeolite synthesis since more than 50 years. The seeds (approximately 1% of the total silica content) were used to accelerate the crystallization – e.g. to avoid the induction period prior to crystal growth which is typical for zeolite crystallization. The seeds, however, led to zeolites of the same framework types as being obtained if no seeds were added.

More recently, seeds (ca. 10% of the silica source) have been used to direct the synthesis towards the crystallization of specific zeolites which had not been obtained without OSDAs before. Pioneering work has been published in a series of papers on this subject by Xiao and coworkers [77, 78]. A comprehensive overview on this synthesis route is presented by Iyoki et al. [79]. The seed-assisted OSDA-free synthesis has some advantages: there is no expensive OSDA required, seeds enhance the crystallization rate, the zeolites do not require calcination to remove the organic compounds, and there is less problematic waste after synthesis.

Beta zeolite has, so far, been investigated most extensively in connection with seed-assisted OSDA-free synthesis. Recently, Zhang et al. described the **seed-d**irected **s**ynthesis of Beta at particular low temperature (120°C). The obtained Beta-SDS120 has an improved quality and is free of impurities. It has a higher thermal stability than the Beta zeolite synthesized at 140°C and also has a higher surface area and micropore volume than Beta-SDS140. It is assumed that Beta-SDS120 has much less framework defects than Beta-SDS140 [80]. The authors of this study were, moreover, able to decrease the amount of Beta seeds down to 1.4%.

The seed-assisted OSDA-free synthesis of zeolite Beta results in products with a very good catalytic performance. The as-made materials possess a high density of active sites suitable for ethylation of benzene as an example. After various post-synthesis treatments (e.g. modifying the Si/Al ratios), beta zeolite is also useful in acylation reactions [81]. The authors state that "the ability to manipulate the framework aluminum content in a very broad range, while maintaining structural integrity, proves that template-free-beta zeolites constitute a powerful toolbox for designing new acid catalysts" [81].

3.6.1.4 Zeolite synthesis by topotactic condensation of hydrous layer silicates

A completely different synthesis route to obtain new zeolite framework types is based on a solid-state reaction using **h**ydrous **l**ayer **s**ilicates (HLSs) as precursors. In comparison to the hydrothermal synthesis, the topotactic conversion of layer silicates into microporous framework materials has been used rarely (Fig. 3.13). Zeolite materials of the MWW framework type were the first successful examples which were described 20–30 years ago [82–86]. Now, there are several high-silica zeolites with different framework topologies which were obtained by a successful topotactic condensation of silicate layers (in the following, the type materials are named): EU-20/EU-20b (framework type **CAS**) [87, 88], CDS-1 (**CDO**) [89], siliceous ferrierite (**FER**) [90], MCM-22 (**MWW**) [84], NU-6(2) (**NSI**) [91], RUB-41 (**RRO**) [92], RUB-24 (**RWR**) [93], guest-free silica sodalite (**SOD**) [94].

A review on the synthesis of zeolites by the topotactic condensation using hydrous layer silicates as precursors was recently published by Marler and Gies [95].

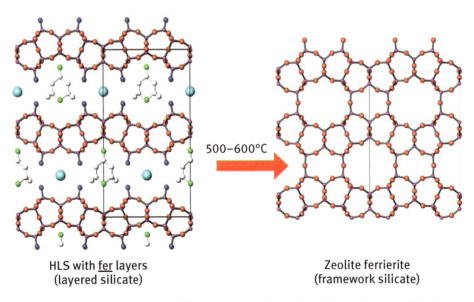

HLS with fer layers
(layered silicate)

Zeolite ferrierite
(framework silicate)

Fig. 3.13: Schematic representation of the topotactic condensation of silicate layers, transforming a HLS into a zeolite.

With respect to zeolite syntheses based on silicate layers as building blocks the ADOR (**A**ssembly-**D**isassembly-**O**rganization-**R**eassembly) and the "Inverse Sigma Transformation" routes are particularly interesting [96, 97]. Here, the starting material is a zeolite which can selectively be disassembled into a layered material. The parent zeolite is a germanosilicate containing germanium at specific sites of the structure. The germanium can be leached out without destroying the silicate layers. The resulting layered silicate can subsequently be organized in different ways (e.g. by intercalation of silicon-containing species between the layers), and finally can be reassembled into zeolites with new topologies by calcination.

In a similar way, Zhao et al. performed a rearrangement of layered building blocks starting with the layered silicate RUB-36 as a precursor containing ferrierite-type silicate layers (fer layers) [98]. While a direct condensation of layers by a mere heating led to a zeolite of the CDO-type, the temporary swelling of the structure with cetyltrimethylammonium cations destroyed the strong interlayer hydrogen bonds between neighboring layers and led to a rearrangement of the layer stacking. After deswelling, the fer layers were condensed by heating (550°C) to form pure silica FER-type zeolite. This concept can probably be applied to other layered silicates and may result in the formation of new zeolite framework types.

The fer layer is in particular interesting since this layer seems to form easily under hydrothermal conditions. Marler et al. described the synthesis of five different hydrous layer silicates (HLSs) containing fer layers (obtained from mixtures of silicic acid, water, and different tetraalkylammonium or tetraalkylphosphonium hydroxides) and their condensation products [99]. The fer layers are well suited for condensation reactions forming zeolite frameworks since the terminal silanol groups of the layers have a distance of at least 7.4 Å. This precludes the formation of **intra**-layer hydrogen bonds but allow for **inter**-layer bonds between silanol groups of neighboring fer layers. The hydrogen bonds predefine the positions where the condensation forming Si-O-Si bridges takes place. **Intra**-layer hydrogen bonds would lead to highly distorted structures in the calcined products.

The conversion of layer silicates – either by the ADOR method or by a direct topotactic condensation – opens up the possibility to obtain zeolite materials which are not accessible by hydrothermal synthesis: So far, topotactic condensation produced four new zeolite framework types (**CDO, NSI, RRO, RWR**) and the ADOR method two (**OKO, PCR**).

The condensation of HLSs forms pure silica framework structures. In order to introduce catalytically active centers, some Si atoms of the layered precursor can be replaced by Al or B atoms. The synthesis of zeolites Al-RUB-41 [100] and B-RUB-41 [101] from layered precursors (Al- and B-RUB-39) prove that the topotactic **condensation** of HLSs can produce microporous materials with some potential to act as catalysts with shape-selective character [100].

3.6.2 Applications

3.6.2.1 Catalysis

Cu-exchanged Beta zeolite is a very good catalyst for selective catalytic reduction (SCR) of NOx with NH_3. SCR is one of the most efficient methods to reduce NOx from exhaust gas streams of diesel engines.

Recently, Cu (or Fe)-exchanged Beta obtained from syntheses without the use of OSDAs (TF-Beta, TF = **t**emplate-**f**ree) has also been tested as SCR catalyst. Compared to conventional Cu-Beta with a typical Si/Al ratio of ca. 19, the Cu-TF-Beta (Si/Al ratio = 4) shows a superior performance at low temperatures (around 150°C). This is beneficial to decrease the total amount of NOx during a conventional driving cycle. The superior low-temperature performance is explained on the basis that isolated Cu cations in TF-Beta can more easily be reduced [102].

Compared to the (monocationic) Cu- or Fe-exchanged TF-Beta catalysts, the Cu(3.0)-Fe(1.3)-TF-Beta catalyst shows even better SCR performance at low-temperatures. Isolated Cu^{2+} and Fe^{3+} ions which are located at the exchange sites are assumed to be the active species for the efficient catalytic reaction at low-temperature. The coexistence of Fe^{3+} and Cu^{2+} improves the dispersion state of the Cu ions compared to pure Cu-TF-Beta. In addition, the Cu(3.0)-Fe(1.3)-TF-Beta exhibits a significant improvement of sulfur resistance. SO_2 being a common component in exhaust gas streams of diesel engines if low-quality fuel is used easily deactivates the catalyst [103].

In summary, it can be concluded that Cu/Fe-exchanged TF-Beta is a very promising catalyst for NH_3-SCR of NO_x.

3.6.2.2 Energy storage

3.6.2.2.1 Heat storage

For the use of renewable energies like storage of solar heat, the development of new and improved materials is necessary. Zeolites are very promising materials for these purposes since they can reversibly be hydrated and dehydrated and possess adsorption heats suitable for such applications (charging temperatures: >180°C) [104].

Among zeolites, Li containing low-silica X zeolite (Li-LSX) with unit cell composition $Li_{96}[Si_{96}Al_{96}O_{384}] \cdot x\ H_2O$ is a promising candidate because of its high water adsorption capacity [105]. To localize the preferred water sorption sites and to optimize the sorption properties, the structure of partially and fully hydrated Li-LSX with water loadings of x = 8, 16, 32, 48, 96, and 270 H_2O/D_2O were analyzed based on neutron and synchrotron powder data [106].

In the water-free state, the Li-cations occupy three different sites in the structure of Li-LSX: two sites in the centers of the 6-ring windows of the small sodalite cage (SI and SII) and one in the large supercage in front of the 4-ring of the sodalite cage (SIII). This Li cation site SIII has a very asymmetric and unfavorable coordination sphere (Fig. 3.14).

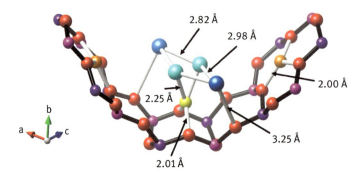

Fig. 3.14: Section of the structure of Li-LSX · 16 H_2O. Si: light purple, Al: dark purple, O: red, Li at site SIII: yellow, Li at site SII: orange, water at site W3: light blue, W4: blue.

The first water molecules added to the pore volume of the zeolite are adsorbed close to site SIII. At slightly higher loadings, the occupancy of Li^+ at SIII decreases with increasing water content. For the completely hydrated material, six different water sites were identified. It is interesting to note that the water adsorption sites are

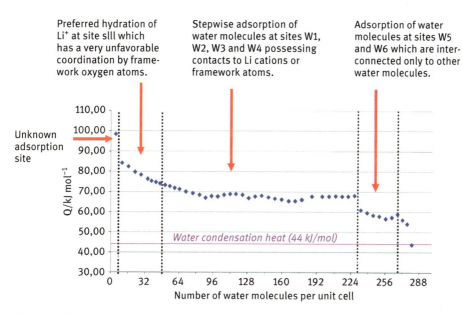

Fig. 3.15: Differential molecular heat of adsorption of Li-LSX/water (Rathmatkariev, unpublished results) with a tentative assignment of water adsorption sites in the structure of Li-LSX.

dominated by the silicate framework and not by the cations. Nearly all water molecules occupy the supercage; only in the fully hydrated stage 2.3 water molecules per unit cell were also located inside the sodalite cage.

Additionally, the differential molar adsorption heat of water/Li-LSX was determined. The energy values vary in several steps between 84 and 54 kJ/mol for the first and last water molecules adsorbed [107]. The comparison of the differential molar adsorption heat with the results of the structure analysis (Fig. 3.15) discloses how to improve the zeolite material to obtain a higher quantity of optimal adsorption sites for the envisaged applications (here, an adsorption heat of ca. 60 kJ/mol). The features to be modified comprise (a) the charge of the silicate framework (i.e. Al/Si ratio), (b) the type of the cation, and (c) impregnations with additional ions (see Fig. 3.16a–c).

The impregnation of a zeolite with additional ions is an interesting approach to modify the hydrophilicity of the material: A commercial zeolite Na-Y (Si/Al = 2.6) was impregnated with NaBr at 700°C under high vacuum according to a procedure of Seidel et al. [108]. Seidel proposed a structure model for NaBr-impregnated Na-Y (Si:Al = 2.4) based on ^{23}Na MAS NMR spectroscopic data that the Br$^-$ ions occupy the sodalite cages and forms [Na$_4$Br] clusters. Now, a structure analysis on Br@Na-Y (Si:Al = 2.6) confirmed this model (Fig. 3.17) [109]. The [Na$_4$Br] clusters which are present in about 90% of the sodalite cages have a positive influence on the desorption temperature of the water (Fig. 3.16c) [109, 110].

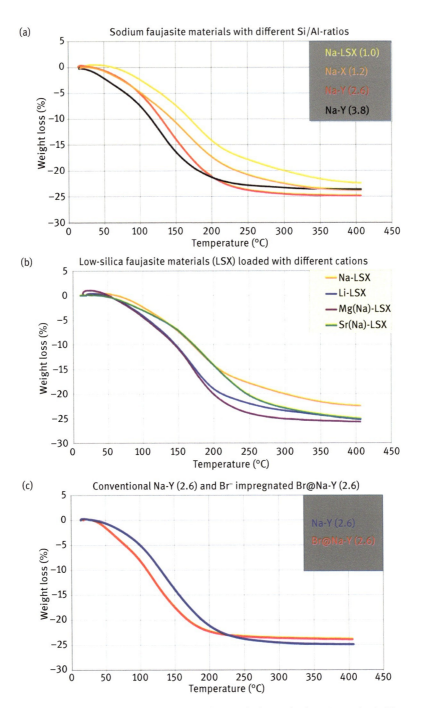

Fig. 3.16: (a) TG curves of Na-LSX (Si/Al = 1), Na-X (1.2) Na-Y (2.6), and Na-Y (3.8). (b) TG curves of low-silica LSX samples Li-LSX, Na-LSX, Mg(Na)-LSY, and Sr(Na)-LSX. (c) TG curves of conventional Na-Y (Si/Al = 2.6) and Br@Na-Y (Si/Al = 2.6).

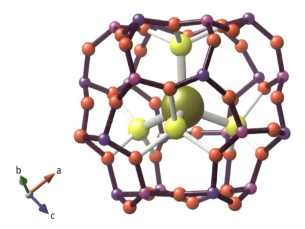

Fig. 3.17: The [Na$_4$Br]$^{3+}$ cluster in the sodalite cages of Br@Na-Y. Na: yellow, Br: olive-green.

3.6.2.2.2 Storage of mechanical energy

A system consisting of a porous solid and a non-wetting liquid has the ability to accumulate, transform, restore, and dissipate energy [111]. Water is a very suitable liquid in such a system: It is inexpensive, non-toxic, has a low viscosity, is easy to obtain, and possesses a high surface tension. The small water molecules are able to access even small micropores. To use water as a non-wetting liquid, a hydrophobic microporous solid is required – all-silica zeolites (zeosils) match this requirement perfectly. During the forced intrusion of water into the pore system, the mechanical energy is converted into an interfacial energy. The bulk water is split into a multitude of molecular clusters, creating new bonds with the (hydrophobic) silica framework and breaking the intermolecular H-bonds in liquid water. When the pressure is released, the system can spontaneously expel the water from the pore of the zeosil. Several water-zeosil systems have been investigated. They show quite diverse behaviors at intrusion-extrusion experiments depending on physicochemical and geometric parameters such as the "degree" of hydrophobic character, pore diameter, pore geometry (cages or channels), and dimensionality of the pore system.

While several channel- and cage-type zeosils show a behavior as molecular springs [112], RUB-41 which has a two-dimensional pore system of 8- and 10-ring channels behaves like a shock absorber of the reversible type [113]. During the intrusion-extrusion experiment, some defect sites of the silanol type are generated, decreasing the hydrophobicity and provoking that not all the water is expelled when releasing the pressure. Silica-ITQ-7 acts as a shock absorber as well, but the phenomenon is non-reversible. A considerable amount of defects is generated. ^{29}Si MAS NMR spectra revealed the breaking of Si-O-Si bridges after the intrusion [114]. In this case, PXRD patterns even showed a modification of the structure at the long-range order. The defective material, however, can be regenerated by calcination at 600°C.

In contrast, calcined silica-sodalite shows hardly any uptake of water, probably due to a very hydrophobic character and very small pore openings (6-rings only) [113].

3.7 Outlook

To our judgment, the use of zeolites will expand in two areas of application in particular: in heat storage and conversion and in environmental catalysis. The driving force to implement new processes in these two areas is the intent to save energy and to reduce the amount of waste products (green chemistry).

Zeolites can be modified in many ways: at first, a particular framework type possessing a distinct pore structure can be chosen from the 232 different framework types that are known [17], the framework composition and polarity can be adjusted to a certain extend during synthesis or post-synthesis treatment, the as-made zeolite can then be loaded with the desired catalytically active cations or nanoparticles, and additionally, it is possible to produce zeolites with a hierarchical pore system comprising, e.g. mesopores for fast diffusion of molecules and micropores to exploit the shape selectivity of the zeolite pores. This flexibility opens up a huge field of research to design zeolites according to the needs.

Heat storage systems based on zeolites and water have a large potential to make use of industrial waste heat of low temperature (ca. >200°C). Presently, hardly any of this waste heat energy generated as a by-product in industry or power-plants is used. Also, the thermochemical storage of solar heat is still in a state of testing. The energy which is made available by solar collectors is typically at an even lower temperature level (ca. 100°C). So far, the use of zeolites which have promising theoretical energy densities of ca. 180 kWh/m^3 suffer from a high charging temperature of zeolite/water systems (>180°C) and a relatively high price. Recent developments concerning a rapid zeolite synthesis by a continuous process using tubes instead of autoclaves and by avoiding costly OSDAs as reaction component will probably reduce the production costs.

Environmental catalysis refers to catalytic processes implemented to reduce the emissions of pollutants (e.g. NOx, SO_2, CO). Due to increasingly strict laws concerning the gas emission of diesel vehicles, improved catalysts are required to remove NOx from the exhaust gas stream. Zeolite materials are likely to meet this requirement, and there is presently an intensive research on zeolites for the use in NH_3-SCR of NOx.

Even if a catalyst is not directly involved in environmentally friendly processes, the use of zeolite catalysts in industrial processes often decreases the required energy, and the shape-selective properties usually lead to less byproducts. This allows to reduce the consumption of energy and to diminish the costs for cleaning up the product.

Renewable resources (plants) and biodegradable waste can be converted to ethanol, for example, as a base chemical for industry using zeolite-catalyzed processes. Ethanol can subsequently be converted into several other essential chemicals again employing zeolite catalysts. It is to be expected that these resources will have a growing impact on the production of base chemicals in the future.

References

[1] Cronstedt AF. Observations and descriptions on an unknown mineral species called zeolites. In: van Ballmoos R, Higgins JB, Treacy HMJ, editors. Proc. 9th Int. Zeolite Conf. Boston: Butterworth-Heinemann; 1993:3–9.
[2] Bosc d'Antic L. Memoire sur la Chabazie. J Hist Nat 1792:181–4.
[3] How DCL. On mordenite, a new mineral from the traps of Nova Scotia. J Chem Soc 1864; 2, 100–4.
[4] Schaller WT. The mordenite-ptilolite group, clinoptilolite a new species. Am Miner 1932;17:128–34.
[5] Rozenberg KA, Sapozhnikov AN, Rastsvetaeva RK, Bolotina NB, Kashaev AA. Crystal structure of a new representative of the cancrinite group with a 12-layer stacking sequence of tetrahedral rings. Crystallogr Rep 2004;4:635–42.
[6] http://www.iza-online.org/natural/.
[7] Gottardi G, Galli E. Natural zeolites. Berlin: Springer; 1985.
[8] Eichorn H. Über die Einwirkung verdünnter Salzlösungen auf Silikate, Poggendorffs. Ann Phys Chem 1858;105:126–33.
[9] Friedel G. Bull Soc Fr Mineral Cristallogr 1896; 19:94–118.
[10] Grandjean F. Comp Rend 1909;149:866.
[11] Fleck RN, Wight CG. Hydrocarbon cracking process and catalyst. US Patent 1961;2962435 A.
[12] http://www.inza.unina.it/.
[13] Natural zeolites. Occurrence, properties, applications. In: David F, Bish L, Ming DW, editors. Reviews of the Mineralogical Society of America. Vol. 45; 2001.
[14] Flanagan DM. Zeolites. US Geological Survey minerals yearbook 2015; November 2016:841–2 [advance release].
[15] Barrer RM. Syntheses and reactions of mordenite. J Chem Soc 1948, 2158–63.
[16] Bennett JM, Marcus BK. The crystal structures of several metal aluminophosphate molecular sieves. Stud Surf Sci Catal 1988;37:269–79.
[17] Baerlocher C, Mc Cusker LB. Database of zeolite structures. http://www.iza-structure.org/databases/.
[18] Foster MD, Treacy MMJ. A database of hypothetical zeolite structures. http://www.hypotheticalzeolites.net.
[19] van Koningsveld H. Compendium of zeolite framework types. Building schemes and type characteristics. Amsterdam: Elsevier; 2007.
[20] Hay RL, Sheppard RA. Occurrences of zeolites in sedimentary rocks. In: David F, Bish L, Ming DW, editors. Natural zeolites: occurrence, properties, applications. Reviews of the Mineralogical Society of America. Vol. 45; 2001:217–234.
[21] Utada M. Zeolites in hydrothermally altered rocks. In: David F, Bish L, Ming DW, editors. Natural zeolites: occurrence, properties, applications. Reviews of the Mineralogical Society of America. Vol. 45; 2001: 305–22.
[22] Damour AA. Description de la faujasite, nouvelle espece minerale. Ann Miner IV Ser I 1842; 395–9.
[23] Gies H, Gerke H, Liebau F. Chemical composition and synthesis of melanophlogite, a clathrate compound of silica. Neues Jahrb Mineral Monatsh 1982;3:119–24.
[24] Gies H. Studies on clathrasils III. Crystal structure of melanophlogite, a natural clathrate. Z Kristallogr 1983;164:247–57.

[25] Ercit TS, van Velthuizen J. Gaultite, a new zeolite-like mineral species from Mont Saint-Hilaire, Quebec, and its crystal structure. Can Mineral 1994;32:855–63.
[26] Walter F. Weinebeneite, $CaBe_3(PO_4)_2(OH)_2 \cdot 4H_2O$, a new mineral species: mineral data and crystal structure. Eur J Mineral 1992;4:1275–83.
[27] Levels of radioactive materials rise near Japanese plant. The Associated Press via New York Times, April 16, 2011.
[28] Breck DW. Zeolite molecular sieves. Malabar (FL): Krieger Publishing; 1984.
[29] Santos KAO, Dantas Neto AA, MCPA Moura, Castro Dantas TN. Separation of xylene isomers through adsorption on microporous materials: a review. Braz J Petrol Gas 2011;5:255–68.
[30] Yoshida S, Ogawa N, Kamioka K, Hirano S, Mori T. Study of zeolite molecular sieves for production of oxygen by using pressure swing adsorption. Adsorption 1999;5:57–61.
[31] Venuto PB, Habib Jr ET. Fluid catalytic cracking with zeolite catalysts. New York: Marcel Dekker; 1979.
[32] Guisnet M, Gilson J-P, editors. Zeolites for cleaner technologies. London: Imperial College Press; 2002.
[33] Degand TF, Chitnis GK, Schipper PH. History of ZSM-5 fluid catalytic cracking additive development at Mobil. Microporous Mesoporous Mater 2000;35–6:245–52.
[34] Bellussi G, Carati A, Clerici MG, Maddinelli G, Millini R. Reactions of titanium silicalite with protic molecules and hydrogen peroxide. J Catal 1992;133:220–30.
[35] Maier-Laxhuber P, Schmidt R, Becky A, Wörz R. Die Anwendung der Zelith/Wasser-Technologie zur Bierkühlung. KI Luft- Kältetech 2002;8:368–70.
[36] Viessmann. Vitosorp, 200-F. http://www.viessmann.de.
[37] Simon M. In: Ramaswamy S, Huang H-J, Ramarao BV, editors. Separation and purification technologies in biorefineries. UK: John Wiley & Sons; 2013.
[38] Jeong JS, Jeon H, Ko KM, Chung B, Choi GW. Production of anhydrous ethanol using various PSA (Pressure Swing Adsorption) processes in pilot plant. Renewable energy 2012;42:41–5.
[39] Rangnekar N, Mittal N, Elyassi B, Caro J, Tsapatsis M. Zeolite membranes – a review and comparison with MOFs. Chem Soc Rev 2015;44:7128–54.
[40] Feng C, Khulbe KC, Matsuura T, Farnood R, Ismail AF. Recent progress in zeolite/zeotype membranes. J Membr Sci Res 2015;1:49–72.
[41] Barrer RM. Hydrothermal chemistry of zeolites. London: Academic Press; 1982: 174–6.
[42] Cooper ER, Andrews CD, Wheatley PS, Webb PB, Wormald P, Morris RE. Ionic liquids and eutectic mixtures as solvent and template in synthesis of zeolite analogues. Nature 2004;430:1012–6.
[43] Wei Y, Marler B, Zhang L, Tian Z, Graetsch H, Gies H. Co-templating ionothermal synthesis and structure characterization of two new 2D layered microporous aluminophosphates. Dalton Trans 2012;41:12408–15.
[44] Mintova S, editor. Verified syntheses of zeolitic materials. 3rd rev. ed. Synthesis Commission of the International Zeolite Association, ISBN: 978-0-692-68539-6, 2016.
[45] Wilson ST. Overview on zeolite synthesis strategies. In: Xu R, Gao Z, Chen J, Yan W, editors. Studies in surface science and catalysis. Vol. 170A. Amsterdam: Elsevier; 2007:3–18.
[46] Gies H, Marler B. The structure controlling role of organic templates for the synthesis of porosils in the system SiO_2/template/H_2O. Zeolites 1992;12:42–9.
[47] Davis ME, Lobo RF. Zeolite and molecular sieve synthesis. Chem Mater 1992;4:756–68.
[48] Lobo RF, Zones SI, Davis ME. Structure direction in zeolite synthesis. J Inclusion Phenom Mol Recog Chem 1995;21:47–78.
[49] Gies H, Marler B. Crystalline microporous silicas as host-guest systems. In: MacNicol DD, Toda F, Bishop R, editors. Comprehensive supramolecular chemistry, Vol. 6: crystal engineering. Oxford: Pergamon Press; 1996: 851–83.

[50] Jackowski A, Zones SI, S-J Hwang, Burton AW. Diquaternary ammonium compounds in zeolite synthesis: cyclic and polycyclic N-heterocycles connected by methylene chains. J Am Chem Soc 2009;131:1092–100.

[51] Barrer RM, Denny J. Hydrothermal chemistry of the silicates. Part IX. Nitrogenous aluminosilicates. J Chem Soc 1961:971–82.

[52] Smolin YI, Shepelev YF, Pomes R, Hoebbel D, Wieker W. Determination of the crystal structure of tetramethylammoniumsilicate 8[N(CH$_3$)$_4$] Si$_8$O$_{20}$·64.8 H$_2$O at T = -100°C. Sov Phys Crystallogr 1979;24L19–23.

[53] Wiebcke M, Hoebbel D. Structural links between zeolite-type and clathrate hydrate-type materials – synthesis and crystal-structure of [NMe$_4$]$_{16}$[Si$_8$O$_{20}$] [OH]$_8$ _116 H$_2$O. J Chem Soc Dalton Trans 1992;16:2451–5.

[54] Ikeda T, Oumi Y, Honda K, Sano T, Momma K, Izumi F. Synthesis and crystal structure of a layered silicate HUS-1 with a halved sodalite-cage topology. Inorg Chem 2011;50:2294–301.

[55] Oberhagemann U, Bayat P, Marler B, Gies H, Rius J. A layer silicate: synthesis and structure of the zeolite precursor RUB-15 – [N(CH$_3$)$_4$]$_8$[Si$_{24}$O$_{52}$(OH)$_4$] _ 20 H$_2$O. Angew Chem Int Ed 1996;35:2869–72.

[56] Marler B, Grünewald-Lüke A, Grabowski S, Gies H. RUB-55, a new hydrous layer silicate with silicate layers known as motives of the sodalite and octadecasil frameworks: synthesis and crystal structure. Z Kristallogr 2012;227:427–37.

[57] Marler B, Gies H. Crystal structure analyses of two TMA silicates with ordered defects: RUB-20, a layered zeolite precursor, and RUB-22, a microporous framework silicate. Z Kristallogr 2015;230(4):243–62.

[58] Ikeda T, Akiyama Y, Izumi F, Kiyozumi Y, Mizukami F, Kodaira T. Crystal structure of a helix layered silicate containing tetramethylammonium ions in sodalite-like cages. Chem Mater 2001;13:1286–95.

[59] Millini R, Carluccio LC, Carati A, Bellussi G, Perego C, Cruciani G, Zanardi S. ERS-12: a new layered tetramethylammonium silicate composed by ferrierite layers. Microporous Mesoporous Mater 2004;74:59–71.

[60] Baerlocher C, Meier WM. Synthesis and crystal structure of tetramethylammonium gismondin. Helv Chim Acta 1970;53:1285–93.

[61] Bellussi G, Millini R, Carati A, Maddinelli A, Gervasini A. Synthesis and comparative characterization of Al, B, Ga, and Fe containing NU-1-type zeolitic frameworks. Zeolites 1990;10:642–9.

[62] Baerlocher C, Meier WM. Synthesis and crystal structure of tetramethylammonium sodalite. Helv Chim Acta 1969;52:1853–60.

[63] Oberhagemann U, Marler B, Topalovic I, Gies H. Rub-10, a boron containing analogon of zeolite Nu-1. In: WeitkampJ, Karge HG, Pfeiffer H, Hölderich W, editors. Studies in surface science and catalysis. Vol. 84. Amsterdam: Elsevier; 1994:435.

[64] Yang X. Synthesis and crystal structure of tetramethylammonium fluoride octadecasil. Mater Res Bull 2006;41:54–66.

[65] Zhao D, Qiu S, Pang W. Zeolites synthesized from nonalkaline media. In: van Ballmoos R, Higgins JB, Treacy HMJ, editors. Proc. 9th Int. Zeolite Conf. Boston: Butterworth-Heinemann; 1993:337–44.

[66] Moliner M, Rey F, Corma A. Towards the rational design of efficient organic, structure-directing agents for zeolite synthesis. Angew Chem Int Ed 2013;52:13880–9.

[67] Pophale R, Daeyaert F, Deem MW. Computational prediction of chemically synthesizable organic structure directing agents for zeolites. J Mater Chem A 2013;1:6750–60.

[68] Chen Y, Su J, Huang Sh, Liang J, X-H Lin, Liao F, Sun J, Wang Y, Lin J, Gies H. PKU-20: a new silicogermanate constructed from sti and asv layers. Microporous Mesoporous Mater 2016;224:384–91.

[69] Yu J. Synthesis of zeolites. In: Čejka J, Bekkum Hv, Corma A, Schüth F, editors. Studies in surface science and catalysis. Vol. 168. Amsterdam: Elsevier; 2007:39–103.
[70] Li C-T. The crystal structure of LiAlSi$_2$O$_6$ III (high-quartz solid solution). Z Kristallogr 1968;127:327–48.
[71] Park S-H, Boysen H, Parise JB. Structural disorder of a new zeolite-like lithosilicate, K$_{2.6}$Li$_{5.4}$[Li$_4$Si$_{16}$O$_{38}$]·4.3H$_2$O. Acta Cryst B62, 2006;42–51.
[72] Kerr IS. Crystal structure of a synthetic lithium zeolite. Z Kristallogr 1974;139:186–95.
[73] Kocman V, Gait RI, Rucklidge J. The crystal structure of bikitaite, Li[AlSi$_2$O$_6$]·H$_2$O. Am Mineral 1974;59:71–8.
[74] Marler B, Grünewald-Lüke A, Gies H. The average structure of RUB-12: a new lithosilicate zeolite. Z Kristallogr 2014;Suppl. 34:118–9.
[75] Park SH, Daniels P, Gies H, RUB-23: a new microporous lithosilicate containing spiro-5 building units. Microporous Mesoporous Mater 2000;37:129–43.
[76] Park S-H, Parise JB, Gies H, Liu H, Grey CP, Toby BH. A new porous lithosilicate with a high ionic conductivity and ion-exchange capacity. J Am Chem Soc 2000;122(44):11023–4.
[77] Song JW, Dai L, Ji Y-Y, Xiao F-Sh. Organic template free synthesis of aluminosilicate zeolite ECR-1. Chem Mater 2006; 18:2775–7.
[78] Meng X-J, Xie B, Xiao F-Sh. Organotemplate-free routes for synthesizing zeolites. Chin J Catal 2009;30:965–71. Chinese.
[79] Iyoki K, Itabashi K, Okubo T. Progress in seed-assisted synthesis of zeolites without using organic structure-directing agents. Microporous Mesoporous Mater 2014; 189:22–30.
[80] Zhang H, Xie B, Meng X-J, Müller U, Yilmaz B, Feyen Ms, Maurer S, Gies H, Tatsumi T, Bao X, Zhang W, De Vos D, Xiao F-Sh. Rational synthesis of beta zeolite with improved quality by decreasing crystallization temperature in organotemplate-free route. Microporous Mesoporous Mater 2013; 180:123–9.
[81] Yilmaz B, Müller U, Feyen M, Maurer S, Zhang H, X-J Meng, Xiao F-Sh, Bao X, Zhang W, Imai H, Yokoi T, Tatsumi T, Gies H, De Baerdemaekerj T, De Vos D. A new catalyst platform: zeolite Beta from template-free synthesis. Catal Sci Technol 2013;3:2580–6.
[82] Puppe L, Weisser J. Crystalline alumosilicate PSH-3 and its process of preparation; 1984. US Patent 4,439,409.
[83] Zones SI. New zeolite SSZ-25; 1987. Eur Patent Appl 231860.
[84] Leonowicz ME, Lawton JA, Lawton SL, Rubin MK. MCM-22: a molecular sieve with two independent multidimensional channel systems. Science 1994;264:1910–3.
[85] Millini R, Perego G, Parker WO, Belussi G, Carluccio L. Layered structure of ERB-1 microporous borosilicate precursor and its intercalation properties towards polar molecules. Microporous Mater 1995;4:221–30.
[86] Camblor MA, Corell C, Corma A, Diaz-Cabanas M-J, Nicolopoulos S, Gonzalez-Calbet JM, Vallet-Regi M. A new microporous polymorph of silica isomorphous to zeolite MCM-22. Chem Mater 1996;8:2415–7.
[87] Blake AJ, Franklin KR, Lowe BM. Preparation and properties of piperazine silicate (EU-19) and a silica polymorph (EU-20). J Chem Soc Dalton Trans 1988;2513–7.
[88] Marler B, Camblor MA, Gies H. The disordered structure of silica zeolite EU-20b, obtained by topotactic condensation of the piperazinium containing layer silicate EU-19. Microporous Mesoporous Mater 2006;90:87–101.
[89] Ikeda T, Akiyama Y, Oumi Y, Kawai A, Mizukami F. The topotactic conversion of a novel layered silicate into a new framework zeolite. Angew Chem Int Ed 2004;43:4892–6.
[90] Schreyeck L, Caullet P, Mougenel JC, Guth J-L, Marler B. PREFER: a new layered (alumino) silicate precursor of FER-type zeolite. Microporous Mater 1996;6:259–71.
[91] Zanardi S, Alberti A, Cruciani G, Corma A, Fornes V, Brunelli M. Crystal structure determination of zeolite Nu-6(2) and its layered precursor Nu-6(1). Angew Chem Int Ed 2004;43:4933–7.

[92] Wang YX, Gies H, Lin JH. Crystal structure of the new layer silicate RUB-39 and its topotactic condensation to a microporous zeolite with framework type RRO. Chem Mater 2007;19:4181–8.

[93] Marler B, Ströter N, Gies H. The structure of the new pure silica zeolite RUB-24, $Si_{32}O_{64}$, obtained by topotactic condensation of the intercalated layer silicate RUB-18. Microporous Mesoporous Mater 2005;83:201–11.

[94] Moteki T, Chaikittisilp W, Shimojima A, Okubo T. Silica sodalite without occluded organic matters by topotactic conversion of lamellar precursor. J Am Chem Soc 2008;130:15780–1.

[95] Marler B, Gies H. Hydrous layer silicates as precursors for zeolites obtained through topotactic condensation: a review. Eur. J. Mineral 2012;24:405–28.

[96] Roth WJ, Nachtigall P, Morris RE, Wheatley PS, Seymour VR, Ashbrook SE, Chlubná P, Grajciar L, Položij M, Zukal A, Shvets O, Cejka J. A family of zeolites with controlled pore size prepared using a top-down method. Nat Chem 2013;5:628–33.

[97] Verheyen E, Joos L, van Havenbergh K, Breynaert E, Kasian N, Gobechiya E, Houthoofd K, Martineau C, Hinterstein M, Taulelle F, van Speybroeck V, Waroquier M, Bals S, van Tendeloo G, Kirschhock CEA, Martens JA. Design of zeolite by inverse sigma transformation. Nature Materials 2012, 11:1059–64.

[98] Zhao Z, Zhang W, Ren P, Han X, Müller U, Yilmaz B, Feyen M, Gies H, Xiao F-Sh, De Vos D, Tatsumi T, Bao X. Insights into the topotactic conversion process from layered silicate RUB-36 to FER-type zeolite by Layer reassembly. Chem Mater 2013;25:840–7.

[99] Marler B, Wang Y, Song J, Gies H. Topotactic condensation of layer silicates with ferrierite-type layers forming porous tectosilicates. Dalton Trans 2014;43:10396–416.

[100] Yilmaz B, Müller U, Tijsebaert B, De Vos D, Xie B, Xiao F-Sh, Gies H, Zhang W, Bao X, Imai H, Tatsumi T. Al-RUB-41: a shape-selective zeolite catalyst from a layered silicate. Chem Commun 2011;47:1812–4.

[101] Grünewald-Lüke A, Gies H, Müller U, Yilmaz B, Imai H, Tatsumi T, Xie B, Xiao F-Sh, Bao X, Zhang W, De Vos D. Layered precursors for new zeolitic materials: synthesis and characterization of B-RUB-39 and its condensation product B-RUB-41. Microporous Mesoporous Mater 2012;147:102–9.

[102] Xu L, Shi C, Zhang Z, Gies H, Xiao F-Sh, De Vos D, Yokoi T, Bao X, Feyen M, Maurer S, Yilmaz B, Müller U, Zhang W. Enhancement of low-temperature activity over Cu-exchanged zeolite beta from organotemplate-free synthesis for the selective catalytic reduction of NOx with NH_3 in exhaust gas streams. Microporous Mesoporous Mater 2014;200:304–10.

[103] Xu L, Shi C, Chen B, Zhao Q, Zhu Y, Gies H, Xiao F-Sh, De Vos D, Yokoi T, Bao X, Kolb U, Feyen M, Maurer S, Moini A, Müller U, Zhang W. Improvement of catalytic activity over Cu-Fe modified Al-rich Beta catalyst for the selective catalytic reduction of NOx with NH_3. Microporous Mesoporous Mater 2016;236:211–7.

[104] Jänchen J, Grimm A, Stach H. Investigation of the storage properties of zeolites and impregnated silica for thermochemical storage of heat. Stud Surf Sci Catal 2001; 135: Full Paper 31-O-04.

[105] Jänchen J, Ackermann D, Stach H, Brösicke W. Studies of the water adsorption on zeolites and modified mesoporous materials for seasonal storage of solar heat. Solar Energy 2004;76:339.

[106] Wozniak A, Marler B, Angermund K, Gies H. Water and cation distribution in fully and partially hydrated Li -LSX zeolite. Chem Mater 2008; 20:5968–76.

[107] Recorded by Rakhmatkariev GU. Tashkent, Usbekistan: Institute of Chemistry, Academy of Science.

[108] Seidel A, Schimiczek B, Tracht U, Boddenberg B. 23Na solid state MAS NMR of sodium halides occluded in zeolites. Solid State Nucl Magn Reson 1997;9:129–41.

[109] Marler B, Müller M, Wozniak A, Gies H. The structure of Br@NaY, $Na_{45}H_{15}[Si_{139}Al_{53}O_{384}Br_7]\cdot$ 247 H_2O, a NaY zeolite impregnated with NaBr [abstract]. In: 25th Annual Conference of the German Crystallographic Society, 2017.

[110] Boddenberg B, Rakhmatkariev GU, Hufnagel S, Salimov Z. A calorimetric and statistical mechanics study of water adsorption in zeolite NaY. Phys Chem Chem Phys 2002;4:4172–80.
[111] Eroshenko V, R-C Regis, Soulard M, Patarin J. Energetics: a new field of applications for hydrophobic zeolites. J Am Chem Soc 2001;123:8129–30.
[112] Tzanis L, Trzpit M, Soulard M, Patarin J. Energetic performance of channel and cage-type zeosils. J Phys Chem C 2012;116:20389–95.
[113] Saada MA, Soulard M, Marler B, Gies H, Patarin J. High pressure water intrusion investigation of pure silica RUB-41 and S-SOD zeolite materials. J Phys Chem C 2011;115:425–30.
[114] Tzanis L, Marler B, Gies H, Patarin J. High-pressure water intrusion investigation of pure silica ITQ-7 zeolite. J Phys Chem C 2013;117:4098–103.

René Gunder, Julien Marquardt, Leonhard Leppin
and Susan Schorr

4 Microstructure analysis of chalcopyrite-type CuInSe$_2$ and kesterite-type Cu$_2$ZnSnSe$_4$ absorber layers in thin film solar cells

4.1 Introduction

Thin film solar cells equipped with polycrystalline compound semiconductors as functional layer for light absorption have continuously been improved in terms of solar energy conversion efficiency, such that they became a competitive alternative to well-established silicon-based solar cells. In 1905, Einstein published a comprehensive, physical description of the photoelectric effect [1] and thus provided the theoretical framework for upcoming research of photovoltaic technologies. The emergence of photovoltaic devices, however, only started about 50 years later, and for several decades, it persisted a niche technology mainly for aerospace applications. Among others, silicon (Si) was known to belong to the group of (extrinsic) elemental semiconductors, and due to its abundance, it was the very first absorber material to be used in solar cells. Triggered by the oil crisis in the 1970s, the research of solar energy conversion technologies finally got a tremendous stimulus. As a result, research not only of silicon-based solar cells but also of other absorber layer materials based on compound semiconductors have been much more extensively endeavored. The latter were also brought into focus in order to address some severe drawbacks of silicon-based solar cells. First of all, the high energy consumption in fabricating single crystal silicon results in a quite long energy amortization time. In addition, the requirements on crystallinity and purity are extremely high while a considerable amount of material is wasted upon slicing silicon wafers. Also, during the growth of silicon single crystals a certain concentration of dopants has to be incorporated in order to induce either extrinsic p-type or n-type conductivity. Despite the energy of the band gap of silicon fitting quite well with the optimal energy determined by the solar spectrum, silicon is an indirect semiconductor whose photonic electron transition from the valence band to the conduction band needs to be assisted by a phononic momentum transfer. This requirement of coincidence between a photon of appropriate energy being absorbed and a phonon transferring impulse to the electron leads to a reduced probability of events of photoelectric charge carrier generation. Correspondingly, the absorber thickness must be augmented in order to compensate the low absorption coefficient. These aforementioned issues, eventually, gave rise to reconsider photovoltaic technologies, being both economical and ecological reasonably applicable in a more widely spread manner. These demands have paved the way for thin film solar

https://doi.org/10.1515/9783110497342-004

cell technologies using compound semiconductors. Those compound semiconductors are intrinsically conductive, and they possess a higher absorption coefficient due to direct electron band transitions (Fig. 4.1).

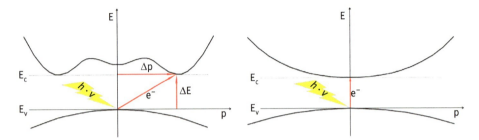

Fig. 4.1: Schematic of the density of states of the valence band and conduction band in (left) indirect and (right) direct semiconductors. The transition of an electron e⁻ from the maximum of the valence band E_v to the minimum of the conduction band E_c in direct semiconductors only depends on the energy of the incident photon $h \cdot v$. In indirect semiconductors, the transition additionally requires a phononic impulse p. Modified after [2].

Two very prominent representatives from such compound semiconductors are the solid solutions Cu(In,Ga)(S,Se)$_2$ and Cu$_2$ZnSn(S,Se)$_4$, which are referred to as chalcopyrite- and kesterite-type compound semiconductors, respectively. Chalcopyrite-type semiconductors are isotypic, with and named after the naturally occurring mineral chalcopyrite (CuFeS$_2$) and are the light-absorbing layers in solar cells since about the 1970s [3]. Even though solar cells based on chalcopyrite-type absorbers have achieved remarkable solar energy conversion efficiencies up to 22.6% [4], the deployment of rare elements in conjunction with the increasing demand for indium not only in photovoltaics has mediated newly emerging photovoltaic technologies based, for instance, on kesterite-type compound semiconductors with efficiencies up to 12.6% [5]. Similar to "chalcopyrites", kesterite-type compound semiconductors are named after the natural mineral kesterite (Cu$_2$ZnSnS$_4$). Incipient research on kesterite-type solar cells was conducted around the mid-1990s [6, 7], which led to fast developments in photovoltaic performance until the present day.

Both chalcopyrites and kesterites crystallize in the tetragonal crystal system with a body-centered unit cell and a four-fold rotational inversion along the c-axis. However, chalcopyrites do have a higher symmetry as they also comprise a two-fold rotational axis along the a-direction as well as a glide plane. Hence, the space group of chalcopyrites is $I\bar{4}2d$ [8] and those of kesterites $I\bar{4}$ [9]. According to the difference in symmetry between chalcopyrites and kesterites, the number of Wyckoff position differs too. In chalcopyrites, three Wyckoff positions (4a, 4b, 8d) are occupied, whereas five Wyckoff positions (2a, 2b, 2c, 2d, 8g) follow from the symmetry of kesterites. The 4a as well as 2a and 2c positions in chalcopyrite and kesterite, respectively, are occupied by a univalent cation (e.g. Cu^+), where the 4a position in chalcopyrite degenerates to 2a and 2c position in kesterite. In contrast, the 4b position in chalcopyrite is occupied by a trivalent cation (e.g. Fe^{3+}), and as symmetry decreases to the kesterite structure, this position is split into two positions, 2b and 2d, being occupied by tetravalent (e.g. Sn^{4+}) and divalent cations (e.g. Zn^{2+}), respectively. This difference in cationic arrangement influences both the anion position, either 8d or 8g, and the tetragonal deformation $\eta = \frac{c}{2a} \neq 1$. The anion position is shifted from the center of the cation tetrahedron. In the chalcopyrite-type structure, the deviation from this ideal position is described by the anion displacement parameter $0.25 - x$, where x is the anion atomic position parameter of the 8d site.

Despite this distinctive structural character, both structures are derived from the diamond-type structure (s.g. $Fd\bar{3}m$) which can be described as two interpenetrating face-centered cubic (fcc) sub-lattices, offset by $\left(\frac{1}{4}, \frac{1}{4}, \frac{1}{4}\right)$, and thereby building a framework of regular, corner-shared tetrahedra [10]. However, with a change in the composition toward multinary compounds, the symmetry decreases progressively both with increasing number of different groups of elements involved and their ordering within the structure. So, in case of binary compounds, either sub-lattices are occupied complementary by the same cationic and anionic species, respectively. By doing so, the symmetry lowers to $F\bar{4}3m$, which is referred to as sphalerite-type structure. The incorporation of a second cation belonging to another element group leads to ordered substitution on the cation site. Thereby, the originally cubic unit cell is expanded to a tetragonal one and anion coordinates then include one independent parameter $\left(x, \frac{1}{4}, \frac{1}{8}\right)$. This is referred to as chalcopyrite structure (s.g. $I\bar{4}2d$) [11]. The number of atomic coordinates further increases in the quaternary stannite structure when three different cations are assembled. The anion position is then characterized by an additional degree of freedom $\left(x, x, \frac{1}{8}\right)$. However, even though containing three different cations as well, it was found that the stannite structure (s.g. $I\bar{4}2m$) further degrades to the kesterite structure (s.g. $I\bar{4}$) due to different structural distribution and partial ordering of Cu^+, Zn^{2+}, and Fe^{2+} in the solid solution stannite-kesterite $Cu_2Fe_{1-x}Zn_xSnS_4$ [9]. Another peculiar feature shown by kesterites is the Cu/Zn disorder among structural sites 2c and 2d [12], and the degree of disorder heavily depends on temperature and equilibration time [13]. In the kesterite structure, eventually, the anion positions are completely independent (x, y, z). The entire structural evolution is emphasized in Fig. 4.2.

A huge advantage both of chalcopyrite- (ternary) and kesterite-type (quaternary) compound semiconductor materials is the ability of their structures to readily facilitate chemical substitution processes, basically involving all cations and anions belonging to the appropriate group of elements. Due to this high chemical variability, the energy of the band gap basically can be adjusted over a quite wide range. Considering the high chemical variability of compound semiconductors, a general notation

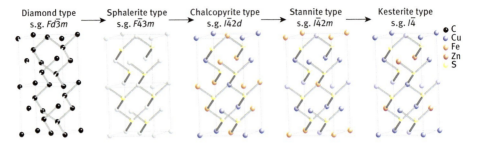

Fig. 4.2: Scheme of the adamantine compound family. The symmetry lowers as the number of different element types increases and ordering of metals occurs. For better visualization, two unit cells are shown, respectively, for cubic diamond and sphalerite-type. Modified after [14].

using capital letters in conjunction with Roman numerals has been established. Each capital letter refers to an isovalent ionic species occupying a specific atomic position in the structure. The Roman numerals denote the number of valence electrons (i.e. element group) of the elements being accommodated by the structure. Subscripted numbers do provide the stoichiometric coefficient. So, in case of ternary chalcopyrite-type compounds, two aliovalent cation positions (A, B) and one anion position (X) is present, which can be written as $A^I B^{III} X^{VI}_2$ [8]. Accordingly, the notation for quaternary kesterite-type compounds is extended by another cationic species (C), yielding $A^I_2 B^{II} C^{IV} X^{VI}_4$. In order to illustrate all possible end-members and their corresponding solid solutions, structurally tolerated elements are compiled in matrices:

(1) ternary chalcopyrites (2) quaternary kesterites

$$\begin{pmatrix} A^I \\ Cu \\ Ag \\ Au \end{pmatrix} \begin{pmatrix} B^{III} \\ Al \\ Ga \\ In \end{pmatrix} \begin{pmatrix} X^{VI}_2 \\ S \\ Se \\ Te \end{pmatrix}_2 \quad \begin{pmatrix} A^I_2 \\ Cu \\ Ag \\ Au \end{pmatrix}_2 \begin{pmatrix} B^{II} \\ Zn \\ Fe \\ Cd \end{pmatrix} \begin{pmatrix} C^{IV} \\ Sn \\ Ge \\ Si \end{pmatrix} \begin{pmatrix} X^{VI}_4 \\ S \\ Se \\ Te \end{pmatrix}_4$$

Being able to design the band gap energy of the absorber layer is of particular importance as the band gap not only has to be optimized with respect to the solar radiation maximum but potentially needs to be adapted to the band gap energies of the other functional layers employed in solar cell devices. Such adjustments of the band gap, aiming to enhance electronic properties in solar cell devices, are referred to as band gap engineering. This is of particular concern when heterojunctions arising from p-n junction interfaces between dissimilar semiconductors are involved. Owing to the unequal band gaps, it is crucial to control the energy bands in semiconductors in order to reasonably tune, for instance, the band gap offset since inappropriate band gap alignments might be detrimental in terms of the charge carrier yield (e.g. through increased number of recombination events), thus deteriorating the performance.

A method to shed some light both on interface and bulk properties is given by the microstructure analysis which is going to be presented herein. Lattice defects evoke (semi-)quantifiable microstrain and usually control the domain size and vice versa [15].

As microstrain and domain size do reflect any kind of lattice defect as well as the degree of crystallinity conclusions can be drawn with respect to the photovoltaic performance since crystal imperfections and even crystallographic orientations influence the mobility and density of charge carriers. This ultimately co-determines the conversion efficiency of solar cells as too many domain boundaries and high microstrain as a result of, for instance, point defects might act as detrimental recombination sites leading to loss in photocurrent. By conducting grazing incidence X-ray diffraction (GIXRD) measurements using different incidence angles, in concert with Beer-Lambert law, it is even possible to spatially assign the observed microstructure within the absorber layer.

A detailed description of the analysis of thin films and on how the presented studies were conducted is provided in the subsequent experimental section. The experimental section is followed by three self-standing reports on recent results from microstructure analysis of chalcogenide absorber layers in thin film solar cells. The first study was titled "Influence of sodium and process temperature on microstructure of CISe thin film absorber layers", which dealt with the impact of sodium on the microstructure of $CuInSe_2$ absorber layers in general. The second report, "Microstructural response to different NaF precursor thicknesses and Cu/In ratios in CISe thin film absorber layers", evaluated the implications for microstructure of $CuInSe_2$ absorber layers when three-stage co-evaporation process is interrupted and sodium content is varied. Eventually, the third study, "Impact of absorber process conditions and substrate material on microstructure of CZTSe thin film absorber layers", gave an impression of some issues the emerging kesterite-type absorbers are struggling with. The focus, however, again is put on the constitution of microstructure both at different process temperatures as well as different substrate materials.

4.2 Experimental

4.2.1 Depth-resolved grazing incidence X-ray diffraction

The grazing incidence setup is the X-ray diffraction technique best suited for depth-resolved thin film characterization. During the measurement, the primary beam is kept at a constant incidence angle ω while the detector moves along the goniometer circle (Fig. 4.3). Hence, this asymmetric geometry enables the depth-resolved investigation of, for instance, composition (indirectly), unit cell metrics, phase content, and microstructure.

The probed sample depth d can be approached using the X-ray linear mass attenuation coefficient μ and the Beer-Lambert law

$$I(d) = I_0 e^{-\mu d}. \qquad (4.1)$$

In a given material, the attenuation length l of monochromatic X-rays defines the path length at which the primary intensity I_0 drops to $1/e$. Due to the fact that the X-rays

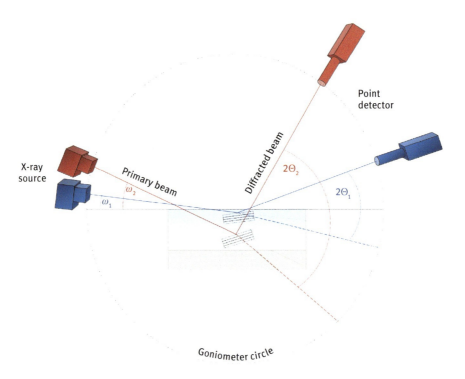

Fig. 4.3: Schematic of the grazing incidence X-ray diffraction geometry (GIXRD) showing two measurements at different incidence angles ω. Depending on the incidence angle, different sample depths can be explored, but the orientation of lattice planes (i.e. scattering vector) being recorded by the detector rotates both as the detector moves and the incidence angle changes. Drawing not to scale.

do not impinge the sample surface perpendicularly, the attenuation length has to be converted in terms of sample depth by simple trigonometry:

$$d(\omega) = l \sin \omega. \tag{4.2}$$

However, because the density and the composition in turn determine the attenuation length, at least the main phase as well as its (approximate) composition must be known beforehand.

4.2.2 Microstructure analysis

The microstructural analysis refers to the quantification of the volume averaged coherently scattering domain size and non-uniform microstrain in crystalline materials. Small domain sizes and microstrain are caused by any kind of local crystallographic defects; thus, they are distinctive from the macrostrain arising from uniformly deformed lattice planes. Point defects like interstitials instead locally distort the three-dimensional periodic arrangement of atoms in a crystal, thus introducing microstrain and in turn may affecting the coherent size of the domains too (e.g. [15]). On the contrary, a poorly crystallized material or an assemblage of sufficiently small grains (e.g. nanocrystals) inherently exhibit small coherence volumes. Whether or not such nanomaterials feature microstrain as well again depends on the presence of crystallographic volume defects within the crystal, but the configuration of either grain boundaries in mutual contact or the epitaxial interfaces and intergrown phases (e.g. through lattice mismatch) also normally influences the microstrain. Lattice mismatch at an epitaxial two-phase junction might be a combined source both for microstrain and macrostrain, which can be readily discriminated in a XRD diffraction pattern. Since macrostrain is caused either by compressive or tensile stress, the lattice plane distances are altered such that peak positions are shifted, while maintaining their shape. Microstructural effects, however, are reflected by a broadening of the diffraction peaks (Fig. 4.4).

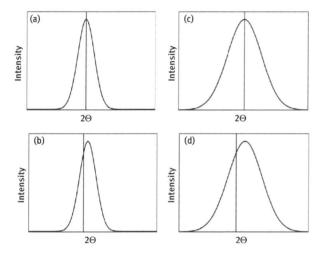

Fig. 4.4: Response of the peak profile in case of no (a) macrostrain and microstrain, (b) compressive macrostrain, (c) microstrain, and (d) combined macrostrain and microstrain.

The microstructure of the $CuInSe_2$ (CISe) and $Cu_2ZnSnSe_4$ (CZTSe) thin film samples have been examined by performing a whole-pattern decomposition of X-ray diffraction data using programs from the *Fullprof Suite* software package [16]. In order to ensure an optimal fit of the peak profile, the diffraction patterns were analyzed by the Le Bail method [17] using the Thompson-Cox-Hastings pseudo-Voigt profile function [18] with the Finger's treatment of the axial divergence [19].

The microstructural information can be extracted from the (isotropic) peak broadening β_{sample} caused by microstrains and/or small average domain sizes. In contrast to full width at half-maximum parameter Γ, the integral breadth β does include information on the integral intensity I_A, that is, the area enveloped by the

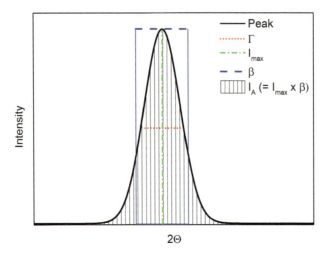

Fig. 4.5: Shape-related characteristics of a diffraction peak. The integral intensity is directly related to phase content by volume, site occupation and preferred orientation. The integral breadth is related to the instrumental resolution and the microstructure of a crystalline material.

peak $\left(\equiv \sum h \cdot v_{X-ray}\right)$. The integral breadth is calculated by the ratio between the integral intensity and Γ and can be represented as the width of a rectangle being as high as I_{max} and including the area of the peak (Fig. 4.5).

In order to describe the shape of a peak properly both Gaussian as well as Lorentzian profile usually must be considered in union. According to the Caglioti (Gaussian) function [20],

$$\Gamma_G^2 = U \tan^2 \Theta_\Gamma + V \tan \Theta_\Gamma + W \quad (4.3)$$

and the Cauchy (Lorentzian) function (e.g. [21]),

$$\Gamma_L = X \tan \Theta_\Gamma + Y/\cos \Theta_\Gamma, \quad (4.4)$$

the width of each peak is calculated at half the maximum intensity for the respective function, given in units of [°].

The calculation of the integral breadth β, by combining Γ_G and Γ_L, is mandatory in order to perform a proper microstructural analysis. However, since the convolution of Gaussian and Lorentzian by applying the Voigt function is mathematically complex, they are treated by the approximating pseudo-Voigt function (e.g. [21])

$$\beta_{pV} = \frac{\pi \Gamma/2}{\eta + (1-\eta)\sqrt{\pi \ln 2}} \text{ with } 0 < \eta < 1, \quad (4.5)$$

where Γ is computed as

$$\Gamma = (\Gamma_G^5 + 2.69269\Gamma_G^4\Gamma_L + 2.42843\Gamma_G^3\Gamma_L^2 + 4.47163\Gamma_G^2\Gamma_L^3 + 0.07842\Gamma_G\Gamma_L^4 + \Gamma_L^5)^{1/5} \quad (4.6)$$

and the mixing parameter η is computed as

$$\eta = 1.36603(\Gamma_L/\Gamma) - 0.47719(\Gamma_L/\Gamma)^2 + 0.11116(\Gamma_L/\Gamma)^3. \quad (4.7)$$

This empirical expression for the pseudo-Voigt approximation is referred to as the Thompson-Cox-Hastings pseudo-Voigt profile function [18]. The Caglioti function (Equation 4.3) was not considered for the microstructure analysis since only the profile parameter U may be influenced by microstrain while V and W exclusively reflect the profile emanating from the instrument. Consequently, only the profile parameters for the Lorentzian profile (i.e. X and Y in Equation 4.4) were used to deduce microstrain and domain size, respectively. The influence of the finite instrumental resolution was previously determined using standard reference material lanthanum hexaboride (NIST SRM-660b for line profile analysis – LaB_6), from which, ultimately, an instrumental resolution function providing Gaussian and Lorentzian could be created.

Once the integral breadth is known and corrected for the instrumental resolution, both microstrain as well as domain size of a sample can be obtained. The isotropic broadening due to small domain size β_{size} is separated from the total sample broadening β_{sample} by the Scherrer equation [22]:

$$\beta_{size} = \lambda_{X-ray}/(D_{<V>} \cos \Theta), \quad (4.8)$$

where $D_{<V>}$ refers to the volume averaged coherently scattering domain size. This means that a particular grain or even a particular crystallite can contain several domains if the periodicity of the three-dimensional arrangement of atoms is interrupted by, for instance, stacking faults and dislocations. So, size broadening is caused by imperfections and, in principle, ranges between the limit cases "fully crystalline" (i.e. completely coherent) and "amorphous" (completely incoherent). That is why the domain size cannot be directly related to the grain or crystallite size, respectively. The domain size obtained from Equation 4.8 does deliver a volume-weighted mean value because the columnar lengths being probed in a crystallite depends on both its shape and orientation. In addition, the resolution limit of a laboratory XRD device normally restricts the detectable broadening β_{size} to $D_{<V>} \leq 100$ nm because larger domain sizes do not contribute significantly to incoherent scattering. The broadening β_{strain} induced by microstrain is extracted by the Wilson equation [23]:

$$\beta_{strain} = 4\bar{\varepsilon} \tan \Theta, \quad (4.9)$$

where $\bar{\varepsilon}$ is treated as the volume averaged microstrain derived from the basic definition of the strain

$$e = \Delta d/d. \quad (4.10)$$

In contrast to the macrostrain, the microstrain emerges from a non-uniform stress field caused by lattice defects, chemical heterogeneities, or lattice mismatch at phase boundaries. Although the sources for microstrain may also affect the domain size, as they control the periodicity, the different broadening effects can be separated in terms of the different angular dependency (cf. Equations 4.8 and 4.9) as well as by the fact that, in contrast to strain broadening, size effects are independent of the order of a reflection.

A graphical separation of size and strain effects is given by the Williamson-Hall analysis (WHA) [24], involving the equations from Scherrer and Wilson (Equations 4.8 and 4.9). The WHA relies on the broadening of each individual Γ_G^{hkl} (Equation 4.3) and its conversion to integral breadths using the pseudo-Voigt function (Equation 4.5) with mixing parameter η obtained from the profile fit. The sample broadening is corrected for the instrumental broadening using the approximation [25].

$$\beta_{sample} = (\beta_{total}^2 - \beta_{XRD}^2)/\beta_{total}. \tag{4.11}$$

The instrumental resolution determined by measuring standard reference material (SRM) LaB$_6$ at exact same conditions as the sample to be investigated is obtained for 2Θ positions at which the SRM does show reflections and can be interpolated for (sample) specific diffraction angles located within the recorded 2Θ range of the SRM (Fig. 4.6).

Each individual integral peak breadth of the sample being analyzed is corrected for finite instrumental resolution at the respective 2Θ position by the approximation given in Equation 4.11, and assuming that the corrected sample broadening represents the sum of size and strain effects of the sample, it can be written as

$$\beta_{sample} = \beta_{size} + \beta_{strain} = \lambda_{X-ray}/(D_{<V>} \cos \Theta) + 4\bar{\varepsilon} \tan \Theta. \tag{4.12}$$

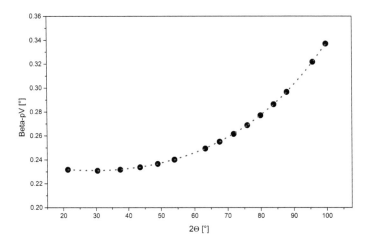

Fig. 4.6: Instrumental broadening β in dependence of diffraction angle 2Θ as obtained from whole pattern decomposition of SRM LaB$_6$ diffraction data using pseudo-Voigt profile function (Equation 4.5).

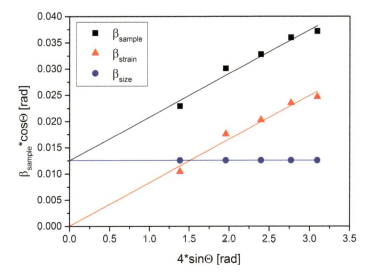

Fig. 4.7: Williamson-Hall plot of a thin polycrystalline molybdenum layer as used as back contact in solar cells. The slope of the plot (black and red line) directly delivers the microstrain while the domain size can be deduced from the y-intercept (black and blue line).

Rearranging Equation 4.12 to the linear equation

$$\beta_{sample} \cos \Theta = \bar{\varepsilon} \cdot 4 \sin \Theta + \lambda_{X-ray}/D_{<V>} \quad (4.13)$$

leads to a separation between strain and size, where the mean strain $\bar{\varepsilon}$ is equal to the slope and the volume-averaged domain size is obtained from the ordinate intercept y_0. In Fig. 4.7, the result of such a WHA is exemplarily shown for a molybdenum back electrode (black symbols/line). Basically, it is sufficient to only plot β_{sample} since it already provides the microstructure information, that is, microstrain = m (slope) ≈ 0.008%, and $D_{<V>} = \lambda/y_0 = 1.54056$ Å$/0.01255 ≈ 123$ Å, but the beauty of a WHP is related to the graphical separation of strain (red symbols/line in Fig. 4.7) and size (blue symbols/line in Fig. 4.7) contribution, respectively.

However, the aforementioned routines of analyzing microstructural effects aim to provide an example of how XRD data can be treated such that additional information from a crystalline sample is extracted, and it is not meant to be detailed in any aspect. The approximation chosen to analyze size and strain effects has to be appropriate for the given conditions and, in accord with a careful interpretation, might be adapted to further enhance the reliability. So, it has to be kept in mind that microstructural analysis can be tricky, in particular in terms of meaningfully interpreting size and strain effects, as it can be difficult to separate them unambiguously from each other. Also, even for an exact same sample, the comparability of microstructure information obtained by different setups or other approaches is likely restricted to qualitative evaluations, although it should result in similar trends. Depth-resolved information extracted by microstructure analysis are best interpreted in conjunction with SEM imaging, either including EDX or compositional depth profiles from glow discharge optical emission spectroscopy (GDOES). For more comprehensive insights into the microstructure analysis, the reader is referred to literature both presenting the various approaches as well as going further into a detailed mathematical and physical discussion about the effects having impact on the peak profile (e.g. [26, 27]).

4.2.2.1 Influence of sodium and process temperature on microstructure of CISe thin film absorber layers

Absorber layers in thin film solar cells are the centerpiece among all the other functional layers building up solar cell devices. While metallic molybdenum is already well established as back electrode, much effort is still put into the search of alternative buffer layer materials in order to substitute highly toxic CdS. Also, the band alignment at the CdS-absorber heterointerface fits better only to a certain solid solution between $CuInSe_2$ and $CuGaSe_2$ (CIGSe) but is rather unfavorable for other compositions or different absorber materials like kesterites, where the p-n interface commonly shows "cliff"-type behavior [28–30]. This kind of interface enhances the recombination of charge carriers and therefore may drastically deteriorates the performance of the solar cell. Despite those efforts aiming to optimize the properties of the heterojunction interface, the key for improving the performance of solar cells is closely connected to a comprehensive understanding of the functionality of the light absorbing layer. However, there is still a considerable lack of understanding which occasionally results in breakthroughs being rather based on fortunate coincidence. By chance, it was found that sodium diffusing from the soda lime glass substrate into the CIGSe absorber layer during thermal treatment significantly improved solar cell performance [31]. Since then, the CIGSe absorber is intentionally doped with sodium, and recently, the possible benefit of other alkali metals (e.g. K, Rb) are intensively studied as well (e.g. [4, 32]). Intentional doping of the absorber layer implies control of the amount of dopants being incorporated. Therefore, the thermally triggered diffusion of sodium from the glass substrate is inhibited by a chromium barrier layer; instead, sodium is supplied either prior to the absorber process using a thin precursor layer of NaF or after the absorber layer deposition by evaporation of NaF onto the absorber (post-deposition treatment). The impact of sodium as well as different process temperatures on microstructure was investigated on two Cu-poor (0.83 < Cu/In < 0.97) $CuInSe_2$ (CISe) thin film sample series produced by a three-stage co-evaporation process. During the first stage, In and Se are evaporated on a molybdenum-coated soda lime glass (SLG) at a substrate temperature of $T_1 = 330°C$. In the second stage, Cu and Se are evaporated onto the In-Se layer at various temperatures $T_2 > T_1$ to react with the In-Se layer and form CISe. At the end of the second stage, the composition passes the stoichiometric point (SP) toward Cu-rich conditions (i.e. Cu/In > 1). This Cu-rich phase plays an important role in the formation of larger crystals and the annihilation of structural defects [33]. In the third stage, In and Se are again evaporated at T_2 to retrogradely cross the stoichiometric point to Cu-poor composition (Fig. 4.8). In this stage, a recrystallization of the CISe layer takes place.

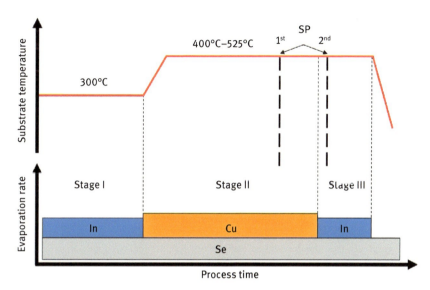

Fig. 4.8: Schematic of the three-stage co-evaporation process.

The two sample series studied are divided into five CISe thin films each, either equipped with a 12 nm NaF precursor layer or without NaF at all. The NaF precursor is deposited on the Mo back contact, prior to absorber deposition. After the three-stage co-evaporation process, the thickness of the absorber layer typically ranges between 1.5 and 2 µm, meaning that grazing incidence angles up to 5° are normally sufficient to probe the absorber layer across its entire thickness.

As microstrains emanate from lattice imperfections, which in turn also limit the coherent domain size, the depth-resolved quantification of the microstrain can be a complementary tool to spatially infer the behavior of charge carriers. In case the concentration of lattice defects exceeds a certain value or the domain sizes are too small, the amount of recombination centers leading to a reduction in photocurrent may increases to an unacceptably high level. This is of particular interest for the vicinity of the heterojunctions where recombination is probably the most critical parameter. In addition, an assessment of the overall absorber quality can be carried out. In general, higher substrate temperatures promote lower microstrains during the growth process of the absorber layer since the recrystallization becomes more enhanced. The substrate temperature, however, cannot be increased arbitrarily for the purpose of good crystallinity. Rather, it has to be a compromise between high crystal quality and substrate temperatures being as low as possible, such that the energy consumption in the course of the absorber growth process is minimized and phase transitions of the substrate material is avoided. The compositional depth profile of the CISe thin film absorber layer after the three-stage co-evaporation process has been obtained by glow discharge optical emission spectroscopy (GDOES) and is shown in Fig. 4.9.

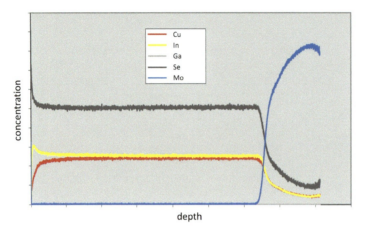

Fig. 4.9: GDOES graph of a CISe thin film without NaF precursor. Apart from the depletion in Cu and enrichment in In at the surface, the composition throughout the absorber is highly homogeneous.

The incorporation of dopants like sodium does have a huge influence on the microstructure (e.g. [34]). Figure 4.10 shows the compiled results obtained for microstrain and lattice parameters of CISe thin films in dependence of the process temperature (T_2) and whether or not a NaF precursor layer was employed. The impact of the process temperature on microstrain can be clearly seen for either series. However, the microstrain values of the samples being not equipped with a NaF precursor layer do not show strong variations with temperature and the progression of the microstrain across the absorber is comparable as the bulk is characterized by lower microstrains than the interfacial regions. Higher microstrains at the surface coincide with a zone being depleted in Cu, which is commonly observed in the uppermost part because Cu is consumed during the superficial formation of secondary Cu phases. The secondary phases, having formed on top the absorber, are removed by etching, leaving behind an absorber surface showing a declining Cu content (cf. Fig. 4.9). Toward the remarkably homogeneous bulk composition, the microstrain decreases and then increases again once the back interface region is reached. The reason for the re-increase is usually connected with the change in the structural environment when the CISe absorber gets in mutual contact with the Mo layer where the resulting lattice mismatch between CISe and Mo introduces microstrains too. This lattice mismatch is also reflected by distinct changes in lattice parameters between the incidence angles 4° and 5° in Fig. 4.9. While in samples without NaF precursor this change is mainly restricted to lattice parameter c (here $c/2$ is used for better comparability with a), the incorporation of Na strongly alters both lattice parameter a and c not only at the back interface but also at the surface. Sodium does have a strong affinity to occupy the vacant Cu (V_{Cu}) positions [35] in the crystal structure, and therefore, the Cu-depleted surface also shows major lattice parameter changes and a different microstrain behavior. The diffusion of sodium from the back interface through the absorber strongly depends on the process temperature (T_2). At lower temperatures, the diffusion of sodium is more sluggish and the spread is more pronounced within the bulk, causing higher overall microstrains. As T_2 increases to 525°C, sodium is more diffusively driven to sites being more favorable for accommodating sodium. Thus, the trend and amount of microstrain are comparable with its analogue without NaF precursor, by, microstrain drops again toward the surface (Fig. 4.10).

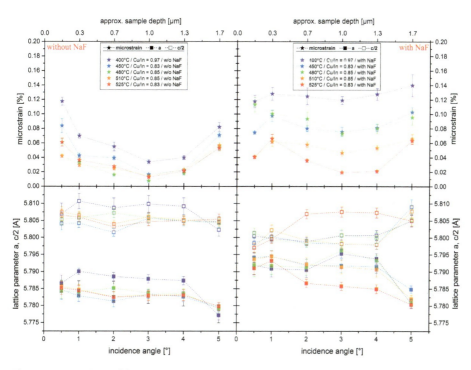

Fig. 4.10: Comparison of the results extracted by the microstructure analysis of CISe thin film absorber layers grown at different temperatures without (left) and with (right) NaF precursor.

The impact on the domain size cannot be fathomed by this method, no matter which configuration (with or without NaF precursor) was investigated. The reason for the incapability in determining the domain size relies on the resolution of conventional X-ray diffractometers, setting the limit of detectable domain sizes around 100 nm and below. However, in SEM, for instance, the effect of sodium on grain growth can be readily observed, as it hinders the growth of larger grains (e.g. [36]). The beneficial influence of sodium is the passivation of grain boundary defects, thus reducing recombination events [37].

4.2.2.2 Microstructural response to different NaF precursor thicknesses and Cu/In ratios in CISe thin film absorber layers

This study was conducted to evaluate the influence of dissimilar NaF precursor thicknesses as well as different compositions on microstructural features of CISe thin film absorber layers. The investigated samples were produced by the three-stage co-evaporation process described above. Although two important differences to the standard process were made: First, the substrate temperature during second stage was significantly lower (T = 370±20°C). Furthermore, the process was interrupted at three different points in time. Thereby, three series were prepared having a Cu-rich (1.22 < Cu/In < 1.26), slightly Cu-rich (1.06 < Cu/In < 1.09), and Cu-poor (0.88 < Cu/In < 0.93) integral composition, respectively. The preparation process of the Cu-rich samples was stopped beyond the first stoichiometric point at the end of second stage and the process of the Cu-poor samples prior to the first stoichiometric point during second stage. Solely, the slightly Cu-rich samples partly underwent the third stage, but the process was interrupted before the samples passed the second stoichiometric point (cf. Fig. 4.8), hence, still retaining their Cu-surplus. Each series consists of seven samples with a distinct sodium content, which was controlled by varying the thickness of a NaF precursor layer (0, 1, 2, 3, 4, 6 and 12 nm) between the Mo back contact and the absorber. The NaF layer was thermally evaporated onto the Mo back contact prior to the CISe evaporation process. To inhibit the diffusion of Na from the soda lime glass substrate into the absorber, a SiN layer between the soda lime glass substrate and the Mo back contact was used.

The absorber layers were pre-characterization by GDOES in order to establish the cross-sectional element distribution. Fig. 4.11 shows the GDOES depth profile of each of the CISe sample equipped with 6-nm NaF precursor. The low process temperature, in particular T_2, is responsible for the poorer homogenization (cf. Fig. 4.8) and likely limits recrystallization significantly. The spike of the Cu concentration (red) at the surface and the simultaneous dent in the In content (blue) indicate the existence of Cu-Se phases at the absorber surface of the two-stage samples. On the contrary, Cu content decreases toward the back interface, whereas In and Se content increases concurrently. The low process temperature also causes Na to be remain more concentrated in the region of the NaF precursor layer between the absorber and the back contact. Information on the thickness of the CISe layer can be deduced from the onset of the Mo signal at about 1.5 µm (Fig. 4.11).

The phase content in dependence of the absorber depth has been investigated by GIXRD. In agreement with the results obtained from GDOES measurements, intensities of diffraction peaks emanating from Cu-Se phases (Cu_2Se, Cu_3Se_2) are highest at low incidence angles (i.e. $\omega = 0.5°$), and diminish or even vanish at higher incidence angles (i.e. $\omega = 5°$). In Fig. 4.12, the most striking differences regarding spatially varying content of Cu-Se phases are indicated by red arrows. Peaks being framed in blue are from the Mo back contact, which start to appear only at higher incidence angles. The feature framed in cyan ($2\theta \approx 26°$, $\omega = 5°$) arises from planar defects in CISe, which are commonly observed in break-off CISe absorbers. For more details on planar

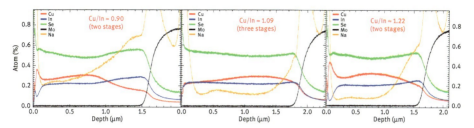

Fig. 4.11: GDOES results of CISe absorbers with 6 nm NaF precursor (left: two-stage Cu-poor, middle: three-stage slightly Cu-rich, right: two-stage Cu-rich). The Na concentration (yellow line) is not to scale. Data and graphs are attributed to Helena Stange, Helmholtz-Zentrum Berlin (HZB).

defects and their analysis in this kind of CISe absorber layers, the reader is referred to Stange et al. [38]. The segregation of secondary phases at the surface of the Cu-rich absorbers can be attributed to the break-off of the three-stage co-evaporation process after the second stage. In the standard process, the evaporation of In and Se during the third stage has this very purpose: to dissolve Cu-Se phases and to form $CuInSe_2$ [39]. The amount of secondary Cu-Se phases in the Cu-rich samples was significantly higher compared to the Cu-poor samples, but Cu-Se phases were still present in the Cu-poor samples. This is due to the low temperatures in the second stage decelerating the formation of the CISe phase and also due to effects caused by the presence of Na as it has been found experimentally by Rudmann et al. [40] and theoretically modeled by Oikkonen et al. [41] that Na impedes the diffusion of Cu and In during the formation of Cu poor CISe absorbers at low temperatures. Oikkonen et al. ascribed this effect mainly to Na occupying Cu vacancies in the CISe lattice. The results obtained for the samples presented here, in fact, support the finding that sodium hinders the formation of CISe, since the amount of secondary phases increases with increasing sodium content. It is remarkable that the Cu-poor 0-nm NaF precursor sample is the only sample showing virtually no secondary phases in neither GDOES nor GIXRD data, thus indicating the important influence of Na on hindering the Cu diffusion in Cu-poor CISe samples at low temperatures. However, the question remains why no indium-rich phases (e.g. In_2Se_3, $CuIn_5Se_8$) were detected by the GIXRD measurements, as could be expected for stoichiometric reasons for overall Cu-poor absorbers with superficially segregated Cu-Se phases and decreasing Cu/In toward the back contact. Possible explanations are that the indium-rich phase is either too thin to produce visible diffraction intensities or that no secondary phase is present but solely very Cu-poor CISe, with a high density of copper vacancies.

The microstructure and unit cell metrics of the CISe absorbers are heavily affected by the Cu/In ratio as the cation ratio determines both the amount of available structural sites tolerating sodium as well as the mode of element distribution and ordering. In the chalcopyrite-type structure, sodium more likely occupies the vacant Cu position (Na_{Cu} antisite) and, less probable, interstitial positions (Na_i) [41]. Sodium also strongly

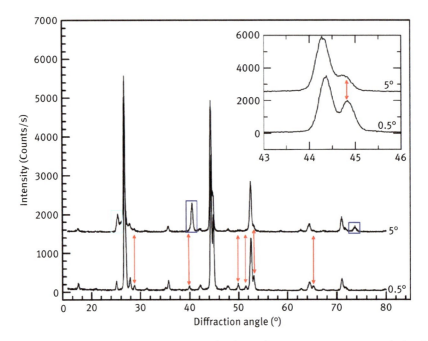

Fig. 4.12: GIXRD patterns of a Cu-poor CISe thin film with 6-nm NaF precursor recorded at different incidence angles to extract information from different depth levels of the absorber layer. The red arrows mark the peaks of the secondary phases. Peaks of secondary phases have higher intensities at ω = 0.5° (lower pattern). Note the increasing shoulder left of the main peak (θ ≈ 26°) marked by a cyan rectangle. This is due to stacking faults as explained in a recent paper by Stange et al. [38]. The peaks framed by blue rectangles are caused by the Mo backcontact. The 5°-pattern is plotted with an y-offset of 1500 counts/s.

modifies the grain growth and interdiffusion of metal ions (i.e. Cu^+, In^{3+}) [42]. Furthermore, it was found that Na preferentially segregates in the vicinity of grain boundaries with rather little amounts being in the interior of the grains [43, 44]. Of course, this difference in Na concentration cannot be resolved by XRD, as it provides volume-averaged information while other techniques do provide higher resolution and visualization (e.g. [45, 46]). However, the behavior of lattice parameters can give valuable insights into the structural responses mediated by the incorporation of sodium. Fig. 4.13 depicts the results with respect to microstrain and lattice parameters for the three CISe sample series. Irrespective of the Cu/In ratio, the break-off samples are more susceptible to changes in unit cell metrics when different NaF precursor thicknesses are used and process temperature is low. Omission of the third stage of the co-evaporation process causes a larger number of remaining structural defects, as two stages are not sufficient to adequately recrystallize the CISe absorber. As a result, the amount of secondary phases is higher and the overall bulk quality and the chemical homogeneity are reduced while the impact of Na is enhanced. This is reflected by stronger variations in microstrains and lattice parameters in dependence of the NaF precursor supply, likewise for Cu-poor and Cu-rich samples being interrupted during second stage. On the other hand, lattice parameters of the three-stage slightly Cu-rich sample series does not vary much with changing NaF precursor thickness, indicating that structural modifications triggered by Na during the growth of CISe are, at least to some extent, revoked by enhanced recrystallization. Larger grains result in reduction of grain boundaries per volume. Taking into account the preference of Na to be concentrated in grain boundary regions the volume-averaged lattice parameters will be less altered by Na and overall microstrains will be lower (Fig. 4.13, middle). However, in consideration of the larger ionic radius of Na$^+$ compared to Cu$^+$ and In^{3+}, it remains enigmatic why unit cell volumes progressively shrink when NaF precursor thickness increases. In the study presented above (cf. e.g. Fig. 4.10), in fact, the employment of a NaF precursor layer generally led to an increase in lattice parameter a and to a decrease in lattice parameter c, causing the tetragonal deformation (not shown) to be closer to one. This possibly means that interdiffusion and disordering in the CISe lattice is much more enhanced when a complete three-stage co-evaporation process is performed.

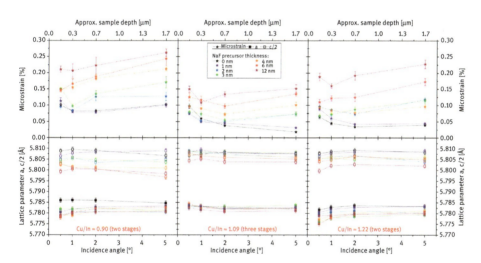

Fig. 4.13: Overview of the results for microstrains and lattice parameters of the three CISe sample series. Both the two-stage Cu-poor samples (left) and the two-stage Cu-rich samples (right) are more off-stoichiometric and inherently contain higher defect concentrations, making the enhanced accommodation of Na in the CISe lattice possible. The three-stage slightly Cu-rich samples (middle) are less vulnerable to changes in lattice parameters because of the progressive annihilation of lattice defects during the third stage.

4.2.2.3 Impact of absorber process conditions and substrate material on microstructure of CZTSe thin film absorber layers

Like in CIGSe solar cell absorber layers, the influence of alkali metals on grain growth and, finally, photovoltaic performance is currently also investigated for kesterite

(CZTSe)-based solar cells [47–51]. However, CZTSe is an emerging technology that is still struggling with various issues of more basic nature. First of all, the quaternary system is more complex and thus way more challenging in terms of phase purity, phase stability as well as structural features. Therefore, the fabrication of kesterite-type absorbers is commonly accompanied by the formation of secondary phases. This is normally also the case for chalcopyrite-type absorber since the composition is always off-stoichiometric. However, chalcopyrites tend to form superficial segregations of Cu-Se which can be readily removed by etchants while kesterites are prone to form not only several secondary phases basically comprising its binary components, but a larger amount is located throughout the entire absorber layer. Segregations of secondary phases on top of the CZTSe absorber layer can be likewise removed, but often, there are agglomerations of secondary phases within the bulk as well as segregations at the back interface which remain inaccessible. In general, this leads to a more heterogeneous and poorer bulk quality and considerably contributes to losses in photocurrent. Another issue arises from the decomposition of CZTSe into its binary components during the thermal treatment when being in contact with the metallic molybdenum back electrode. This decomposition reaction next to the molybdenum back contact releases selenium which commonly results in the pervasive formation of a thick $MoSe_2$ layer between CZTSe absorber and Mo back contact [52]:

$$2Cu_2ZnSnSe_4 \rightarrow 2Cu_2Se + 2ZnSe + 2SnSe + Se_2 \quad (4.13)$$

$$Mo + Se_2 \rightarrow MoSe_2 \quad (4.14)$$

Another issue arises from the disorder of Cu^+ and Zn^{2+} among the 2c and 2d structural sites, leading to spatial band gap fluctuations which presumably contribute critically to the V_{oc} deficit [53]. The amount of Cu-Zn disorder depends on the extent of off-stoichiometry and yet more on the cooling history of the kesterite-type compound (e.g. [13]).

The study presented here is dealing with four CZTSe thin film absorber samples prepared by a direct current magnetron sputtering process (DC-Sp) of metallic precursors and a subsequent annealing in selenium atmosphere. The DC-Sp process is physical vapor deposition (PVD), where atoms are ejected (i.e. sputtered) from a target by ion bombardment and eventually condense onto a substrate. The ions are generated by glow discharge when the electrical potential ΔU applied between two electrodes is sufficient to establish an electric current. By collisions of the charge carriers (electrons) with the sputtering gas, the gas atoms are ionized. The resulting cascade of secondary electrons then maintains and amplifies the plasma. However, if the trajectory of the electrons is only determined by collisional deflections and the electric field, such a configuration simply results in a closed circuit (i.e. diode), finally counteracting the potential. That is why the plasma has to be confined by a magnetic field which additionally increases the probability of ionization events resulting in denser plasma in the vicinity of the target [54]. In order to avoid contamination issues of the deposited film, inert gases are used [54, 55]. The CZTSe absorbers were sputtered on different substrate materials already coated with a molybdenum back contact by using argon plasma. As substrate materials, stainless steel (austenitic, ferritic) and soda lime glass were used. In case of austenitic and ferritic steel substrates, a chromium barrier layer was deployed between the substrate and the back contact to preclude the diffusion of iron and other contaminants into the back contact. For any substrate material, a thin (~10-nm) ZnO layer deposited onto the back contact has also been employed in order to prevent the decomposition of CZTSe in mutual contact with metallic Mo and to reduce the lattice mismatch between the CZTSe absorber and the Mo back contact [55]. The metals contained in kesterite were supplied by corresponding targets and were deposited sequentially at ambient temperature. In this first stage, the sputtering sequence of the metals is Cu/Sn/Cu/Zn. In the second stage, the metallic stack is treated by a reactive annealing process under selenium atmosphere, thereby CZTSe absorber layer starts to form [55]. The final integral composition of the grown thin film (CZTSe and secondary phases) is located in the Cu-poor (0.75 < Cu/(Zn+Sn) < 0.81) and Zn-rich (1.14 < Zn/Sn < 1.25) regime, as it delivers the best performing solar cells. The superficial formation of ZnSe and CuSe secondary phases was removed by etching using $KMnO_4+Na_2S$ [55]. So, the kesterite absorber is supposed to represent the uppermost layer of the thin film samples, which ensures an analysis being not interfered by overlying phases.

GIXRD patterns were recorded for each of the four CZTSe thin film samples using various incidence angles of the primary beam for depth-resolved characterization (Fig. 4.14, left). Every diffraction pattern show distinct features typical for different absorber depths. At low incidence angles, the irradiated sample volume is smallest; thus, intensities are lower. Toward higher incidence angles, the 110 and 211 reflections of the Mo back contact appears at about 40.5° and 73.7° diffraction angles, respectively. Another feature also visible at higher incidence angles is a broad peak at about 32° 2θ, which is attributable to 100/101 Bragg peaks of $MoSe_2$, meaning the formation of this phase has occurred despite using ZnO intermediate layer. Although there is no hint on CZTSe having decomposed, the formation of $MoSe_2$ is likely promoted during reactive annealing treatment under Se atmosphere. Fig. 4.14 (right) shows the results obtained for microstructure and lattice parameters. In general, the progression of microstrain through the absorber is comparable with those observed for CISe thin films, yet more distinctive. Both interface regions are characterized by higher microstrains while microstrain drops toward the mid-region. GDOES measurements demonstrated that the surface region is considerably depleted in Cu and slightly depleted in Zn, which were partly consumed upon the formation of CuSe and ZnSe secondary phases at the surface. The increase in microstrain is due to lattice mismatch between CZTSe absorber and ZnO intermediate layer, which is in part altered by the Mo lattice underneath. This distortional lattice mismatch is also evident from the behavior of lattice parameters between 4° and 5° incidence angles, that is, near the back interface (Fig. 4.14, bottom right). The important role of process temperature with respect to microstrain can be seen when comparing the microstrains between those samples being likewise deposited onto austenitic steel substrates (denoted as a-steel in Fig. 4.14) since microstrains are significantly higher if the CZTSe sample has been annealed at only 450°C in a one-step process. The ferritic steel substrate (denoted as f-steel in Fig. 4.14) promotes somewhat lower microstrains compared to austenitic steel under very same conditions (i.e. 550°C, two annealing steps). The overall lowest microstrains are observed when using the soda lime glass substrate. Apart from the expected increase in microstrain at each interface region, the bulk absorber exhibits constantly low microstrains and lattice parameters change monotonously until the back interface is approached.

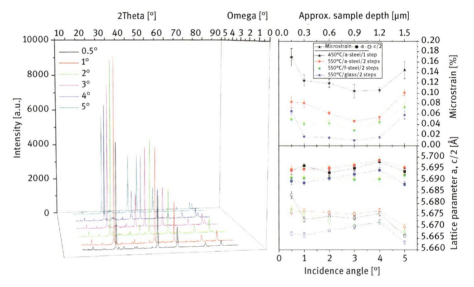

Fig. 4.14: Three-dimensional plot of GIXRD patterns recorded for the CZTSe sample on glass substrate (left). The amount of microstrain (upper right) and lattice parameters (lower right) is influenced by the substrate material being used (i.e. differential thermal expansion coefficients) as well as the process parameters.

Microstrains of the CZTSe absorber deposited on soda lime glass substrate have also been depicted by an "idealized" Williamson-Hall plot (Fig. 4.15, left). Domain sizes could not be detected; hence, all graphs meet at the origin of coordinates while the microstrain related to the different incidence angles measured is given by the slope of each graph (cf. Fig. 4.7 and Equation 4.13). Fig. 4.15 (right) compiles the results for microstrains and lattice parameters of the CZTSe absorber sample on soda lime glass. As aforementioned, the increased microstrain at the surface is a consequence of the larger amount of lattice defects introduced by declining Cu and Zn concentration. On the other hand, microstrain in the vicinity of the back interface is basically controlled by yet another mechanism. Distortional lattice mismatch and general interface properties do play a key role in affecting microstructure at the bottom part of the absorber, but element concentrations normally also changes toward the back interface.

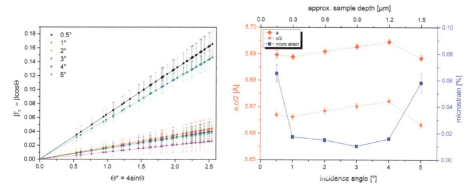

Fig. 4.15: (Left) Microstrains obtained using TCH-pV profile function as they would look in a Williamson-Hall plot. The microstrain can be directly inferred by calculating the slope. Domain sizes are given by the y-intersect (here equal to zero because domain sizes are too large and therefore inaccessible). (Right) The same microstrain values plotted together with lattice parameters vs. incidence angle (sample depth).

4.3 Summary

The three studies exemplarily presented in this chapter demonstrate the potential of microstructure analysis. Research aiming at improved power conversion efficiencies in solar cell devices requires the adjustment of plenty of parameters interacting among each other, until the best configuration is found. Unfortunately, the recipe providing the best overall conditions and that leads to optimal exploitation of solar energy is not obvious in any case. Normally, breakthroughs in solar cell device performance are due to coincidence and trial and error. Nowadays, however, extensive empirical experiences have already been made, and using the concept of trial and error seems to be not sufficient anymore. Therefore, theoretical considerations, device modeling and simulation as well as ab initio calculations became much more essential tools within the scope of systematically exploring properties and setups being beneficial for solar cell performance. However, the actual effect of the various approaches, in particular their repercussion on specific solar cell parameters, remains very complex. The microstructure analysis of functional layers building up thin film solar cells does help to fundamentally understand the impact of different process adjustments being applied on very small scales, which ultimately can be assembled to a general view. In addition to XRD, microstructural features can also be examined on different length scales using other techniques like electron backscatter diffraction (EBSD), Raman microspectroscopy, and high-resolution scanning transmission electron microscopy

(HR-STEM), allowing for detailed mappings of microstrain distribution [45] as well as comprehensive studies with respect to electronic properties [46].

References

[1] Einstein A. The photoelectric effect. Ann Phys 1905;17(132):4.
[2] Würfel P. Semiconductors. In: Physics of solar cells. Wiley-VCH, Weinheim; 2007:37–79.
[3] Suntola T, Antson J. Method for producing compound thin films; 1977. Google Patents US Patent 4,058,430.
[4] Jackson P, Wuerz R, Hariskos D, Lotter E, Witte W, Powalla M. Effects of heavy alkali elements in Cu (In, Ga) Se$_2$ solar cells with efficiencies up to 22.6%. Phys Status Solidi RRL 2016;10(8):583–6.
[5] Wang W, Winkler MT, Gunawan O, Gokmen T, Todorov TK, Zhu Y, Mitzi DB. Device characteristics of CZTSSe thin-film solar cells with 12.6% efficiency. Adv Energy Mater 2014;4(7).
[6] Nakayama N, Ito K. Sprayed films of stannite Cu$_2$ZnSnS$_4$. Appl Surf Sci 1996;92:171–5.
[7] Katagiri H, Sasaguchi N, Hando S, Hoshino S, Ohashi J, Yokota T. Preparation and evaluation of Cu$_2$ ZnSnS$_4$ thin films by sulfurization of E·B evaporated precursors. Solar Energy Mater Solar Cells 1997;49(1):407–14.
[8] Kühn G, Neumann A. AlBIIICVI2-Halbleiter mit Chalcopyritstruktur. Zeitschr Chem 1987;27(6):197–206.
[9] Hall SR, Szymanski JT, Stewart JM. Kesterite, Cu$_2$(Zn,Fe)SnS$_4$, and stannite, Cu$_2$(Fe,Zn)SnS$_4$, structurally similar but distinct minerals. Can Mineral 1978;16(2):131–7.
[10] Pamplin B. The adamantine family of compounds. Prog Crystal Growth Charact 1980;3(2):179–92.
[11] Jaffe J, Zunger A. Theory of the band-gap anomaly in ABC 2 chalcopyrite semiconductors. Phys Rev B 1984;29(4):1882.
[12] Schorr S, Hoebler H-J, Tovar M. A neutron diffraction study of the stannite-kesterite solid solution series. Eur J Mineral 2007;19(1):65–73.
[13] Többens DM, Gurieva G, Levcenko S, Unold T, Schorr S. Temperature dependency of Cu/Zn ordering in CZTSe kesterites determined by anomalous diffraction. Phys Status Solidi b 2016;253(10):1890–7.
[14] Schorr S. Structural aspects of adamantine like multinary chalcogenides. Thin Solid Films 2007;515(15):5985–91.
[15] Balzar D, Ledbetter H. Microstrains and domain sizes in Bi-Cu-O superconductors: an X-ray diffraction peak-broadening study. J Mater Sci Lett 1992;11(21):1419–20.
[16] Rodriguez-Carvajal, Juan. "FULLPROF version 3.0.0." Laboratorie Leon Brillouin, CEA-CNRS (2003).
[17] Le Bail A, Durot H, Fourquet JL. Ab initio structure determination of LiSbWO$_6$ by X-ray powder diffraction. Mater Res Bull 1988;(23):447.
[18] Thompson P, Cox D, Hastings J. Rietveld refinement of Debye-Scherrer synchrotron X-ray data from Al$_2$O$_3$. J Appl Crystallogr 1987;20(2):79–83.
[19] Finger LW, Cox DE, Jephcoat AP. A correction for powder diffraction peak asymmetry due to axial divergence. J Appl Crystallogr 1994;27(6):892–900.
[20] Caglioti G, Paoletti FP, Ricci FP. Choice of collimators for a crystal spectrometer for neutron diffraction. Nucl Instrum 1958;3(4):223–8.

[21] Young RA. The Rietveld method. Oxford University Press, Oxford, UK; 1995.
[22] Scherrer P. Estimation of the size and internal structure of colloidal particles using X-rays. Nachr Ges Wiss Göttingen 1918;2:96–100.
[23] Stokes A, Wilson A. The diffraction of X rays by distorted crystal aggregates – I. Proc Phys Soc 1944;56(3):174.
[24] Williamson G, Hall W. X-ray line broadening from filed aluminium and wolfram. Acta Metall 1953;1(1):22–31.
[25] Halder NC, Wagner CNJ. "Separation of particle size and lattice strain in integral breadth measurements." Acta Crystallographica 20.2 (1966): 312-313.
[26] Snyder RL, Bunge HJ, Fiala J. Defect and microstructure analysis by diffraction. Oxford University Press, Oxford, UK; 1990.
[27] Dinnebier RE. Powder diffraction: theory and practice. Royal Society of Chemistry, Cambridge, UK; 2008.
[28] Hengel I, Neisser A, Klenk R, Lux-Steiner MC. Current transport in $CuInS_2$: Ga/Cds/Zno-solar cells. Thin Solid Films 2000;361:458–62.
[29] Ito K, et al. Theoretical model and device performance of $CuInS_2$ thin film solar cell. Jpn J Appl Phys 2000;39(1R):126.
[30] Klenk R. Characterisation and modelling of chalcopyrite solar cells. Thin Solid Films 2001;387(1):135–40.
[31] Rudmann D, da Cunha AF, Kaelin M, Haug FJ. Effects of Na on the growth of Cu(In,Ga) Se_2 thin films and solar cells. In: MRS Proceedings. Cambridge University Press, Cambridge, UK; 2003.
[32] Chirilă A, Reinhard P, Pianezzi F, Bloesch P, Uhl AR, Fella C, Kranz L, Keller D, Gretener C, Hagendorfer H, Jaeger D, Erni R, Nishiwaki S, Buecheler S, Tiwari AN. Potassium-induced surface modification of Cu(In,Ga) Se_2 thin films for high-efficiency solar cells. Nat Mater 2013;12(12):1107–11.
[33] Mainz R, Simsek Sanli E, Stange H, Azulay D, Brunken S, Greiner D, Hajaj S, Heinemann MD, Kaufmann CA, Klaus M, Ramasse QM, Rodriguez-Alvarez H, Weber A, Balberg I, Millo O, van Aken PA, Abou-Ras D. Annihilation of structural defects in chalcogenide absorber films for high-efficiency solar cells. Energy Environ Sci 2016;9(5):1818–27.
[34] Stephan C, Greiner D, Schorr S, Kaufmann CA. The influence of sodium on the point defect characteristics in off stoichiometric $CuInSe_2$. J Phys Chem Solids 2016;98:309–15.
[35] Caballero R, Kaufmann CA, Efimova V, Rissom T, Hoffmann V, Schock HW. Investigation of Cu(In,Ga)Se_2 thin-film formation during the multi-stage co-evaporation process. Prog Photovolt Res Appl 2013;21(1):30–46.
[36] Bodegard M, Stolt L, Hedstrom J. The influence of sodium on the grain structure of $CuInSe_2$ films for photovoltaic applications. In: 12th European Photovoltaic Solar Energy Conference, 1994.
[37] Rudmann D, da Cunha AF, Kaelin M, Kurdesau F, Zogg H, Tiwari AN. Efficiency enhancement of Cu(In,Ga)Se_2 solar cells due to post-deposition Na incorporation. Appl Phys Lett 2004;84(7):1129–31.
[38] Stange H, Brunken S, Hempel H, Rodriguez-Alvarez H, Schäfer N, Greiner D, Scheu A, Lauche J, Kaufmann CA, Abou-Ras D, Mainz R. Effect of Na presence during $CuInSe_2$ growth on stacking fault annihilation and electronic properties. Appl Phys Lett 2015;107(15):152103.
[39] Gabor AM, Tuttle JR, Albin DS, Contreras MA, Noufi R. High-efficiency $CuIn_xGa_{1-x}Se_2$ solar cells made from $(In_x, Ga_{1-x})_2Se_3$ precursor films. Appl Phys Lett 1994;65(2):198–200.
[40] Rudmann D, Brémaud D, da Cunha AF, Bilger G, Strohm A, Kaelin M, Zogg H, Tiwari AN. Sodium incorporation strategies for CIGS growth at different temperatures. Thin Solid Films 2005;480:55–60.

[41] Oikkonen LE, Ganchenkova MG, Seitsonen AP, Nieminen RM. Effect of sodium incorporation into $CuInSe_2$ from first principles. J Appl Phys 2013;114(8):083503.
[42] Rudmann D, Bilger G, Kaelin M, Haug FJ, Zogg H, Tiwari AN. Effects of NaF coevaporation on structural properties of $Cu(In,Ga)Se_2$ thin films. Thin Solid Films 2003;431:37–40.
[43] Kronik L, Cahen D, Schock HW. Effects of sodium on polycrystalline $Cu(In,Ga)Se_2$ and its solar cell performance. Adv Mater 1998;10(1):31–6.
[44] Cojocaru-Mirédin O, Choi P, Wuerz R, Raabe D. Atomic-scale distribution of impurities in CuInSe 2-based thin-film solar cells. Ultramicroscopy 2011;111(6):552–6.
[45] Schäfer N, Wilkinson AJ, Schmid T, Winkelmann A, Chahine GA, Schülli TU, Rissom T, Marquardt J, Schorr S, Abou-Ras D. Microstrain distribution mapping on $CuInSe_2$ thin films by means of electron backscatter diffraction, X-ray diffraction, and Raman microspectroscopy. Ultramicroscopy 2016;169:89–97.
[46] Abou-Ras D, Schmidt SS, Schäfer N, Kavalakkatt J, Rissom T, Unold T, Mainz R, Weber A, Kirchartz T, Simsek Sanli E, van Aken PA, Ramasse QM, Kleebe HJ, Azulay D, Balberg I, Millo O, Cojocaru-Mirédin O, Barragan-Yani D, Albe K, Haarstrich J, Ronning C. Compositional and electrical properties of line and planar defects in $Cu(In,Ga) Se_2$ thin films for solar cells – a review. Phys Status Solidi RRL 2016;10(5):363–375.
[47] Prabhakar T, Jampana A. Effect of sodium diffusion on the structural and electrical properties of Cu_2ZnSnS_4 thin films. Solar Energy Mater Solar Cells 2011;95(3):1001–4.
[48] Li JV, Kuciauskas D, Young MR, Repins IL. Effects of sodium incorporation in Co-evaporated $Cu_2ZnSnSe_4$ thin-film solar cells. Appl Phys Lett 2013;102(16):163905.
[49] Tong Z, Yan C, Su Z, Zeng F, Yang J, Li Y, Jiang L, Lai Y, Liu F. Effects of potassium doping on solution processed kesterite Cu_2ZnSnS_4 thin film solar cells. Appl Phys Lett 2014;105(22):223903.
[50] Xin H, Vorpahl SM, Collord AD, Braly IL, Uhl AR, Krueger BW, Ginger DS, Hillhouse HW. Lithium-doping inverts the nanoscale electric field at the grain boundaries in $Cu_2ZnSn(S,Se)_4$ and increases photovoltaic efficiency. Phys Chem Chem Phys 2015;17(37):23859–66.
[51] Hsieh YT, Han Q, Jiang C, Song TB, Chen H, Meng L, Zhou H, Yang Y. Efficiency enhancement of $Cu_2ZnSn(S,Se)_4$ solar cells via alkali metals doping. Adv Energy Mater 2016;6:1502386.
[52] Liu X, Feng Y, Cui H, Liu F, Hao X, Conibeer G, Mitzi DB, Green M. The current status and future prospects of kesterite solar cells: a brief review. Prog Photovolt Res Appl 2016;24:879–98.
[53] Kelly P, Arnell R. Magnetron sputtering: a review of recent developments and applications. Vacuum 2000;56(3):159–72.
[54] López-Marino S, Neuschitzer M, Sánchez Y, Fairbrother A, Espindola-Rodriguez M, López-García J, Placidi M, Calvo-Barrio L, Pérez-Rodriguez A, Saucedo E. Earth-abundant absorber based solar cells onto low weight stainless steel substrate. Solar Energy Mater Solar Cells 2014;130:347–53.
[55] López-Marino S, Placidi M, Pérez-Tomás A, Llobet J, Izquierdo-Roca V, Fontané X, Fairbrother A, Espíndola-Rodríguez M, Sylla D, Pérez-Rodríguez A, Saucedo E. Inhibiting the absorber/Mo-back contact decomposition reaction in $Cu_2ZnSnSe_4$ solar cells: the role of a ZnO intermediate nanolayer. J Mater Chem A 2013;1(29):8338–43.

Peter Majewski, Karyn Jarvis and Behnam Akhavan
5 Surface-engineered silica via plasma polymer deposition

5.1 Plasma polymerization

Silica, in its various crystallographic modifications, is one of the most abundant mineral in the earth's crust and on the surface. In the form of quartz sand and sand stone, it is widely used in construction for millennia. In modern times, the application of quartz sand is also considered for innovative technologies, for example as raw material for the fabrication of silicon used for semiconductors and as simple quartz sand in energy storage systems for capturing solar energy [1]. Silica is also manufactured in vast quantities as amorphous silica or silica gel and used in a range of applications, such as addition to elastomers to increase their strength and durability, in paints and coatings for its white color and also to increase the scratch resistivity of the coatings, as absorbents for chromatography and de-humidifier, in water treatment and food processing, and many more applications [2]. Often, silica gel is applied in a surface functionalized form that allows to change the surface charge of the particles or to trigger specific chemical reactions on the surface to bond the silica particles to specific compounds. Surface functionalization can be achieved through various techniques, such as by using silanes with specific moieties [3], or polymers deposited through wet chemistry approaches or by using plasma polymerization.

It is well documented that surfaces exposed to low-pressure glow discharges of organic vapors are coated with a solid polymer film. This phenomenon is now known as "plasma polymerization" [4]. Goodman used a plasma polymer film in parts of nuclear batteries in 1960 [5]. Since then, plasma polymerization has been employed for a large number of applications, and significant research was conducted in the synthesis, application, and characterization of plasma polymer films [6–12].

Surface modification through plasma polymerization provides a number of advantages. Plasma polymerization is a substrate-independent process by which polymer films can be deposited onto almost all solid materials including metals, ceramics, polymers, and even low-cost natural materials. The selection of such substrates is independent of their shape, surface chemistry, and geometry [8]. Plasma polymer films only influence the surface of a substrate without altering the bulk properties [13–17]. Plasma polymerization is also a solvent-free and environmentally friendly process producing virtually no waste. Plasma polymer films are chemically and physically stable, homogeneous, void-free, highly cross-linked, and often highly adherent to the substrate [18, 19]. Such unique properties of plasma polymer films have led to their extensive applications in the production of biocompatible [20, 21], anti-bacterial [22, 23], adhesive [24], super-hydrophobic [25, 26], anti-fouling [6], and barrier coatings [27, 28].

https://doi.org/10.1515/9783110497342-005

Figure 5.1 shows a schematic drawing of a rotary plasma generator which was extensively used for the preparation of plasma polymer films on silica particles as described below. The 13.56-MHz RF reactor was usually run at various RF power, chamber base pressures, and flow rate of degassed monomers depending on the desired chemistry plasma polymer film. The chamber rotation speed during polymerization was kept to 14 rpm.

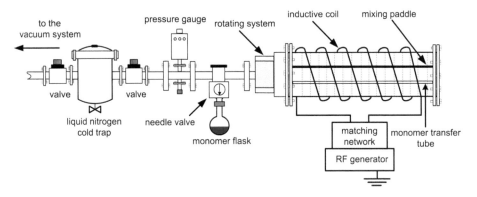

Fig. 5.1: Schematic illustration of a rotating plasma polymerization reactor.

5.2 Polymer plasma synthesis of amine-functionalized silica particles

The efficiency of amine functionalization of silica particles for the removal of organic contaminants and pathogens in water has been well demonstrated [29–33]. The use of plasma polymer deposition techniques provides a synthesis method that uses much less chemical ingredients such as commonly used 3-aminopropyl trimethoxysilane [29–33] and is much more time efficient.

Plasma polymerization of allylamine has been demonstrated to be a successful method for the formation of amine-functionalized quartz particles. Allylamine plasma-polymerized films can be prepared within 60 minutes at 25 W of plasma energy exhibiting a linear relationship between polymerization time and amine concentration and a homogeneous distribution of nitrogen species on quartz particles (Fig. 5.2) [14].

X-ray photoelectron spectroscopy (XPS) surface elemental compositions of uncoated and allylamine plasma polymer-coated silica particles are shown in Fig. 5.3. It can be observed that with an increase in polymerization time, the atomic concentrations of carbon and nitrogen increase, while those of oxygen and silicon decrease. The atomic concentrations of carbon and nitrogen increase with

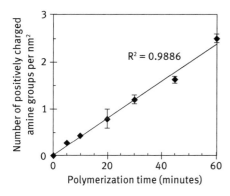

Fig. 5.2: Number of positively charged adsorption sites per square nanometer as a function of polymerization time for allylamine plasma-polymerized quartz particles.

polymerization time as more allylamine fragments are deposited on the surface. Silicon and oxygen concentrations decrease with polymerization time as the polymer layer gets thicker; thus, the signal from the underlying quartz particles is obscured. Increasing the polymerization time past 45 minutes does not appear to result in further modification of the atomic concentration of the particle surface. For polymerization times of 45 and 60 minutes, silicon makes no contribution to the surface atomic concentration. For a Mg K_α source, XPS has a silicon sampling depth of 8.4 nm [34]. It can be therefore concluded that a polymerization time of 45 minutes results in a polymer layer that is at least 8.4 nm thick. Oxygen is often present in the XPS spectra of plasma-polymerized surfaces either from residual oxygen inside the plasma chamber or from post-deposition reactions with atmospheric oxygen once samples are exposed to air [35].

Time of flight – secondary ion mass spectrometry (ToF-SIMS) studies also show that the allylamine coating already occurs after ~5 minutes of polymerization, which is followed by an increase of the thickness of the coating (Fig. 5.4) [14].

According to the linear increase of the amine concentration on the surface of the particles, a steady shift of the isoelectric point (IEP) toward higher pH values can be

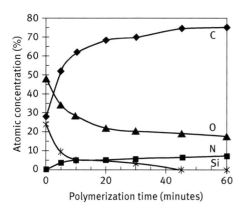

Fig. 5.3: XPS atomic survey concentrations of allylamine coated quartz particles as a function of polymerization time.

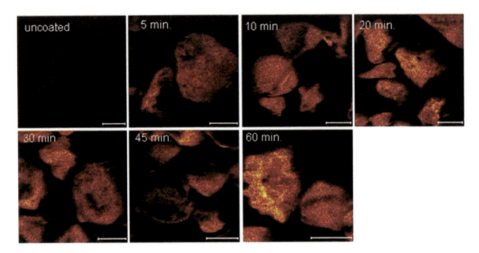

Fig. 5.4: CN⁻ ion maps from ToF-SIMS for uncoated and allylamine plasma-polymerized quartz particles (scale bar: 100 μm). CN⁻ species has been selected because it represents a tracer ion for allylamine coating. Although a relatively uniform distribution of CN⁻ species was observed for all polymerization times, its density and thickness appear to increase with polymerization time.

observed (Fig. 5.5) [14]. Uncoated quartz particles have an IEP of 5.1 due to the surface SiO⁻ groups. Plasma polymerization for 5 minutes increases the IEP to 6.0. Increasing the polymerization time further gradually increases the IEP, which reaches a maximum of 7.5 for a polymerization time of 60 minutes. The chemistry of the amines is dominated by the ability of the lone pair on the nitrogen atom to capture protons forming NH_3^+-groups. This phenomenon is the reason for the positive charge of the

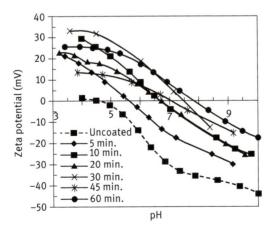

Fig. 5.5: Zeta potential of uncoated and allylamine plasma-polymerized quartz particles as a function of pH for various polymerization times.

allylamine coated particles at pH values below ~7. Contrary to that, at pH values above ~7, the hydrogen of the amines is sufficiently acidic to undergo deprotonation forming negatively charged NH⁻ -groups, explaining the negative surface charge of the particles at such pH values.

5.3 Polymer plasma synthesis of sulfonate-functionalized silica particles

Thiophene has been demonstrated to be a suitable precursor monomer for the formation of sulfur-functionalized silica particles via polymer plasma deposition. The use of thiophene provides a versatile method for the functionalization of silica particles with a range of different sulfur-containing moieties (Fig. 5.6).

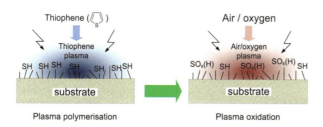

Fig. 5.6: Schematic drawing of the formation of a thiophene film and its subsequent oxidation via either air or oxygen plasma treatment to form SO_xH moieties.

Plasma polymerization of thiophene for approximately 30 minutes results in the deposition of a sulfur-rich coating on silica particles. The thickness of this plasma polymer coating is expected to be at least 10 nm as indicated by the absence of silicon signals in the XPS survey spectrum (Fig. 5.7) [33–35]. Even after short polymerization times of only few minutes, sulfur signals can be detected on the surface of the silica particles followed by a steady increase in sulfur concentration.

The spatial distribution of SH⁻ counts for the uncoated and plasma-polymerized thiophene (PPT) coated quartz silica particles are shown in Fig. 5.8. As observed, no SH⁻ counts were detected for the uncoated quartz silica particles, whereas an abrupt increase of these counts was clearly visible after 2 minutes of polymerization. Further increasing the polymerization time did not show any significant increase of the distribution density of SH⁻ ions. This may indicate that a homogeneous layer of PPT can be achieved after only ~2 minutes of polymerization and longer polymerization times result in an increase of the thickness of the films as indicated by the XPS analysis.

Planar silicon substrates were coated along with the silica particles and were applied for thickness measurements using ellipsometry. The increase of the

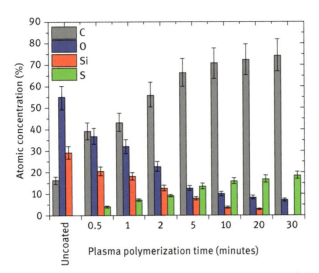

Fig. 5.7: XPS survey spectra atomic concentrations of uncoated and PPT-coated quartz silica particles as a function of plasma polymerization time (plasma-specific energy = 0.08 kJ/cm³). The disappearance of silicon at the polymerization time of 30 minutes indicates a film thickness of more than 8.4 nm.

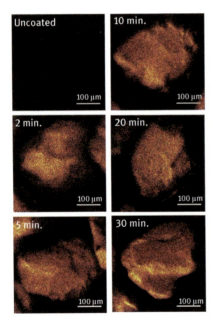

Fig. 5.8: SH⁻ ion distribution maps from ToF-SIMS for uncoated and PPT-coated silica particles (plasma-specific energy = 0.08 kJ/cm³).

thiophene film thickness has again been demonstrated to be in linear dependency to the polymerization time [36–38] (Fig. 5.9).

Plasma treatment of thiophene films via either air or oxygen plasmas can be applied to transform the SH functionality provided by the thiophene film into SO_xH

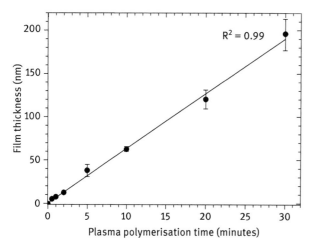

Fig. 5.9: Ellipsometry thicknesses of PPT coatings deposited onto planar silicon substrates as a function of plasma polymerization time (plasma-specific energy = 0.08 kJ/cm³).

functionalities. Oxidation of polymers via oxidative plasmas is always accompanied by a degree of ablation [36–38] (Fig. 5.10). The predominance of one of these two processes over the other is predominately controlled by the processing parameters such as chemistry of the gas, input specific energy, and treatment time [39]. The kinetic energy of the ions arriving at the surface is highly dependent on external parameters of the plasma polymerization process including the gas flow rate and input RF power [40]. The power-to-flow rate ratio (W/F), which shows the available energy per unit volume of air/oxygen, directly correlates with the energy of ions attacking the surface [39].

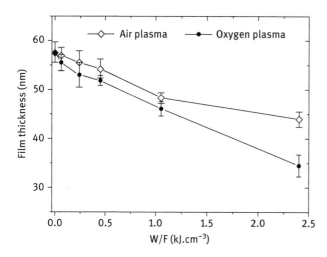

Fig. 5.10: Film thickness of PPT films treated by air or oxygen plasmas as a function of plasma-specific energy (plasma treatment time = 2 minutes).

The concentration of oxygen atoms at high input specific energies is considerably lower for air plasma compared to oxygen plasma. Such a difference may be explained by the higher oxidation kinetics of oxygen plasma which results in an immediate oxidation of the ablated surface. As observed, by increasing the plasma-specific energy, the thickness of the thiophene film decreases for both air and oxygen plasmas, thus suggesting significant etching of the films. At W/F values greater than 0.45 kJ/cm^3, the etching of thiophene film becomes more significant. For example, the thickness reduction of the film upon oxygen plasma treatment increases from less than 3% for W/F = 0.06 kJ/cm^3 to more than 40% for W/F = 2.4 kJ/cm^3. It appears that at high W/F ratios, the balance of oxidation and ablation processes tends more toward ablation. Such behavior can be explained by a greater number of ions that possess sufficient energy to etch the thiophene film before it becomes significantly oxidized.

The surface-incorporated sulfur-oxygen moieties can be identified via ToF-SIMS – normalized negative counts shown in Fig. 5.11 [36–38]. It is observed that air/oxygen

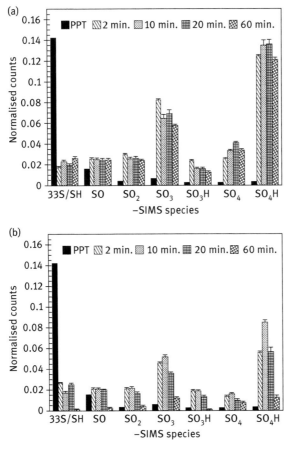

Fig. 5.11: Average normalized negative SIMS counts of sulfur-oxygen species from ToF-SIMS for PPT films treated by (a) air and (b) oxygen (95% confidence intervals – N ≥ 9, plasma-specific energy = 0.24 kJ/cm^3)

plasma treatment of PPT film results in a significant decrease of thiol (SH) counts and an increase in oxidized sulfur fragments. Consistent with XPS results, sulfur atoms in untreated PPT film are mainly in the form of thiol (SH) and sulfoxide (SO) groups. As expected, sulfur-containing functionalities at high oxidation states are not significantly observed for untreated PPT film. On the contrary, a significant contribution of $-SO_x(H)$ species is observed for air/oxygen plasma treated PPT films even at short polymerization times. The main oxidized sulfur-containing species are at high oxidation states that include SO_2, SO_3, SO_3H, SO_4, and SO_4H.

According to the increase of SO_xH groups on the surface of the silica particles, a decrease of the IEP of the particles can be observed, which is due to the strong deprotonation behavior of SO_xH moieties. Such strongly negatively charged surfaces can be applied for the removal of metal cations from water. The use of silica particles coated with oxidized thiophene films for the removal of copper and zinc from water has recently been demonstrated [41].

Tab. 5.1: Maximum adsorption capacity, maximum removal efficiency, and optimum removal time for plasma polymer-functionalized silica particles and a number of other copper and zinc adsorbents reported in the literature [35].

Adsorbent	Adsorbate	Maximum adsorption capacity (mg/g)	Maximum removal efficiency (%)	Optimum removal time	Solution pH
Plasma polymer-functionalized silica particles	Cu	25.0	>96.7	60 minutes	5.5
	Zn	27.4	>96.7	60 minutes	5.5
Polyvinyl alcohol/EDTA resin	Zn	38.6	~75	90 minutes	5
Graphene oxide/Fe_3O_4 composites	Cu	18.3	~55	10 hours	5.3
Aminated electrospun polyacrylonitrile microfiber	Cu	~24	~60	12 hours	4
Aminated electrospun polyacrylonitrile nanofiber	Cu	~36	~90	12 hours	4
Bentonite	Zn	~20	~80	6 hours	4
MnO_2-coated magnetic nanocomposites	Cu	Not reported	~98	30 minutes	6.3
	Zn		~80	30 minutes	6.3
Fly ash	Zn	Not reported	~98	3 hours	6

5.4 Plasma polymer synthesis of hydrophobic films on silica particles

In recent years, functionalized hydrophobic materials have attracted considerable interest as oil-removal agents. Plasma polymerization has been applied as a novel method to develop hydrophobic and oleophilic particles for water purification. Plasma-polymerized 1,7-octadiene (PPOD) was deposited onto silica particles [42].

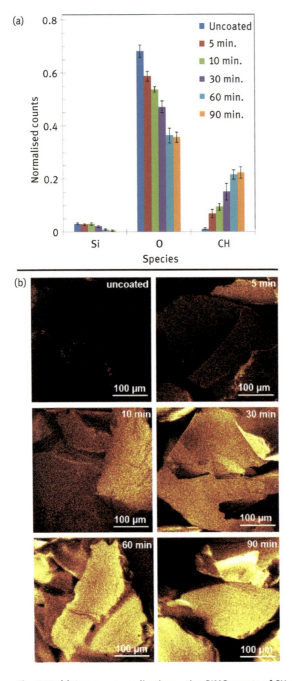

Fig. 5.12: (a) Average normalized negative SIMS counts of CH, Si, and O from ToF-SIMS for uncoated and PPOD-coated silica particles (95% confidence intervals, N ≥ 9), (b) CH⁻ ion distribution maps from ToF-SIMS for uncoated and PPOD-coated silica particles indicating homogeneous PPOD films after ~60 minutes of polymerization.

It was shown that within ~60 minutes of polymerization, homogeneous films with a thickness of approximately 10 nm can be prepared on silica particles (Fig. 5.12), resulting in significant enhancement of the hydrophobic character (Fig. 5.13).

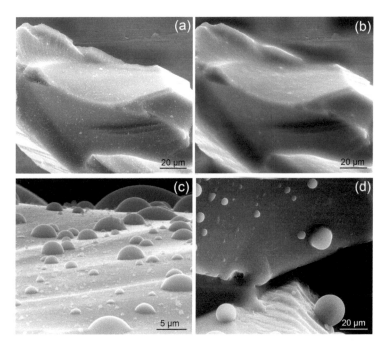

Fig. 5.13: Environmental scanning electron microscopy (ESEM) images of (a) dry uncoated silica particle, (b) wetted uncoated silica particle, (c) silica particle coated at 10 minutes, and (d) silica particle coated at 60 minutes.

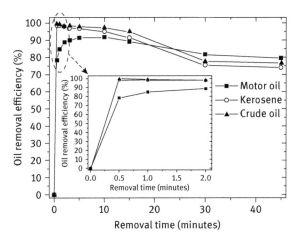

Fig. 5.14: Oil removal efficiency for PPOD-coated particles as a function of removal time (Ci = 20 g/L, particle mass = 40 g/L, plasma polymerization time = 45 minutes).

The effectiveness of PPOD-coated particles for the removal of hydrophobic matter from water was demonstrated by adsorption of motor oil, kerosene, and crude oil [43]. The PPOD-coated particles have shown to have excellent oil removal efficiency (Fig. 5.14). These particles were capable of removing 99.0–99.5% of high viscosity motor oil in 10 minutes, while more than 99.5% of low-viscosity crude oil and kerosene was adsorbed in less than 30 seconds. Plasma polymerization has shown to be a promising approach to produce a new class of materials for fast, facile, and efficient oil removal.

Tab. 5.2: Adsorption capacity, removal efficiency, and removal time for plasma polymer-coated silica particles and a number of other oil adsorbents reported in the literature [37].

Absorbent	Absorbate	Adsorption capacity (g/g)	Removal efficiency (%)	Removal time
Plasma polymer-coated silica particles	Motor oil	0.33	>99	10 minutes
	Crude oil	0.5	>99	15 seconds
	Kerosene	0.5	>99	15 seconds
Tanned wastes	Motor oil	5	98.6	10 minutes
Sugar cane bagasse	Gasoline	~1.2	94.3	60 minutes
Perlite	Natural petroleum-contaminated water	0.27	91	100 minutes
Walnut shell	Standard mineral oil	0.8	83	6 hours
Ionic liquid treated yellow horn shell	Crude oil	0.6	Not reported	3 hours
Macro-porous butyl rubber	Crude oil	23	Not reported	5 minutes
Magnetic macro-porous carbon nanotubes	Diesel oil	56	Not reported	30 minutes

References

[1] Schlipf D, Schicktanz P, Maier H, Schneider G. Using sand and other small grained materials as heat storage medium in a packed bed HTTESS. Energy Proc 2015;69:1029–38.

[2] Waddell WH, Evans LR. Silica – amorphous silica. In: Kirk-Othmer encyclopedia of chemical technology. John Wiley & Sons; New York, USA, 2001.

[3] Majewski P. Synthesis and application of surface engineered silica (SES). In: Der T, editor. Nanotechnology for water purification. Central Eur J Chem 2013:209–34, Montreal, Canada.

[4] Yasuda H. Glow discharge polymerization. J Polym Sci 1981;16:199–293.

[5] Goodman J. The formation of thin polymer films in the gas discharge. J Polym Sci 1960;44:551–2.

[6] Favia P, d'Agostino R. Plasma treatments and plasma deposition of polymers for biomedical applications. Surf Coat Technol 1998;98:1102–6.

[7] Biederman H. Polymer films prepared by plasma polymerization and their potential application. Vacuum 1987;37:367–73.

[8] Chan CM, Hiraoka H, Ko TM. Polymer surface modification by plasmas and photons. Surf Sci Rep 1996;24:3–54.

[9] Biederman H, Slavínská D. Plasma polymer films and their future prospects. Surf Coat Technol 2000;125:371–6.
[10] Osada Y. Plasma polymerization and plasma treatment of polymers. Review 1988;30:1922–41.
[11] Yasuda H, Gazicki M. Biomedical applications of plasma polymerization and plasma treatment of polymer surfaces. Biomaterials 1982;3:68–77.
[12] Michelmore A, Steele DA, Whittle JD, Bradley JW, Short RD. Nanoscale deposition of chemically functionalised films via plasma polymerization. RSC Adv 2013;3:13540–57.
[13] Jarvis KL, Majewski P. Influence of film stability and aging of plasma polymerized allylamine coated quartz particles on humic acid removal. ACS Appl Mater Interfaces 2013;5(15):7315–22.
[14] Jarvis KL, Majewski P. Plasma polymerized allylamine coated quartz particles for humic acid removal. J Colloid Interface Sci 2012;380:150–8.
[15] Jarvis KL, Majewski P. Removal of acid orange 7 dye from water via plasma polymerized allylamine coated quartz particles. Water Air Soil Pollut 2014;225:2227–36.
[16] Jarvis KL, Majewski P. Influence of particle mass and flow rate on plasma polymerized allylamine coated quartz particles for humic acid removal. Plasma Process Polym 2015;12(1):42–50.
[17] Jarvis KL, Majewski P. Optimizing the surface chemistry of plasma polymerized allylamine coated particles for humic acid removal. Plasma Process Polym 2016;13(8):802–13.
[18] Vandencasteele N, Reniers F. Plasma-modified polymer surfaces: characterization using XPS. J Electron Spectrosc Relat Phenom 2010;178(9):394–408.
[19] Easton CD, Jacob MV, Shanks RA. Fabrication and characterisation of polymer thin-films derived from cineole using radio frequency plasma polymerization. Polymer 2009;50:3465–9.
[20] Farhat S, Gilliam M, Rabago-Smith M, Baran C, Walter N, Zand A. Polymer coatings for biomedical applications using atmospheric pressure plasma. Surf Coat Technol 2014;241:123–9.
[21] Choi C, Hwang I, Cho YL, Han SY, Jo DH, Jung D, Moon DW, Kim EJ, Jeon CS, Kim JH, Chung TD, Lee TG. Fabrication and characterization of plasma-polymerized poly(ethylene glycol) film with superior biocompatibility. ACS Appl Mater Interfaces 2013;5:697–702.
[22] Qureshi AT, Terrell L, Monroe WT, Dasa V, Janes ME, JGimble JM, Hayes DJ. Antimicrobial biocompatible bioscaffolds for orthopaedic implants. J Tissue Eng Regen Med 2014;8:386–95.
[23] Kumar V, Jolivalt C, Pulpytel J, Jafari R, Arefi-Khonsari F. Development of silver nanoparticle loaded antibacterial polymer mesh using plasma polymerization process. J Biomed Mater Res A 2013;101:1121–32.
[24] Moreno-Couranjou M, Blondiaux N, Pugin R, Le Houerou V, Gauthier C, Kroner E, Choquet P. Bio-inspired nanopatterned polymer adhesive: a novel elaboration method and performance study. Plasma Process Polym 2014;11:647–54.
[25] Woodward I, Schofield WCE, Roucoules V, Badyal JPS. Super-hydrophobic surfaces produced by plasma fluorination of polybutadiene films. Langmuir 2003;19:3432–8.
[26] Irzh A, Ghindes L, Gedanken A. Rapid deposition of transparent super-hydrophobic layers on various surfaces using microwave plasma. ACS Appl Mater Interfaces 2011;3:4566–72.
[27] Lin Y, Yasuda H, Miyama M, Yasuda T. Water barrier characteristics of plasma polymers of perfluorocarbons. J Polym Sci Pol Chem 1996;34:1843–51.
[28] Bahroun K, Behm H, Mitschker F, Awakowicz P, Dahlmann R, Hopmann C. Influence of layer type and order on barrier properties of multilayer PECVD barrier coatings. J Phys D Appl Phys 2014;47:1–8.
[29] Majewski P. Removal of organic matter in water by functionalised self-assembled monolayers on silica. Separ Purif Technol 2007;57:283–8.
[30] de los Reyes M, Keegan A, Monis P, Majewski P. Removal of pathogens by functionalised self-assembled monolayers. J Water Supply Res Technol Aqua 2008;57(2):93–100.
[31] Majewski P, Chan CP. Removal of Cryptosporidium parvum by silica coated with functionalized self-assembled monolayers. J Nanosci Nanotech 2008;8:1–4.

[32] Majewski P; Keegan A. Surface properties and water treatment capacity of surface engineered silica coated with 3-(2-aminoethyl) aminopropyltrimethoxysilane. Appl Surf Sci 2012;258:2454–8.

[33] Majewski P, Luong J; Stretton K. The application of surface engineered silica for the treatment of sugar containing wastewater. Water Sci Technol 2012;65:46–52.

[34] Briggs D, Surface analysis of polymers by XPS and static SIMS. In: Clarke DR, Suresh S, Ward IM, editors. Cambridge solid state science series. Cambridge University Press, Cambridge, UK; 1998.

[35] Wang X, Wang J, Yang Z, Leng Y, Sun H, Huang N. Structural characterization and mechanical properties of functionalized pulsed-plasma polymerized allylamine film. Surf Coat Technol 2010;204:3047.

[36] Akhavan B, Jarvis K, Majewski P, Development of negatively charged particulate surfaces through a dry plasma-assisted approach. RSC Adv 2015;5:12910–21.

[37] Akhavan B, Jarvis K, Majewski P, Plasma polymerization of sulfur-rich and water-stable coatings on silica particles. Surf Coat Technol 2015;264:72–9.

[38] Akhavan B, Jarvis K, Majewski P, Development of oxidized sulfur polymer films through a combination of plasma polymerization and oxidative plasma treatment. Langmuir 2014;30:1444–54.

[39] Hegemann D, Körner E, Guimond S. Plasma polymerization of acrylic acid revisited. Plasma Process Polym 2009;6:246–54.

[40] Coen MC, Keller B, Groening P, Schlapbach L, Functionalization of graphite, glassy carbon, and polymer surfaces with highly oxidized sulfur species by plasma treatments. J Appl Phys 2002;92:5077–83.

[41] Akhavan B, Jarvis K, Majewski P. Plasma polymer-functionalized silica particles for heavy metals removal. Appl Mater Interfaces 2015;7:4265–74.

[42] Akhavan B, Jarvis K, Majewski P. Tuning the hydrophobicity of plasma polymer coated silica particles. Powder Technol 2013;249:403–411.

[43] Akhavan B, Jarvis K, Majewski P. Hydrophobic plasma polymer coated silica particles for petroleum hydrocarbon removal. ACS Appl Mater Interfaces 2013;17:8563–71.

Wolfgang W. Schmahl, Korbinian Schiebel, Peter M. Kadletz,
Xiaofei Yin and Markus Hoelzel

6 Crystallographic symmetry analysis in NiTi shape memory alloys

6.1 Introduction

Shape memory alloys are fascinating materials which show a displacive phase transformation from a paraelastic to a ferroelastic state [1–3]. The transition is diffusionless, thermodynamically first-order, and involves a shear of the unit cell. In metallurgy, such transitions are termed "martensitic". If the volume change at the transition is sufficiently small, the material does not suffer from irrecoverable (plastic) strains and a shape change related to the phase transition and/or change in the transformation-twin-microstructure related to the phase transition can be recovered. This is the shape memory effect. There are two fundamental application modes of shape memory alloys [1–3]:

(i) Superelasticity or "pseudoelasticity": Here the martensitic transition temperature is below the application temperature, and in the absence of applied stresses, the phase state of the material is that of the symmetric parent-structure ("austenite"). The phase transition to the sheared derivative structure ("martensite") occurs upon the application of (non-isotropic) stress. Beyond the initial elastic regime, the strain response to the applied stress is entirely related to the shear deformation associated with the martensitic phase transition. When the stress is released, the stress-induced martensite becomes thermodynamically unstable and the material regains the stable austenite structure, while the original shape is recovered (Fig. 6.1).

(ii) Shape memory or "pseudoplasticity". Here the martensitic transition temperature is above the application temperature, and in the absence of applied stresses the phase state of the material is that of the distorted derivative structure (martensite). In this state, the material has a twinned microstructure. The twin states are related by symmetry elements of the austenite phase which are lost upon the phase transition to martensite. When polycrystalline material is cooled free of stress from the symmetric austenite state to the martensite state, the twins will form an accommodation structure to minimize the overall shape change of the austenite grains within the polycrystalline fabric. Upon the application of (non-isotropic) stress, the twin states will adjust to the macroscopic strain by shifting the domain walls between the twins. This "detwinning" mechanism produces the macroscopic strain. When the stress is released, the material stays in the "detwinned" state, as the martensite phase is thermodynamically stable and has little reason to transform in any way. However, when the material is heated until the austenite becomes thermodynamically stable, the twinning/detwinning-related macroscopic strains disappear upon transition to austenite, and the original shape is recovered. By cooling without external stresses, the material falls back into the original twinned martensite structure, keeping the original shape (Fig. 6.1).

The binary alloy NiTi is a popular shape memory alloy used in engineering, as it combines recoverable shape changes of up to 8% with high strength, high actuator forces related to the martensitic phase change, excellent resistance to functional fatigue, and a tunable martensitic transformation temperature as a function of the Ni/Ti ratio [4]. The present chapter thus focuses on the B2 austenite to B19' martensite phase transition in NiTi.

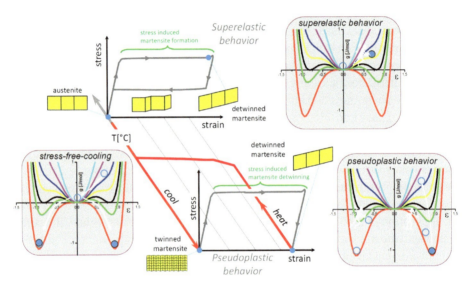

Fig. 6.1: Stress-strain and thermodynamic characteristics of superelasticity and shape memory. The superelastic (top) and shape memory (bottom) stress-strain curves are related by the temperature axis. The three plots showing Landau isotherms (free energy density g vs. strain ε) indicate the thermodynamic aspects. The black free energy isotherm corresponds to the equilibrium transition temperature (T_{tr}) in a stress-free situation. Austenite corresponds to the free energy minimum at $\varepsilon=0$; martensite corresponds to the free energy minima at $\varepsilon \neq 0$. Superelastic behaviour occurs on an isotherm well above T_{tr} (e.g. here the yellow isotherm), where martensite is formed and stabilized by applied stress (symbolized by the little white man pushing the system "up hill"). The martensite formation leads to a shape-change and this martensite transforms back to austenite when the stress is released. Pseudoplastic behaviour occurs at temperatures below T_{tr} (e.g. here the red isotherm). When cooling down from high temperature in stress-free situation below, equal amounts of symmetry-equivalent martensite twin variants (minima at positive and negative ε) are formed; application of stress moves the system into one twin variant. The original twinned state is regained by heating well above T_{tr} (here the blue isotherm) followed by cooling in stress-free situation.

Symmetry analysis of crystals had been the domain of mineralogists interested in classification and morphology of minerals, and later, from the advent of crystal structure analysis [5, 6] of crystallographers, who use symmetry as a tool to determine and describe a crystal structure with an appropriate minimum of free variables. With the exploitation of ferroic, i.e. ferromagnetic, ferroelectric, and/or ferroelastic, substances as functional materials which show strong responses to applied external fields, symmetry analysis of the ferroic order parameters became an important tool to understand this functionality. For these substances, typically, a high-symmetry parent phase exists (paramagnetic, paraelectric, or paraelastic), from which a symmetry-breaking long-range correlation develops in the crystal structure below a critical temperature (the Curie temperature). It is the order parameter, i.e. the magnitude or amplitude of this correlation process, that is conjugate to the external field and shows a characteristic temperature dependence of its restoring force. In a famous article, Landau [7, 8] proposed a macroscopic thermodynamic theory of phase transitions, which is based on the expression of the thermodynamic potential as a power series expansion in terms of order parameter, temperature, and field. The power of the Landau approach was not immediately recognized at the time. First, the minimalistic approach of Landau, who concentrated on the generalities of structural phase transitions rather than on details of any particular substance, failed to explain quantitatively the ferromagnetic-order parameter behavior of many systems known at the time. Second, many scientists are more satisfied to have an atomistic explanation at hand for a process, rather than a macroscopic thermodynamic approach. Many atomistic models have been developed to explain or predict the spontaneous development of structural order and its macroscopic field and thermal dependence. However, they usually relied on certain simplifications and specific assumptions about each specific system to make them solvable. For example, Weiss [9], considering magnetic systems, and Bragg and Williams [10], considering ordering in a solid solution, used a *mean field* concept to approximate the influence of the material's structure on the energy of a single spin (or atom) inside the structure. Their quantitative predictions with this hybrid approach between macroscopic and atomistic concepts were of only partial success either. The elegant Ising [11] model, which considers only two discrete states for each atomistic site and coupling between nearest neighbors only, was finally solved by Onsager [12] for the two-dimensional square lattice structure. It took almost three decades until the so-called renormalization theory was developed by Wilson [13, 14], which essentially scales the mean field with the correlation length of the order parameter, that a successful general theory of critical phenomena became available.

Although originally introduced by Landau [7, 8], it was not until Devonshire's [15–17] publications on ferroelectric $BaTiO_3$ that, what we call Landau theory today, became widely recognized as a powerful empirical tool to look at structural phase transitions in ferroic materials, and other systems, including minerals [18, 19].

6.2 Elementary Landau theory

Taking the thermodynamic approach of Landau theory in this chapter, we describe the structural state of the system by an order parameter (or several order parameters if needed). The order parameter is zero in the parent phase and describes the amplitude of the deviation of the structure of the derivative phase (here the martensite) from that of the parent phase (here the austenite). In the thermodynamic sense the order parameter Q is the derivative of the thermodynamic potential density g with respect to the conjugate applied field of strength h ($g = G/V$, where G is the Gibbs function and V is the volume of the system):

$$Q = \left(-\frac{\partial g}{\partial h}\right)_T \tag{6.1}$$

Landau's fundamental assumption is that the thermodynamic potential can be expressed as a power series expansion in Q. If it is sufficient to consider a single one-dimensional order parameter, this would read

$$g = g_0 + \sum_{n=1}^{m} \frac{1}{n} a_n Q^n, \tag{6.2}$$

where g_0 represents the free-energy density of the parent phase (paraphase) with $Q = 0$. Further, according to Equation 1, we identify the coefficient of the first-order term with the applied field strength: $a_1 = -h$. To model the occurrence of spontaneous order in the absence of an external field below a critical temperature, Landau introduced a temperature dependence in the very suitable most minimalistic way: by expanding the leading term for $h = -a_1 = 0$, i.e. a_2, in terms of temperature T as

$$a_2 = \alpha(T - T_0) + \ldots \tag{6.3}$$

(similarly, other coefficients may be expanded, or the expansion may include pressure- or composition-dependence of the coefficients a_n). To discuss some fundamental concepts, let us assume that it is sufficient to consider terms up to $n = 6$ in Equation 6.2, and that the system is symmetric with respect to Q such that $g(Q) = g(-Q)$. We thus arrive at:

$$g = g_0 - hQ + \frac{1}{2}a_2 Q^2 + \frac{1}{4}a_4 Q^4 + \frac{1}{6}a_6 Q^6 + \ldots \tag{6.4}$$

Falk [20] discussed the free energy of shape memory alloys within the framework of Equation 6.4. Shape memory metals belong to the *ferroelastic* class of materials. They possess a high-symmetry phase, which is the stable state at high temperature, and they develop a symmetry-breaking spontaneous shear-strain below a critical temperature. Further, as a requirement for *ferroic* materials, the shear strain can be switched between different "opposite" states. Correspondingly, Falk [20] identified the order parameter Q with the symmetry-breaking (shear) strain, and Equation 4 allows two "opposite" states, $+Q$ and $-Q$. Moreover, the paraelastic-ferroelastic phase transition is a first-order phase transition for all known shape memory systems. This means that all first-order derivatives of the free energy (such as order parameter, entropy, or elastic strains) are discontinuous at the transition, as opposed to a second-order phase transition, where these derivatives vary continuously during the transition and the discontinuities are found in the second-order derivatives (such as heat capacity, thermal expansion, etc.). In metallurgy, diffusionless, first-order paraelastic-ferroelastic phase transitions are termed *martensitic* phase transitions in analogy to the cubic-tetragonal phase transition from austenite (fcc iron) to martensite (bct iron). The requirement of a discontinuous transition is also met by Equation 6.4 if a_4 is negative (see below). Note that the highest-order coefficient, here a_6, always needs to be positive such that the free energy always increases for large $|Q|$ and thus essentially forms a potential well.

The stable (or metastable) states of the system described by (6.4) are associated with the equilibrium values $\langle Q \rangle$ for which the system is in a (local) minimum of the thermodynamic potential:

$$\left(\frac{\partial g}{\partial Q}\right)_T = 0, \quad \left(\frac{\partial^2 g}{\partial Q^2}\right)_T > 0 \qquad (6.5)$$

In absence of a field, the requirements of Equation 6.5 are met by the following two types of solutions: one solution describes the undistorted ("disordered") high-symmetry parent phase,

$$\langle Q \rangle_{pp} = 0, \qquad (6.6)$$

whereas the other solution describes the two symmetry-equivalent but "opposite" sheared ("ordered") states of the low-symmetry derivative phase (ferrophase, fp):

$$\langle Q \rangle_{fp} = \pm \left(\frac{-a_4 + \sqrt{a_4^2 - 4\alpha a_6 (T - T_0)}}{2a_6} \right)^{1/2}. \qquad (6.7)$$

The temperature dependence resides in the coefficient of the quadratic term as in (6.3).

If both the a_4 and a_6 coefficients of the 4th-order and 6th-order terms in (6.4) are positive, the system of equations describes a second-order phase transition with $\langle Q \rangle$ evolving continuously to zero from lower temperatures up to T_0, which is the transition temperature in this case. If a_4 is negative (and a_6 positive), the system describes a first-order phase transition occurring in equilibrium and in the absence of a field at the transition temperature

$$T_{tr} = T_0 + \frac{3a_4^2}{16\alpha a_6}. \qquad (6.8)$$

Equation 6.8 is found from the equivalence of the free energies of the two equilibrium states $g(\langle Q \rangle_{pp}) = g(\langle Q \rangle_{fp})$ at T_{tr}. This behavior can be seen from the free-energy isotherms (g vs. Q curves) plotted in Fig. 6.1 for a sequence of temperatures. If the system is in equilibrium, the order parameter jumps between $\langle Q \rangle_{pp} = 0$ and $\langle Q \rangle_{fp} = \langle Q(T = T_{tr}) \rangle_{fp}$ at the transition temperature. This jump $\langle \Delta Q(T = T_{tr}) \rangle$ is

$$\langle \Delta Q \rangle_{T=T_{tr}} = \sqrt{\frac{-3a_4}{4a_6}}. \tag{6.9}$$

A second temperature, T_{Mf}, which is characteristic for this scenario, is the temperature up to which the martensitic (sheared) phase can exist in metastable form. It is the temperature where the free-energy minimum at $\langle Q \rangle_{fp}$ becomes a saddle-point:

$$T_{Mf} = T_0 + \frac{4a_4^2}{16\alpha a_6} = T_0 + \frac{4}{3}(T_{tr} - T_0). \tag{6.10}$$

Note that the lowest temperature at which the austenitic (unsheared) phase can exist in metastable form is T_0, where the minimum at $\langle Q \rangle_{pp} = 0$ turns into maximum for temperatures lower than T_0 (Figs. 6.1 and 6.2).

The excess free-energy of the equilibrium ferrophase compared to the equilibrium paraphase in the field-free state is

$$\langle \Delta g(T) \rangle = g(T, \langle Q \rangle_{fp}) - g_0(T)$$
$$= \frac{1}{2}\alpha(T - T_0)\langle Q \rangle_{fp}^2 + \frac{1}{4}a_4 \langle Q \rangle_{fp}^4 + \frac{1}{6}a_6 \langle Q \rangle_{fp}^6 + \ldots \tag{6.11}$$

Around $T = T_{tr}$, this equation can be written as a power series expansion with leading terms:

$$\langle \Delta g(T) \rangle = \frac{3\alpha a_4(T - T_{tr})}{8} - \frac{\alpha^2(T - T_{tr})^2}{2a_4} - \frac{4\alpha^3 a_6(T - T_{tr})^3}{3a_4^3} - \ldots \tag{6.12}$$

In a system with volume V the entropy reduction (excess entropy) associated with the order parameter becomes

$$\Delta S = -\frac{V \partial \langle \Delta g(T) \rangle}{\partial T} = -V\frac{\alpha}{2}\langle Q \rangle_{fp}^2, \tag{6.13}$$

which, at the transition temperature, involves the entropy step

$$\Delta S_{tr} = V\frac{3\alpha a_4}{8a_6}, \tag{6.14}$$

which corresponds to a latent heat of transition of

$$\Delta H_{tr} = T_{tr}\Delta S_{tr} = V\frac{3\alpha a_4 T_{tr}}{8a_6}. \tag{6.15}$$

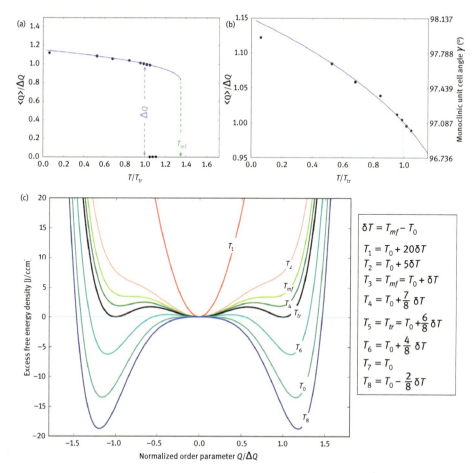

Fig. 6.2: Order parameter behavior (a, b) and free-energy isotherms (c) as described by equations (6.4)–(6.10) with the parameters $a_2 = 0.08897$ (J/ccm), $T_0 = -16.35$ K, $a_4 = -159.56$ (J/ccm), $c = 158.11$ (J/ccm). The monoclinic unit cell γ angle as determined from neutron diffraction was taken as a measure of the order parameter ($\langle Q \rangle_{obs} = \tan[\gamma - 90°]$). (a) Measured (black dots) and calculated (continuous line) order parameter evolution, (b) zoom into (a), (c) free-energy isotherms.

Adjusting the Landau coefficients to the transition enthalpy of 25.2 J/g which we measured by DSC [4] and to monoclinic shear-angle as measured by neutron diffraction as a function of temperature on the SPODI diffractometer at FRM-II [21], we can determine the free-energy functional and order parameter behavior as displayed in Fig. 6.2 for $Ni_{50.14}Ti_{49.86}$. Here we used the tangent of the monoclinic shear angle of the unit cell, $\tau = \gamma_{mon} - 90°$, as a measure of the thermodynamic-order parameter. The obtained value for T_0 is just below 0 K, meaning that the cubic B2 phase possesses a local free-energy minimum at all relevant temperatures.

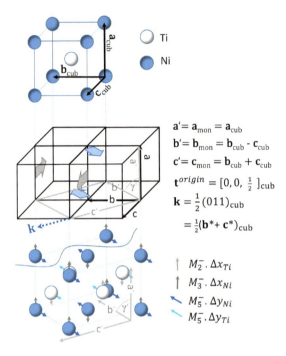

Fig. 6.3: Crystallographic relationship between B2 austenite and B19′ monoclinic martensite of NiTi, shown for one selected domain. Broad arrows indicate the monoclinic (blue) and orthorhombic (gray) Γ_5^+ shear modes. The atomic displacements associated with the phase transition are indicated by normal arrows, and they are transverse with respect to the distortion wave vector $\bar{k} = \frac{1}{2}(011)_{cub}$ (M-point of the Brillouin zone, a corresponding cosine wave is indicated). The displacements in x_{mon} direction belong to the M_2^- irreducible representation for Ti and M_3^- for Ni. The M_5^- irreducible representation displaces both Ti and Ni with different amplitudes in y_{mon} direction.

The correspondence of the austenite parent structure and the martensite derivative structure is shown in Fig. 6.3 along with the relationship between the unit cells and the atomic displacements occurring at the phase transition. Table 6.1 compares the two structures numerically.

Figures 6.4 and 6.5 displays the thermal behavior of the lattice parameters and the atomic displacements away from the positions in the symmetric structure that occur in association with the martensitic phase transition in the material, respectively. Table 6.1 lists the structural parameters at room temperature (293 K).

While Falk's 2-4-6 free-energy polynomial produces some essential features of the martensitic phase transition, the situation in a real system such as NiTi [1, 22, 23] is more complicated, as the transition involves several order parameters of different symmetry and more than two ferroelastic twin states. The full symmetry relationship between the austenite parent phase and the martensite derivative phase needs to be considered.

Tab. 6.1: Crystal structure data of B2 austenite (space group $Pm\bar{3}m$) and B19' martensite (space group $P2_1/m$, setting: unique axis c) for monoclinic B19' $Ni_{50.14}Ti_{49.86}$ at room temperature and for B2 austenite at 340 K as obtained by neutron diffraction.

Unit cell dimensions (standard deviations referring to the last digit)						
B19' martensite (space group $P2_1/m$)					B2 austenite (space group $Pm\bar{3}m$)	
a/Å	b/Å	c/Å	γ/°		a/Å	
2.8940(2)	4.6399(3)	4.1247(2)	97.252(8)		3.0200(1)	
$[100]_{cub}$	$[01-1]_{cub}$	$[011]_{cub}$				
Fractional atomic coordinates						
	B19' martensite			B2 austenite		
	x	y	z	x	y	z
Ni	$0 + \Delta x$	$¼ - \Delta y$	¾	0	0	0
Ni'	$0 - \Delta x$	$¾ - \Delta y$	¼	0	1	0
Ni: $\Delta x, \Delta y$	0.0309(8)	-0.0729(4)				
Ti	$½ + \Delta x$	$¼ - \Delta y$	¼	½	½	½
Ti'	$½ - \Delta x$	$¾ + \Delta y$	¾	½	3/2	½
Ti: $\Delta x, \Delta y$	0.0718(14)	0.0327(10)				

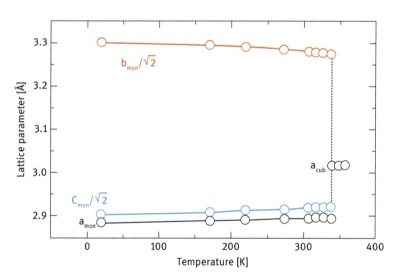

Fig. 6.4: Evolution of the lattice parameters through the cubic ↔ monoclinic phase transition of $Ni_{50.14}Ti_{49.86}$. See Fig. 6.2 for the monoclinic unit cell angle γ.

Let $\{G_0\}$ be the symmetry group of the parent phase and $\{G\}$ be that of the derivative phase. Both groups are related by the set equation

$$\{G_0\} = \{G\} \oplus \sum_{i}^{N} \{S_i \odot \{G\}\}, \quad (6.16)$$

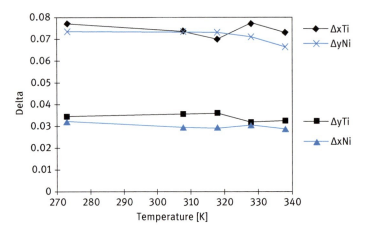

Fig. 6.5: Evolution of the displacement of the atomic positions in $Ni_{50.14}Ti_{49.86}$ B19' ($P2_1/m$) martensite away from the positions in the parent austenite phase (see Tab.6.1 for definition of the displacements).

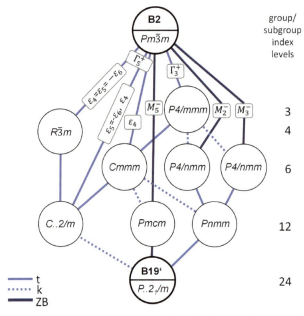

Fig. 6.6: Space-group symmetry relationship between B2 austenite (space group $Pm\bar{3}m$) and B19' martensite (space group $P2_1/m$) in relation to the different symmetry-breaking distortion modes occurring at the phase transition. The *translationengleiche* (t) and *klassengleiche* (k) group-subgroup relationships are marked as light continuous (t) and dotted (k) lines, respectively. Dark connecting lines mark the Brillouin zone-boundary modes which break translational as well as point group symmetry. *Pmcm* is space group #51 (standard setting *Pmma*); *Pnnm* is space group #59 (standard setting *Pmmn*). The two isomorphic *P4/nmm* space groups differ in origin: M_2^- mode: $[0, -1, 1/2]_{cub}$, M_3^- $[0, 1/2, 1/2]_{cub}$.

where S_i are the symmetry elements contained in $\{G_0\}$ which are not in $\{G\}$, the operator symbol $\oplus$ indicates the formation of the union of sets, and $\{S_i \circ \{G\}\}$ is the set obtained by multiplication of all symmetry elements in $\{G\}$ with S_i. $\{G\}$ and each so called coset $\{S_i \circ \{G\}\}$ represents the symmetry group of a domain of the derivative structure; the different domains are symmetry-equivalent under the operations S_i. The summation in 6.16 runs over all non-overlapping cosets of $\{G\}$ in $\{G_0\}$. Their number, N, which is called the index of the group/subgroup relationship is 24 for the present case.

Shchyglo et al. [23] investigated the symmetry properties of the strains leading from the cubic B2-type austenite parent phase (space group $Pm\bar{3}m$; point group $m\bar{3}m$, order 48) to the monoclinic B19'-type martensite derivative phase (space group $P2_1/m$, point group $2/m$, order 4). Note that the strain does not explain the doubling of the unit cell between austenite and martensite (Fig. 6.2). The number of symmetry-equivalent ferroelastically switchable twin states derived from symmetry loss from $m\bar{3}m$ to $2/m$,

$$I = \frac{\text{order of point group of parent phase}}{\text{order of point group of derivative phase}}, \tag{6.17}$$

is thus 48/4 = 12 in the present case. For each of these 12 twin states two antiphase-domains exist due to the doubling of the unit cell, corresponding to displacement waves with positive or negative amplitude, respectively.

To obtain a complete symmetry analysis, considering the full space group description is possible with the very useful software utility ISODISTORT [24, 25], which separates the overall structure distortion into orthogonal symmetry-adapted modes. The metric distortion (strain) of the material, which is related to the macroscopic mechanical shape memory effect, is the same for all unit cells in a domain. These distortion modes propagate with a wavelength of infinity, corresponding to a wave vector of zero, also called the Γ point of reciprocal space (Γ point of the Brillouin zone, [26]). Figure 6.4 displays the group-subgroup table of the symmetry relationship between both structures. The doubling of the unit cell corresponds to transverse distortion modes propagating with a wave vector of the {½, ½, 0} family of reciprocal lattice points, which are labeled M-points [26, 27].

6.3 M-point transverse distortion modes

For the martensite domain selected for the drawing in Fig. 6.3, the atomic displacements are associated with a commensurate transverse displacement wave of wave-vector $\mathbf{k} = \left(0 \; ½ \; ½\right)_{cub}$. The displacements are indicated by the simple arrows in Fig. 6.3. They lead to a doubling of the cubic unit cell in the direction of its (100)-face diagonal, and produce a new unit cell with dimensions

$\mathbf{a}_{mon} = \mathbf{a}_{cub}$, $\mathbf{b}_{mon} = \mathbf{b}_{cub} - \mathbf{c}_{cub}$, $\mathbf{c}_{mon} = \mathbf{b}_{cub} + \mathbf{c}_{cub}$. To comply with the standard setting of space group description as tabulated in the International Tables of crystallography, the origin of the description of the martensite structure in space group $P112_1/m$ is situated in $\left[0, 0, \frac{1}{2}\right]_{cub}$. The lattice correspondence expressed in the **P**, **Q** transformation matrix notation of Arnold's [28] article in the *International Tables for Crystallography* for the representative domain selected in Fig. 6.3 becomes

$$\mathbf{P} = \begin{pmatrix} 1 & 0 & 0 \\ 0 & 1 & 1 \\ 0 & -1 & 1 \end{pmatrix}, \quad \mathbf{t}^{origin} = \begin{pmatrix} 0 \\ 0 \\ 1/2 \end{pmatrix}_{cub}, \quad \mathbf{Q} = \mathbf{P}^{-1} = \begin{pmatrix} 1 & 0 & 0 \\ 0 & 1/2 & 1/2 \\ 0 & -1/2 & 1/2 \end{pmatrix} \quad (6.18)$$

$$\mathbf{x}_{mon} = \mathbf{Q}\left(\mathbf{x}_{cub} + \mathbf{t}^{origin}_{cub} + \{\Delta\mathbf{x}^{modes}_{cub}\}\right). \quad (6.19)$$

where $\mathbf{x}_{mon}$ and $\mathbf{x}_{cub}$ is the atomic position coordinate vector in the cubic and the monoclinic unit cell, respectively. The vector in the curly brackets in Equation 19 represents the sum of displacements caused by all distortional modes of the phase transition for the particular atom located at $\mathbf{x}_{cub}$. Let $\mathbf{u}_{cub}(j)$ be the displacement vector for the considered atom in the displacement pattern which is caused by mode j with wave vector $\mathbf{k}(j)$, and amplitude $Q(j)$ (no relation to the **Q**-transformation matrix of eqns. 18, 19), then the total displacement becomes

$$\{\Delta\mathbf{x}^{modes}_{cub}\} = \sum_{Modes\,j} Q(j)\, \mathbf{u}_{cub}(j)\, \cos(\mathbf{k}_{cub}(j)\cdot\mathbf{x}_{cub}). \quad (6.20)$$

For the NiTi B2-B19′ phase transition, there are four active modes breaking the translational symmetry, all with wave vector $\mathbf{k} = (0\ \frac{1}{2}\ \frac{1}{2})_{cub}$ for the representative domain. Table 6.2 lists the four modes along with the symmetry properties, displacement pattern, and amplitudes; Fig. 6.3 attempts a corresponding graphical display. The modes M_2^- and M_3^- correspond to three dimensional irreducible representations, while M_5^- is 6-dimensional. The number of domain variants for each mode is two times the dimension.

Tab. 6.2: M-point displacive modes for the B2-B19′ phase transition, $\mathbf{k} = (0, \frac{1}{2}, \frac{1}{2})$.

Mode	Direction	Subgroup	Δx^{mode}_{mon}				Value$_{mon}$	Amplitude (Å)
M_2^-	(0, 0, a)	P4/nmm	Ti	Δx	0	0	0.0718(14)	0.217(4)
M_3^-	(0, 0, a)	P4/mm	Ni	Δx	0	0	0.0309(8)	0.093(2)
M_5^-	(0, 0, 0, 0, a, −a)	Pmma	Ti	0	$-\Delta y$	0	0.0327(10)	0.140(4)
			Ni	0	Δy	0	−0.0729(4)	0.312(2)

6.4 Γ-point strain distortion modes

The ISODISTORT mode analysis identifies the following strain modes (active Γ-point irreducible representations), with strain tensors referring to the cubic reference frame:

Complete Strain $\qquad\qquad\qquad \Gamma_1^+ \quad + \quad \Gamma_3^+ \quad + \quad \Gamma_5^+(m) \quad + \quad \Gamma_5^+(o)$

$$\begin{pmatrix} v+t & -m & m \\ -m & v-t/2 & o \\ m & o & v-t/2 \end{pmatrix} = \begin{pmatrix} v & 0 & 0 \\ 0 & v & 0 \\ 0 & 0 & v \end{pmatrix} + \begin{pmatrix} t & 0 & 0 \\ 0 & -t/2 & 0 \\ 0 & 0 & -t/2 \end{pmatrix} + \begin{pmatrix} 0 & -m & m \\ -m & 0 & 0 \\ m & 0 & 0 \end{pmatrix} + \begin{pmatrix} 0 & 0 & 0 \\ 0 & 0 & o \\ 0 & o & 0 \end{pmatrix}$$

Space group $\qquad\qquad\qquad Pm\bar{3}m \qquad\qquad P4/mmm \qquad\qquad\qquad C2/m$

Order parameter direction $\qquad\qquad\qquad\quad (t, t\sqrt{3}) \qquad\qquad (m, o, -m)$ (6.21)

These strain modes have already been discussed by Shchyglo et al. [23]. The Γ_1^+ mode is the non-symmetry-breaking volume strain. As expected for a shape memory material, this strain component is very small and amounts to $v = 0.00212(5) \sim 0.2\,\%$. Γ_3^+ produces a tetragonal distortion, where the tetragonal c-axis corresponds to the direction in which the repeat unit is maintained, i.e. the $\bar{a}$-axis direction in the representative domain in Fig. 6.3. This axis is shortened while the axes in the plane perpendicular to it are extended. The irreducible representation of this mode is two-dimensional, it allows three domain states, depending on whether the cubic **a**, **b**, or **c**- axis is chosen as the short axis of the monoclinic cell. Γ_5^+ is a three-dimensional irreducible representation, where the B19' cell is generated by the direction $(m, o, -m)$ (refer to Tab. 6.3 for more information). The shear components denoted as m produce the deviation of the monoclinic angle of the B19' martensite unit cell from 90°. This shear deformation occurs such that one plane in the {100} family of planes of the austenite unit cell represents the shear plane and the shear occurs in a <110>-type direction on that {100} plane. In Fig. 6.3, it is shown as a blue flat arrow, which shears in the $[0\bar{1}1]_{cub}$ direction on the $(100)_{cub}$ plane. Table 6.3 lists the combinations of shear planes and shear directions, which lead to the 12 metric states between which ferroelastic switching ("twinning" or "detwinning") is possible under applied non-isotropic stress. Note that for each metric (ferroelastic) state, the M-point modulation produces two antiphase-domain states (corresponding to a positive or a negative mode amplitude, respectively), such that 24 domains result in total. The shear component denoted as o produces an orthorhombic strain, which corresponds to the relative difference in length between the monoclinic **b**- and **c**-axes of the reference domain, i.e. the axes perpendicular to the cubic direction which is shortened by the tetragonal Γ_3^+ strain. It is indicated by a gray flat arrow in Fig. 6.3.

Tab. 6.3: Strain tensors corresponding to the 12 switchable ferroelastic metric states of B19' martensite.

Shear direction	$\vec{a}_{mon} = \vec{a}_{cub}$ shear plane (100)	Shear direction	$\vec{a}_{mon} = \vec{b}_{cub}$ shear plane (010)	Shear direction	$\vec{a}_{mon} = \vec{c}_{cub}$ shear plane (001)
$[0\bar{1}1]$	$\begin{pmatrix} t & -m & m \\ -m & -t/2 & o \\ m & o & -t/2 \end{pmatrix}$	$[10\bar{1}]$	$\begin{pmatrix} -t/2 & m & o \\ m & t & -m \\ o & -m & -t/2 \end{pmatrix}$	$[\bar{1}10]$	$\begin{pmatrix} -t/2 & o & -m \\ o & -t/2 & m \\ -m & -m & t \end{pmatrix}$
$[01\bar{1}]$	$\begin{pmatrix} t & m & -m \\ m & -t/2 & o \\ -m & o & -t/2 \end{pmatrix}$	$[\bar{1}01]$	$\begin{pmatrix} -t/2 & -m & o \\ -m & t & m \\ o & m & -t/2 \end{pmatrix}$	$[1\bar{1}0]$	$\begin{pmatrix} -t/2 & o & m \\ o & -t/2 & -m \\ m & -m & t \end{pmatrix}$
$[011]$	$\begin{pmatrix} t & m & m \\ m & -t/2 & -o \\ m & -o & -t/2 \end{pmatrix}$	$[101]$	$\begin{pmatrix} -t/2 & m & -o \\ m & t & m \\ -o & m & -t/2 \end{pmatrix}$	$[110]$	$\begin{pmatrix} -t/2 & -o & m \\ -o & -t/2 & m \\ m & m & t \end{pmatrix}$
$[0\bar{1}\bar{1}]$	$\begin{pmatrix} t & -m & -m \\ -m & -t/2 & -o \\ -m & -o & -t/2 \end{pmatrix}$	$[\bar{1}0\bar{1}]$	$\begin{pmatrix} -t/2 & -m & -o \\ -m & t & -m \\ -o & -m & -t/2 \end{pmatrix}$	$[\bar{1}\bar{1}0]$	$\begin{pmatrix} -t/2 & -o & -m \\ -o & -t/2 & -m \\ -m & -m & t \end{pmatrix}$

The indicated shear plane and direction refer to the monoclinic strain Γ_5^+ with component m.

The amplitudes of all strain modes at room temperature are as follows (with limits of uncertainty of ca. $5*10^{-5}$):

$$\begin{pmatrix} v+t & -m & m \\ -m & v-t/2 & o \\ m & o & v-t/2 \end{pmatrix} = \begin{pmatrix} 0.00212 - 0.04600 & -0.04554 & 0.04554 \\ -0.04554 & 0.00212 + 0.02300 & -0.05936 \\ 0.04554 & -0.05936 & 0.00212 + 0.02300 \end{pmatrix}$$

(6.22)

6.5 Mode coupling and invariants in Landau free-energy polynomial

We have seen that symmetry analysis indicates that the symmetry-breaking process at the B2-B19' phase transition involves a total of 5 orthogonal symmetry-breaking modes with 7 independent components. An 8th parameter is the non-symmetry breaking volume change. Correspondingly, the crystallographic structure description features 4 degrees of freedom in the atomic coordinates and another 4 degrees of freedom for the metric of the monoclinic unit cell. The Landau free-energy is thus a much more complex story than discussed above as we deal with a landscape of 8 nominally independent "order parameters".

Tab. 6.4: Summary of polynomials of degree 2, 3, and 4 in order parameter component amplitudes which are allowed by symmetry at the B2⟷B19' phase transition.
This table notably shows the allowed coupling terms in the Landau free-energy expansion between the mode amplitudes.

Modes and components	Degree 2	Γ_5^+ (o, m, m)	Γ_3^+ t	M_2^-, M_3^- q, r	M_5^- s	Γ_5^+ (o, m, m)	Γ_3^+ t	M_2^-, M_3^- q, r	M_5^- s
		Degree 3				Degree 4			
Γ_5^+ (o, m, m)	$2m^2 + o^2$	$m^2 o$	$tm^2 - to^2$	–	–	m^4, $m^2 o^2$, o^4	$t^2 m^2$, $t^2 o^2$	$q^2 m^2$, $q^2 o^2$, $r^2 m^2$, $r^2 o^2$	$s^2 o^2$, $s^2 m^2$
Γ_3^+ t	t^2		t^3	$t^2 q$, $t^2 r$	$t^2 s$		t^4	$t^2 q^2$, $t^2 r^2$	$t^2 s^2$
M_2^-, M_3^- q, r	q^2, r^2			–	–			q^4, r^4	$s^2 q^2$, $s^2 r^2$
M_5^- s	s^2				–				s^4

ISODISTORT produces a list of all polynomial terms in the independent order parameters (up to a certain order of the polynomial) which are invariant under the symmetry of the parent phase (here $Pm\bar{3}m$). In Tab. 6.4, we summarized the terms up to the 4th degree to demonstrate the essential coupling mechanisms between the modes. Note the presence of a third-order term in the tetragonal distortion, t^3, which forbids a continuous (2nd order) phase transition in the present case.

From the total symmetry reduction from $Pm\bar{3}m$ to $P112_1/m$ and lattice correspondence (1 0 0, 0 1 –1, 0 1 1) with index 24, it is clear that only 24 out of all conceivable variations of these eight order parameter components lead to a global minimum in the free-energy. And, at least for the investigated alloy composition, only two phases, B2 austenite and B19' martensite are experimentally observed and thus form experimentally identifiable minima in the 8-dimensional free-energy landscape. (However, there are potentially more local minima with yet more active irreducible representations [29]; the corresponding phases are unstable compared to B19' and B2 for the investigated composition.) At the present state it is not possible to evaluate numerical values for all conceivable coupling coefficients. Nevertheless the observed data allow some important conclusions on the system.

Only a combination of the Γ_5^+ mode and the M-point modes is able to produce the $P112_1/m$ symmetry. An important observation in this system is that the atomic displacements Δx^{mode} (Eq. 6.19) show very little to no temperature dependence, and they essentially jump to a constant value at the cubic to monoclinic transition (Fig. 6.5). The Γ_5^- mode, however, shows the typical temperature dependence of an order parameter softening toward the transition temperature (Fig. 6.2). Let us thus concentrate on the coupling between the ferroelastic distortion and the displacement wave. If a phase transition involves the breaking of translational symmetry in a way that leads to a change in point group which increases the number of metrical degrees of freedom, the transition is typically also associated with a metric distortion ε (Γ-point distortion) of the unit cell (e.g. [30–33]). If positive and negative amplitude $Q(j)$ of the displacement wave lead to the same unit cell (same ε), the lowest-order free-energy coupling term is of the linear-quadratic form [$\varepsilon Q^2(j)$], as it is independent of the sign of $Q(j)$. A result of this coupling, the (secondary) strain-order parameter follows the square of the displacive (primary) order parameter as $\varepsilon \propto Q^2(j)$ (see e.g. [30–33]). This implies a flatter temperature dependence of ε than that of $Q(j)$; in the NiTi system, however, the opposite is the case: the atomic displacements corresponding to $Q(j)$ appear independent of temperature in the monoclinic phase while the monoclinic shear angle displays a typical temperature dependence of a structural order parameter driving the phase transition.

This leads us to the following conclusion: The practical independence of the M-mode amplitudes suggests that the coefficients for the M-modes in the Landau free-energy are fairly independent of temperature, while the transition is driven by the monoclinic strain. This must correspond to the temperature-dependence of the coefficient of the quadratic term of the strain-order-parameter in the Landau free-energy expansion. Moreover, the structure produced by the M-point displacive modes alone (i.e. with no monoclinic strain) forms, at least, a local minimum in the free-energy at all relevant temperatures. In other words, it is an alternative structure to the B2 structure, and the system jumps discontinuously into this alternative M-point distorted structure when the temperature gets low enough to benefit from a strain distortion. This occures as the temperature-dependent coefficient of the quadratic term in strain becomes more and more negative on cooling.

6.6 Model of biquadratic order parameter coupling

The lowest-order symmetry-permitted coupling term between monoclinic strain and displacement wave amplitudes in the present case is biquadratic (Tab. 6.4). A tractable Landau free-energy expansion with biquadratic coupling between an atomic displacement mode, Q, and a strain mode, e, was discussed by Gufan and Larin [34], Salje and Devarajan [35], and Schmahl et al. [32, 33] and would read as

$$\begin{aligned} g = g_0 &+ \frac{1}{2}\alpha_Q(T - T_Q)Q^2 + \frac{1}{4}b_Q Q^4 + \frac{1}{6}c_Q Q^6 + \ldots \\ &+ \frac{1}{2}\alpha_e(T - T_e)e^2 + \frac{1}{4}b_e e^4 + \ldots \\ &+ \frac{1}{2}\lambda e^2 Q^2 + \ldots \end{aligned} \qquad (6.23)$$

Tab. 6.5: Four possible phases arising from biquadratic coupling of two order parameters, e, and Q.

	Symbol	$\langle e \rangle$	$\langle Q \rangle$
Parent structure	0	0	0
Derivative structures			
	I	$\langle e(T) \rangle \neq 0$	0
	II		$\langle Q(T) \rangle \neq 0$
	III	$\langle e_c(T) \rangle \neq 0$	$\langle Q_c(T) \rangle \neq 0$

(this potential describes one domain of B19′ only). The equilibrium conditions $[(\partial g/\partial e) = 0, (\partial g/\partial Q) = 0]$ permit four different phases or thermal regimes which are summarized in Tab. 6.5.

The equilibrium order parameters are given by the following equations:

(i) quasi-independent order parameter behavior. In a temperature range where only one of the order parameters is active (i.e. below the critical onset temperature T_{tr} of one of the order parameters but above that of the other), we can have either phase I or phase II, depending essentially on the values of T_e and T_Q, respectively. The corresponding equilibrium order parameters are:

Phase I:

$$\langle e \rangle^2 = -(\alpha_e/b_e)(T - T_e), \; \langle Q \rangle = 0 \tag{6.24}$$

with transition temperature phase 0 → phase I at T_e;

Phase II:

$$\langle Q \rangle^2 = (2c_Q)^{-1}\left(-b_Q + \sqrt{b_Q^2 - 4\alpha_Q c_Q(T - T_Q)}\right), \; \langle e \rangle = 0 \tag{6.25}$$

with transition temperature phase 0 → phase II at

$$T_{tr} = T_Q \qquad\qquad \text{for } b_Q \geq 0 \tag{6.25.1}$$

$$= T_Q - 3b_Q^2/16a_Q c_Q \qquad \text{for } b_Q < 0 \tag{6.25.2}$$

(ii) simultaneously active coupled order parameters (phase III)

Phase III forms when the temperature is low enough to activate both order parameters. To calculate the order parameter behavior for phase III, we use the condition $(\partial g/\partial e) = 0$ which gives

$$\langle e_c \rangle^2 = -\left(\frac{\alpha_e}{b_e}\right)(T - T_e) - \left(\frac{\lambda}{b_e}\right)\langle Q_c \rangle^2$$

$$= \langle e(\text{phase I}) \rangle - \left(\frac{\lambda}{b_e}\right)\langle Q_c \rangle^2, \tag{6.26}$$

where the subscript c of the order parameters indicates the coupled regime (phase III). Note that the first contribution to $\langle e_c \rangle$ in Equation 6.26 is the quasi-independent

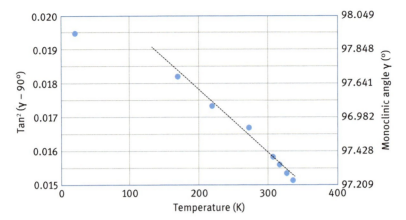

Fig. 6.7: The square of the monoclinic shear order parameter of the B2→B19' transition in NiTi is nearly linear in temperature, but off-set from zero at the transition temperature (~ 323 K).

behavior of phase I, to which a contribution quadratic in $\langle Q_c \rangle$ is added. If $\langle Q_c \rangle$ is rather independent of temperature, the behavior of $\langle e_c \rangle$ is approximately linear in temperature (Equation 6.26), but in contrast to a classical second-order Landau order parameter, it will not continuously run to zero with increasing temperature, as it is off-set by the term in $\langle Q_c \rangle^2$. Such a behavior is actually what we seem to observe, as the monoclinic shear angle does behave quite like this (Fig. 6.7).

From Equation 6.26, it is obvious that for $\lambda < 0$ both order parameters enhance each other as the systems gains stability by simultaneously activating both order parameters. This is the case for NiTi. (For $\lambda > 0$, both order parameters are competing, and within regime of phase III, as one order parameter increases, it will suppress the competing second order parameter, a scenario described in [32, 33]). Equation 6.26 may now be used to eliminate either e or Q from the free-energy expression

$$g = g_0 + \frac{1}{2}\alpha_Q(T - T_Q)Q_c^2 + \frac{1}{4}b_Q Q_c^4 + \frac{1}{6}c_Q Q_c^6 + \ldots$$
$$+ \frac{1}{2}\frac{\lambda \alpha_e}{b_e}(T - T_e)Q_c^2 - \frac{\lambda^2}{4b_e}Q_c^4$$
$$+ \frac{\alpha_e^2(T - T_e)^2}{4b_e} \tag{6.27}$$

The last term in Equation 6.27 gives the free-energy contribution similar to the quasi-independent behavior of <e>. The remaining new terms give a renormalization of the quadratic and the 4th-order terms in Q, such that Equation 6.27 may be rewritten as

$$g = g_0 + \frac{1}{2}A(T - T_0^{III})Q_c^2 + \frac{1}{4}BQ_c^4 + \frac{1}{6}CQ_c^6 + \ldots \tag{6.28}$$

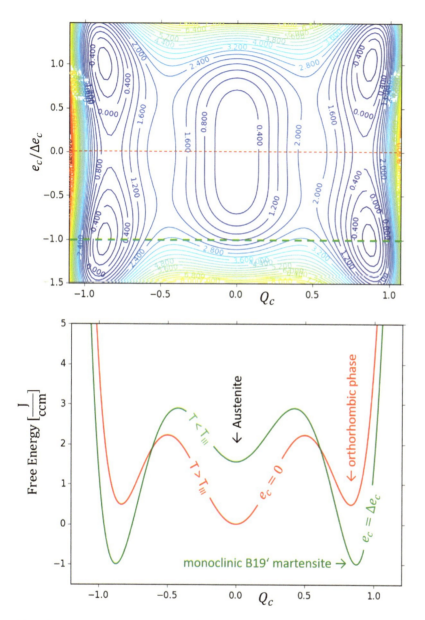

Fig. 6.8: Free-energy landscape for the model of coupled order parameters e_c and Q_c near the martensitic transition temperature. Top: Contour plot, Bottom: Section through the contour plot at the red and green transects. It is the coupling term $\lambda Q_c^2 e_c^2$ with $\lambda < 0$ that switches the system from the austenite ($\langle e_c \rangle = 0$, $\langle Q_c \rangle = 0$) to the martensite ($\langle e_c \rangle \neq 0$, $\langle Q_c \rangle \neq 0$) state when the threshold temperature T_{III} is reached.

with

$$A = \alpha_Q - \lambda \alpha_e/b_e$$
$$B = b_Q - \lambda^2/b_e$$
$$C = c$$
$$T_0^{III} = \frac{\alpha_Q}{A}\left(T_Q - \frac{\lambda \alpha_e T_e}{\alpha_Q b_e}\right) \qquad (6.29)$$

The circumstance that Equation 6.28 looks like a single-order-parameter model is related to the fact that biquadratic coupling of order parameters is always symmetry-allowed. With the renormalized coefficients and characteristic temperature given in Equation 6.29 for phase III the equilibrium value for the coupled order parameter $\langle Q_c \rangle$ in phase III thus becomes

$$\langle Q_c \rangle^2 = (2C)^{-1}\left(-B + \sqrt{B^2 - 4AC(T - T_0^{III})}\right), \qquad (6.30)$$

while that of $\langle e_c \rangle$ can be obtained from Equation 6.26 using Equation 6.29.

The martensitic phase transition in NiTi with the investigated composition occurs directly from B2 austenite (phase 0) to B19' martensite (phase III) but discontinuously ($B < 0$) with a first order step at the transition temperature

$$T_{III} = T_0^{III} - 3B^2/16AC \qquad (6.31)$$

Figure 6.8 depicts the free-energy landscape of this scenario. We postulate that the M-point displacements describe a structure with *Pmma* symmetry (Fig. 6.6) that forms a local minimum in free energy (Fig. 6.8, bottom, red line) which is fairly independent of temperature (a_Q of Equation 6.23 is small, T_Q is very high). When the system is able to gain energy from the ferroelastic distortion e_c as the corresponding quadratic term $a_0(T-T_e)$ gets more and more negative with decreasing temperature, the system jumps from the free-energy minimum corresponding to the austenite state ($e_c = 0$, $Q_c = 0$) to the free-energy minimum corresponding to the martensite state ($\langle e_c \rangle = \Delta e_c$, $\langle Q_c \rangle = \Delta Q_c$). Here ΔQ_c is the constant value of the M-point-order parameter and Δe_c is defined according to Equation 6.26 for the jump occurring at T_{III}. This scenario explains why we have a martensitic transition with a "soft" behavior of the monoclinic shear angle, while the atomic positions in the B19' martensite structure jump strongly at the phase transition but remain practically unaffected by temperature within in the martensite phase.

Acknowledgment

We thank the Deutsche Forschungsgemeinschaft for funding and Mrs. Stefanie Hoser for technical assistance.

References

[1] Otsuka K, Wayman CM. Shape memory materials. Cambridge University Press; 1999.
[2] Van Humbeeck J. Shape memory alloys: a material and a technology. Adv Eng Mater 2001;3:837–50.
[3] Jani JM, Leary M, Subic A, Gibson MA. A review of shape memory alloy research, applications and opportunities. Mater Design 2014;56:1078–113.
[4] Frenzel J, George EP, Dlouhy A, Somsen C, Wagner M-X, Eggeler G. Influence of Ni on martensitic phase transformations in NiTi shape memory alloys. Acta Mater 2010;58:3444–58.
[5] Laue Mv. Eine quantitative Prüfung der Theorie für die Interferenzerscheinungen bei Röntgenstrahlen. Ann Phy 1913;346:989–1002.
[6] Bragg WL. The structure of some crystals as indicated by their diffraction of X-rays. Proc R Soc Lond A 1913;89:248–77.
[7] Landau LD. On the theory of phase transitions I. Zh Eksp Teor Fiz 1937;7:19.
[8] Landau LD. On the theory of phase transitions II. Zh Eksp Teor Fiz 1937;7:627.
[9] Weiss P. L'hypothèse du champ moléculaire et la propriété ferromagnétique. J Phys Theor Appl 1907;6:661–90.
[10] Bragg WL, Williams E. The effect of thermal agitation on atomic arrangement in alloys. Proc R Soc Lond A 1934;145:699–730.
[11] Ising E. Beitrag zur Theorie des Ferromagnetismus. Z Phys A 1925;31:253–8.
[12] Onsager L. Crystal statistics. I. A two-dimensional model with an order-disorder transition. Phys Rev 1944;65:117.
[13] Wilson KG. Renormalization group and critical phenomena. I. Renormalization group and the Kadanoff scaling picture. Phys Rev B 1971;4:3174.
[14] Wilson KG. Renormalization group and critical phenomena. II. Phase-space cell analysis of critical behavior. Phys Rev B 1971;4:3184.
[15] Devonshire AF. XCVI. Theory of barium titanate. Lond Edinb Dubl Philos Mag 1949;40:1040–63.
[16] Devonshire AF. CIX. Theory of barium titanate – part II. Lond Edinb Dubl Philos Mag 1951;42:1065–79.
[17] Devonshire A. Theory of ferroelectrics. Adv Phys 1954;3:85–130.
[18] Salje EKH. Phase transitions in ferroelastic and co-elastic crystals. 2nd ed. Cambridge: Cambridge University Press; 1993.
[19] Salje EKH. Application of Landau theory for the analysis of phase transitions in minerals. Phys Rep 1992;215:49–99.
[20] Falk F. Ginzburg-Landau theory of static domain walls in shape-memory alloys. Z Phy B 1983;51:177–85.
[21] Hoelzel M, Senyshyn A, Juenke N, Boysen H, Schmahl W, Fuess H. High-resolution neutron powder diffractometer SPODI at research reactor FRM II. Nucl Instrum Meth A 2012;667:32–7.
[22] Wagner MFX, Windl W. Lattice stability, elastic constants and macroscopic moduli of NiTi martensites from first principles. Acta Mater 2008;56:6232–45.
[23] Shchyglo O, Salman U, Finel A. Martensitic phase transformations in Ni-Ti–based shape memory alloys: The Landau theory. Acta Mater 2012;60:6784–92.
[24] Stokes HT, Campbell BJ, Hatch DM. ISOTROPY software suite. Department of Physics and Astronomy, Brigham Young University; Provo, Utah, USA 2013.
[25] Campbell BJ, Stokes HT, Tanner DE, Hatch DM. ISODISPLACE: a web-based tool for exploring structural distortions. J Appl Crystallogr 2006;39:607–14.
[26] Aroyo MI, Perez-Mato JM, Capillas C, et al. Bilbao crystallographic server: I. Databases and crystallographic computing programs. Z Kristallogr 2006;221:15–27.
[27] Stokes HT, Campbell BJ, Cordes R. Tabulation of irreducible representations of the crystallographic space groups and their superspace extensions. Acta Crystallogr A 2013;69:388–95.

[28] Arnold H. Transformations of symmetry operations (motions). In: Hahn T, editor. International tables for crystallography volume A: space-group symmetry. Springer, Dordrecht, The Netherlands; 2006:86–9.
[29] Khalil-Allafi J, Schmahl WW, Reinecke T. Order parameter evolution and Landau free energy coefficients for the B2↔R-phase transition in a NiTi shape memory alloy. Smart Mater Struct 2005;14:S192–S6.
[30] Schmahl WW, Swainson I, Dove M, Graeme-Barber A. Landau free energy and order parameter behaviour of the α/β phase transition in cristobalite. Z Kristallogr 1992;201:125–46.
[31] Schmahl WW. Diffraction intensities as thermodynamic parameters-orientational ordering in $NaNO_3$. Z Kristallogr 1988:231–3.
[32] Schmahl WW, Salje E. X-ray diffraction study of the orientational order/disorder transition in $NaNO_3$: evidence for order parameter coupling. Phys Chem Miner 1989;16:790–8.
[33] Schmahl WW. Landau-model for the λ-transition in $NaNO_3$. Ferroelectrics 1990;107:271–4.
[34] Gufan YM, Larin E. Theory of phase transitions described by two order parameters. Sov Phys Solid State 1980;22:270–5.
[35] Salje E, Devarajan V. Phase transitions in systems with strain-induced coupling between two order parameters. Phase Transit 1986;6:235–47.

Part II: **Environmental mineralogy**

S. Heuss-Aßbichler and M. John
7 Gold, silver, and copper in the geosphere and anthroposphere: can industrial wastewater act as an anthropogenic resource?

7.1 Introduction

Ores are of tremendous importance for human being and the exploration to find new ore deposits is an urgent task. A review of the different types of ore deposits, for example, gold, silver, or copper, shows that they are distributed in different regions all over the world in a wide range of geological formations. They may occur highly concentrated or finely disseminated in the host rock. The formation of ore deposits depends on the specific geological environment. Various factors like change in temperature, pressure, redox and pH conditions, or biological activities may influence the ore formation. It is common sense that fluids are essential for the dissolution and transport of the metals [1–2], whereby there are still many open questions (e.g. [2, 3]). These processes are also of particular interest to understand the impacts of mine waste weathering [4].

In ancient times, metals were used for tools, weapon, money, and jewelry. With the development of civilization, the application area of metals changed continuously. In the modern society, there is an increasing need for metals with special structures and clearly defined morphology. Especially nanoparticles with specific properties have become integral parts of the technology. This trend implicates sophisticated synthesis methods. In general, the synthesis methods are complex and energy intensive.

At the same time, the huge demand for critical elements opens up the view on anthropogenic residues as potential secondary resources. Currently, various waste flows are tested if they provide an option for recovering metals. In general, wastewater and effluents with metals in the range of milligrams to grams per liter are regarded as too low concentrations for cost-efficient metal recovery. Most of these effluents are treated with lime to co-precipitate the metals with $Ca(OH)_2$, and the resulting neutralization sludge are disposed in landfills, which means dissipation of these metals. Moreover, in many countries, these heavy metal–containing effluents are discharged into the environment without any treatment, causing a serious source of contamination and leading to serious harm to the health of the people in this area. Therefore, heavy metal pollution is still one of the most serious environmental problems today [5].

There are many questions: Is it possible to precipitate metal particles efficiently from solution under low-temperature conditions? Can this procedure be adapted to recover metals from waste water? Are the phase assemblages obtained during these processes comparable with those observed in natural systems, and if so, can they help us to understand better the processes leading to various types of ore formations and vice versa?

7.2 Gold, silver, and copper in the geosphere

The average crustal abundance for gold and silver is in the range of 1–70 ppb, and for copper, about 50 ppm. In some ore deposits, we observe concentrations around 5 g/t, which means enrichment factors in the range of 1000 times and more. In hydrothermal solutions, the concentration of gold and silver are thought to range from 1 to 1 ppm, and those associated with copper, lead, and zinc ores in the range of 1–1000 ppm (or mg/L) metal [2]. Formation of ore minerals is the result of complex spatial and temporal interaction of rock, fluid, and/or melt [1]. Various parameter like temperature, pressure, chemistry of the host rock and fluid including pH and redox potential control the solubility, transport, and precipitation of the metal species [1–2, 4]. For example, ligands like chloride (Cl^-) or bisulfide (HS^-) increase the solubility of most metals in solution [2]. The driving force for many ore deposit formations are, for instance, changed reaction conditions, e.g. decrease in temperature or change in the physical properties or composition of the fluid, or mixing of two fluids with distinct temperature, pH, and redox characteristics [1].

A special type of ore accumulation represent supergene ore deposits which are formed by interaction of primary ore with very saline solutions or sulfuric acid combined with oxidation and leaching conditions within the surface near alteration zone [1, 4]. Within the unsaturated zone, the dissolution of the primary ore minerals is initiated by the oxidation of sulfides, and the metal ions like copper, zinc, silver, gold, nickel, or uranium are transported to deeper, less altered regions with higher pH and reducing conditions. The remaining leached cap rock, also called gossan, consists of quartz and Fe-(hydro)oxides such as goethite, hematite, ferrihydrite ($Fe_{10}O_{14}(OH)_2$), jarosite ($KFe_3[(OH)_6/(SO_4)_2]$), or schwertmannite ($Fe_8O_8(OH)_6(SO_4) \cdot nH_2O$), whereas within the subjacent oxide zone, ore minerals coexist with magnetite or sulfites [1, 4]. Further factors that have to be considered are microorganisms like sulfate-reducing bacteria (SRB). They can cause high H_2S concentrations including reducing conditions in the environment and thus accelerate the kinetics of ore formation [4]. However, adsorption processes are also important. Sorption of metals from solutions highly depends on prevailing pH conditions, and it may be the first step to incorporate metals in ore minerals (e.g. [6]).

7.2.1 Gold (Au)

Gold has been mined since the third millennium BC. It exists preferentially as native (zero-valent) element mineral (Fig. 7.1a) or electrum, an Au-Ag alloy. In primary deposits, gold may also occur with sulfide minerals such as pyrite, chalcopyrite, arsenopyrite, and sulfosalts. In supergene ore deposits, gold up to 2 mm and of extremely high purity are observed. It is often associated with crystalline and amorphous iron oxides. Dissolution of primary ore is the main process how even traces of gold are mobilized and precipitate later as nuggets [4, 7–8]. Three different processes are thought to evoke the precipitation of gold from a solution: decrease in activity of Cl^- or SO_4^{2-}, increase of pH, or reducing conditions by decrease in f_{O_2} [8]. Under acidic conditions, gold is preferentially adsorbed onto phases such as pyrite (FeS_2) and pyrrhotite ($Fe_{(1-x)}S$; $x = 0$–0.2) [9]. In a second step, electron transfer between the metal ion and mineral substrate is suggested to promote the formation of gold clusters on the mineral surface [10–11]. Microbial activity is also discussed as agent for mobilization and precipitation of gold (e.g. [12]).

7.2.2 Silver (Ag)

The history of silver mining goes back to 5000 BC. It can occur as native element mineral (Fig. 7.1b), such as the silver wires from Kongsberg (Norway), or alloyed with gold. The most common primary silver ores are argentite or acanthite (Ag_2S), proustite ($3Ag_2S \cdot As_2S_3$) and pyrargyrite ($Ag_2S \cdot Sb_2S_3$). Most of the silver in ores, however, is associated with sulfur, arsenic, antimony or chlorine [13].

Silver-rich mines are often associated with lead mines. About 75% of silver is obtained from ores with major metal being either lead, copper or zinc. In supergene silver deposits, the amount of acid and Fe^{3+}-producing sulfide minerals and acid-neutralizing gangue minerals are decisive [13]. Increasing Eh and decreasing pH promotes the solubility of Ag^+. Saline groundwater fosters the precipitation of chlorargyrite ($AgCl$), also called "cerargyrite" or "horn silver", whereas the formation of native silver preferentially occurs at the iron redox front, where the reduction of Ag^+ to Ag^0 is accompanied by the oxidation of Fe^{2+} to Fe^{3+} [13]. In neutral to alkaline environment, silver is frequently associated with various secondary minerals like iron- or manganese-(hydro)oxides, jarosite, or cerussite ($PbCO_3$).

7.2.3 Copper (Cu)

The main copper ore minerals in primary deposits are sulfides, e.g. bornite (Cu_5FeS_4), chalcopyrite ($CuFeS_2$), enargite (Cu_3AsS_4). About 70% of the copper inventory worldwide provide porphyry copper deposits. Supergene enrichment of copper is associated with secondary minerals, including native copper (Fig. 7.1c), oxides like tenorite (CuO) and cuprite (Cu_2O), sulfates like brochantite ($Cu_4SO_4(OH)_6$), chalcanthite ($CuSO_4 \cdot 5H_2O$), and antlerite ($Cu_3SO_4(OH)_4$), carbonates like malachite ($Cu_2CO_3(OH)_2$) and azurite ($Cu_3(CO_3)_2(OH)_2$), hydroxy-chlorides like atacamite ($Cu_2Cl(OH)_3$),

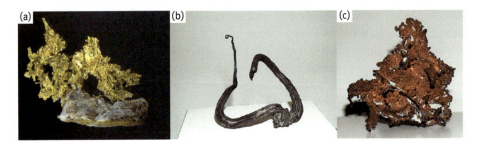

Fig. 7.1: (a) Native gold, from Eagle's Nest Mine, Michigan Bluff, CA, USA. (b) Native silver, from Kongsberg, Norway. (c) Native copper, from Ajo, AZ, USA. Photos: Mineralogische Staatssammlung München.

phosphates like turquoise $(CuAl_6(PO_4)_4(OH)_8 \cdot 4H_2O)$ and silicates like chrysocolla $(Cu_{2-x}Al_x)H_{2-x}Si_2O_5(OH)_4 \cdot n\ H_2O)$ [4]. It occurs in association with other supergene minerals such as hematite (Fe_2O_3) and goethite (α-FeOOH). Under more reduced conditions, beneath the oxidized zone, copper may replace iron in sulfides, producing secondary chalcocide (Cu_2S) and covellite (CuS) [4].

Minerals like delafossite $(CuFeO_2)$ are very rare and were found, e.g. in Bissbee, Arizona, USA [14] and the Besshi ore deposit, central Shikoku, Japan [15]. Cuprospinel or copper ferrite $(CuFe_2O_4)$ occurs naturally in Baie Verte, Newfoundland, Canada.

7.3 Gold, silver, and copper in the anthroposphere

7.3.1 Synthesis of gold, silver, and copper phases from aqueous solutions

Gold, silver, and copper are essential in a wide field of industrial applications. Recently, the demand for nanoparticles increased enormously due to their specific physical and chemical properties. Much effort is invested to develop new techniques to produce nanoparticles with specific design and properties [16]. There are various partly energy-intensive methods to synthesize gold, silver, or copper-(iron)oxide nanoparticles, including both chemical methods (e.g. thermal decomposition, hydrolysis, co-precipitation, chemical reduction, hydrothermal reaction, sol-gel procedures) and physical procedures (e.g. vapor deposition, grinding, laser ablation) [17–25]. Focus is on a high level of homogeneity of the particles with specific size, shape, and surface properties. Successful techniques are hydrothermal procedures at medium to high temperatures and saturation vapor pressure, for example [26–31]. Often, additives such as glycerol or polyvinyl-pyrrolidone are added as a reducing agent to stabilize a specific oxidation state of ions in the solution [26–27].

In comparison, there are only very few synthesis experiments done at low temperatures ≤100°C. We observed that at low temperatures, precipitation of metals occurs readily by addition of a defined amount of iron to the aqueous solution. The reaction is often initiated by iron-rich, process-active, metastable phases. Studies show that magnetite can be formed by various intermediate phases like iron-(hydro)oxides or green rust (GR), a Fe^{2+}-Fe^{3+}–double-layered hydroxy salt, which are not preserved in the final assemblage (e.g. [32–34]). Correspondingly, different metal (M)-oxides, metal (M)-ferrites (MFe_2O_4), delafossite ($CuFeO_2$) and even native metals are generated. Our studies showed that under alkaline reaction conditions, ferrihydrite ($Fe_{10}O_{14}(OH)_2$) and green rust (GR) can act as precursor phases [35–38]. In the following section, we would like to focus on reaction paths implying these reactive intermediate phases.

7.3.1.1 Transformation of metastable phases to more stable iron-oxides

Many synthesis routes are described to gain GR: (a) partial reduction of lepidocrocite (γ-FeOOH) or ferrihydrite ($Fe_{10}O_{14}(OH)_2$) [39], (b) the reaction of γ-FeOOH with Fe^{2+}_{aq} [40], and (c) co-precipitation from Fe^{2+}_{aq} and Fe^{3+}_{aq} containing solutions under anoxic conditions [41–45]. The key factor is to ensure the presence of Fe^{2+} in the solution due to the preference of Fe^{2+}_{aq} for oxidation [46]. An alternative approach is to precipitate $Fe(OH)_2$ from Fe^{2+}_{aq}-containing solutions and then the partial oxidization of Fe^{2+} to achieve GR [47–50]. The last step is promoted by adding SO_4^{2-} [51].

GR is a very reactive phase and transforms readily with changed reaction condition (e.g. pH, temperature, oxidazing agent, kinetics) [52–54]. Furthermore, GR acts as strong reducing agent during oxidation [35, 37–38, 55]. The oxidation rate is an important parameter controlling the formation of the desired reaction products. Magnetite Fe_3O_4 occurs spontaneously without oxidants [40, 56–58]. At low oxidizing conditions, however, various iron oxide-hydroxide phases like goethite (α-FeOOH), akaganeite (β-FeO(OH,Cl)), lepidocrocite (γ-FeOOH), ferroxyhyte (δ-FeOOH) as well as schwertmannite ($Fe^{3+}_{16}[O_{16}|(OH)_{9-12}(SO_4)_{3.5-2}]$) may grow [47, 52, 54, 59–61]. GR can also oxidize to maghemite (γ-Fe_2O_3) or hematite (α-Fe_2O_3) [62, 63]. The formation of ferrihydrite, the most poorly ordered type of iron hydroxide, and its further transformation to goethite is of special interest because of its abundance in natural systems [53, 64]. It is favored if the oxidation process is too rapid to dissolve ferrihydrite [53].

The question is, which phase assemblages can be precipitated in aqueous gold, silver, or copper systems and can we provoke the direct synthesis of gold, silver, and copper-(iron)-oxides or co-precipitate native (zero-valent) metals from aqueous solutions? In the following, we present case studies showing different reaction products obtained depending on the initial metal concentration and adjusted M/Fe-ratio, respectively.

7.3.1.2 Gold (Au)
Case 1: co-precipitation of gold with ferrihydrite and/or goethite

Greffie and co-workers [7] performed experiments with $AuCl_4^-$ solutions adding $Fe(NO_3)_3 \cdot 9\,H_2O$ with various Au/Fe ratios at different pH values. They obtained gold with a particle size up to 60 nm together with ferrihydrite (5–14 nm) or well-crystallized goethite. We used $Fe(SO_4) \cdot 7\,H_2O$ as educt and observed the precipitation of a mixture of μm-sized Au^0 and nano-sized goethite at alkaline pH values.

Case 2: separate formation of native gold (Au^0) and magnetite

Our investigations clearly showed that it is also possible to gain Au^0 separately from the Fe-containing reaction solution. After the addition of Fe^{2+} to the gold-containing solution, the first precipitates exclusively contain single-phase rosette-shaped Au^0 with an average size of 1.5 to 4 μm. After separation of these primary precipitates from the Fe-containing reaction solution and adjusting the pH ≥ 10, iron precipitates as 5- to 20-nm magnetite crystals. Small amounts of gold observed on the surface of magnetite indicate either incorporation, co-precipitation, or adsorption of gold on the surface of magnetite.

7.3.1.3 Silver (Ag)
Case 1: formation of native silver Ag^0 and magnetite

The reduction of Ag^{1+}, Au^{3+}, Cu^{2+}, and Hg^{2+} by formation of GR and subsequent transformation of GR to magnetite ($Fe^{2+}Fe^{3+}_2O_4$) was observed by O'Loughlin and coworkers [33]. Our experiments prove that in the presence of Fe^{2+} in solutions with initial silver concentrations of more than 5 g/L, part of the Ag^{1+} ions can readily be reduced to Ag^0. We gained silver as an Ag^0 foil directly on a glass substrate [38, 55]. At lower concentrations during alkalization, Ag^{1+} ions can also be reduced to Ag^0 with simultaneous oxidation of the iron precursor phase [38, 55]. In contrast to the gold system, the separation of the primary silver (Ag^0) precipitates is not necessary in order to gain magnetite as iron-bearing phase. The oxidation of GR directly to Fe_3O_4 is obviously promoted by electron exchange in the presence of Ag^{1+} ions [55].

Case 2: composite nanoparticles as silver core shell (Ag^0@magnetite (Fe_3O_4) and Ag-delafossite ($AgFeO_2$))

We observed that in case of low concentrated silver solutions (1 g/L), and after the precipitation of GR and Ag^0, two further scenarios of GR oxidation are possible [38]: In the first scenario, the oxidization of GR to magnetite is accompanied by the reduction of small amounts of adsorbed Ag^{1+} to Ag^0, forming a kind of layer around Fe_3O_4. The second scenario comprises the intergrowth of Ag-delafossite ($AgFeO_2$) and magnetite (Fe_3O_4) nanoparticles partially surrounded by a layer of Ag^0. We suggest that a local excess of Ag^{1+} ions adsorbed on single GR particle promotes additional growth of delafossite [38].

Case 3: formation of Ag-Delafossite ($AgFeO_2$)

The formation of pure $AgFeO_2$ with delafossite structure is also possible using low concentrated (1 g/L) silver solutions. In this case, iron was added in oxidized form (Fe^{3+}) and nitrate was used as reactant [55]. After a reaction time of 10 minutes, thin hexagonal plates of AgFeO$_2$ up to 50 nm diameter were observed. It occurs in both polytypes, rhombohedral 3R delafossite and hexagonal 2H delafossite.

7.3.1.4 Copper (Cu)

In comparison to gold and silver, the copper system is very complex and multifaceted since a huge variety of phase assemblages is possible. Génin and co-workers [51, 65] showed that the ratio of OH$^-$/Fe^{2+} strongly influences the formation and oxidation of GR. Our results show that in the presence of Cu^{2+} ions, not only the absolute amount of added NaOH (= ratio of OH$^-$/Fe^{2+}) but also the concentration of the added NaOH solution is crucial [38]. The addition of low concentrated NaOH solutions (<32% NaOH) promotes the formation of ferrite as the main phase. In contrast, we obtain delafossite structures (CuFeO$_2$) by adding highly concentrated NaOH solution (32% NaOH).

Case 1: solid solution of copper ferrite (CuFe$_2$O$_4$)-magnetite (Fe$_3$O$_4$) and co-precipitation of either tenorite (CuO), cuprite (Cu$_2$O) or native copper (Cu^0)

In all experiments, we observed that GR and Cu$_2$O precipitate first [36–38]. The reduction of Cu^{2+} ions to their monovalent state (Cu$_2$O) is especially promoted in initially copper-rich solutions. Additional NaOH supply for at least 5 minutes and subsequent aging for more than 1 hour result in the transformation of GR to magnetite. At more oxidizing conditions, ferrihydrite is obtained.

In highly alkaline solutions, Cu$_2$O is metastable and decomposes readily to CuO and/or Cu^0 [35, 38], depending on oxygen fugacity: In solutions with initially high amounts of Cu^{2+} ions (concentrations >5 g/L), during aging, Cu^0 is formed on the expense of Cu$_2$O. In contrast, in solutions with low Cu^{2+} concentrations (<5 g/L) or a rather low Fe proportion (Fe/Cu of ≤1/1), Cu$_2$O preferentially oxidizes to CuO. In the latter case, CuO precipitates simultaneously with GR during the initial alkalization process. This is obviously the effect of the low absolute amount of GR in the solution and hence lower demand of oxygen needed for the transformation of GR. However, later, during aging, CuO is partially reduced to Cu$_2$O.

Several studies with solutions containing maximum 10 mol% Cu^{2+} stated the reduction of Cu^{2+} to Cu^0 by the oxidation of GR to goethite (α-FeOOH) [66–67] and magnetite [33]. Some of our experiments show that, depending on the reaction and aging conditions, ferrihydrite can spontaneously transform to goethite [38]. Tamaura and co-workers [68] pointed out that the formation of ferrite is promoted by adsorption of divalent heavy metals on the surface of FeOOH. A similar mechanism of Cu^{2+} adsorption on the surface of ferrihydrite and goethite may also explain the formation of copper ferrite (CuFe$_2$O$_4$) in our system.

Chaves and co-workers [69], however, showed that divalent heavy metals can substitute Fe^{2+} in GR. The incorporation of Cu^{2+} ions directly into GR (= Cu-GR) during the initial alkalization process seems to promote the formation of copper ferrite–magnetite solid solution [38]. The solid solution seems to be restricted as we observed maximal 45 mol% copper ferrite end-member [Cu$_{0.45}$Fe$_{0.55}$]$^{2+}$ Fe$^{3+}_2$O$_4$.

Case 2: formation of Delafossite ($CuFeO_2$)

Various studies deal with the synthesis of delafossite [17–20, 26–28, 31]. However, we could present the precipitation of delafossite via GR as a precursor for the first time [36] and described the reaction mechanism of ferrihydrite and Cu_2O leading to the crystallization of delafossite under alkaline conditions [36–37], called (low-temperature) Lt-delafossite process. Three steps are involved: During the first step, the precipitation of GR and Cu_2O requires a sufficient amount of OH^- which is achieved by adding a highly concentrated NaOH solution (32% NaOH). During the second step, the oxidization of GR and formation of the intermediate reaction phase ferrihydrite is accompanied by the reduction of Cu^{2+}. This leads to the precipitation of Cu_2O as a further intermediate compound. During the last step, the high pH value in the solution promotes the growth of delafossite at the expense of the unstable phases ferrihydrite and Cu_2O (Fig. 7.2).

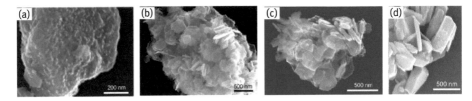

Fig. 7.2: SEM pictures of the delafossite formation: (a) first delafossite crystals after 3 hours of aging, (b) delafossite crystals in nanocrystalline matrix after 6 hours, (c) almost complete transformation of the matrix to delafossite due to further aging, (d) thick delafossite crystals after 7 days of aging.

The formation of delafossite can be influenced in various ways: The reaction rate increases with increasing aging temperature, reaction pH, and, in particular, NaOH concentrations in the solution. A deviation from the optimum Cu/Fe ratio of 1/1 leads to the growth of either ferrihydrite or Cu_2O in addition to copper ferrite.

Case 3: low-Fe approach – doped Cu-oxide (CuO or Cu_2O)

It is also possible to obtain Fe-doped CuO if a low amount of Fe is added to the solution [38]. CuO precipitates as single phase in aqueous solutions with 1–3 g/L Cu (Cu/Fe ratios higher than 1/0.1) at 40°C and pH 11. At lower Cu/Fe ratios, traces of co-precipitated ferrihydrite may occur and at higher ratios (Cu/Fe ratio 1/≥0.5) Cu_2O precipitates in addition to CuO.

7.3.2 Wastewater: an anthropogenic resource for gold, silver, and copper?

Industrial effluents often contain metals in sulfuric acid solutions at low pH conditions and hence are comparable with natural solutions forming supergene metal deposits. There are several methods to remove metals from wastewater, e.g. chemical precipitation, flotation, adsorption, ion exchange, electrochemical deposition [5, 70]. The role of microorganisms in nature inspired the development of bioremediation technologies to remove heavy metals by SRB (e.g. [71]). However, precipitation of metals as hydroxides at basic conditions is the most widely used method [70]. The application of lime for co-precipitation is a cheap and convenient method. The high volume of the neutralization sludge and low concentrations of the metals, however, make the recovery of these metals uneconomic, and as a result, the sludge is mostly deposited at high costs in landfills. Therefore, the main question is, can we evoke the precipitation of the metals as oxides or even as zero-valent metals without producing voluminous hydroxides?

First, studies were performed with the so-called ferrite process. The aim was to remove low concentrations of bivalent cations up to 1 g/L from laboratory wastewater [72–74]. The focus of all these studies was to achieve the incorporation of divalent non-Fe metal ions into the ferrite MFe_2O_4 with spinel structure [75–79]. In studies, the formation of copper ferrite, for example, was described without co-precipitated phases [78, 80]. Recently, our group developed an environmentally friendly and energy-saving approach, called "SPOP" (specific product–oriented precipitation). This technique enables to recover noble and transition metals from industrial wastewater with the main goal to precipitate valuable secondary raw materials as oxide components or even zero-valent elements [35, 37–38, 55]. We tested wastewaters from the electroplating industry and catalyst productions, for example, and obtained purification rates of ≥99.98% up to 100% for gold, silver, and copper [35, 37–38, 55]. Figure 7.3 shows

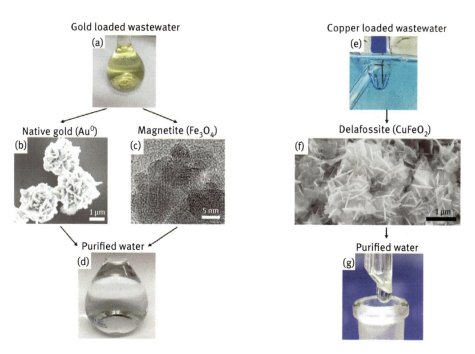

Fig. 7.3: (a) Gold-containing wastewater before treatment. (b) SEM image of zero-valent gold rosettes. (c) TEM image of magnetite. (d) Purified gold wastewater after treatment. (e) Copper containing wastewater before treatment. (f) SEM image of delafossite ($CuFeO_2$). (g) purified copper wastewater after treatment.

the gold and copper wastewater before and after treatment with our SPOP technology and the corresponding precipitated products.

Our results indicate that the phase assemblage of the residues strongly depends on the reaction conditions, including metal type, initial metal concentration, Fe content, reaction temperature, pH, as well as alkalization and partially aging conditions. By adapting specific reaction parameters from our synthesis procedures (see Section 3.1), we are able to produce residues free of hydroxide components [35, 37–38, 55, 81]. The benefit of our SPOP procedure is an extremely low volume of recovered residues compared to other treatment techniques, with both the density of the precipitated material and the overall concentration of the target metal having the highest amount. Additional elements mostly present in wastewater from electroplating industries such as Cr, Ni or Zn can also be removed simultaneously. They are incorporated in the structures of the ferrite, delafossite, or metal-oxide components [35, 37–38, 55, 81].

7.4 What do/did we learn?

The detailed mineralogical characterization of reaction products obtained after the application of the ferrite process to treat copper containing wastewater showed that the assemblage of reactions products can be controlled by adjustment of boundary conditions. This was the starting point for the development of SPOP as a sophisticated technique: It enables to purify effectively even highly contaminated rinsing water with heavy metals >25 g/L in compliance with the legal requirements for environmental regulations. Furthermore, this new approach makes the recovery of the valuable metals and metalloids possible and hence opens the door to regard wastewater as a potential resource for secondary raw materials. Moreover, this concept presents an efficient method to synthesize nanocrystalline metal oxides, metal ferrites (MFe_2O_4), delafossite ($MFeO_2$), and native metals via precipitation at low temperatures (20–70°C). Last but not the least, it is an effective method to prevent environmental pollution and it is a very good example of an efficient circular economy.

The phases obtained in the experiments show very similar assemblages as observed in some supergene metal deposits. An important contribution is the reaction triggered by the intermediate phases. Phases like ferrihydrite or goethite survive in natural environment. However, for the formation of the ore deposits, the active contribution of metastable products like GR, observed in soil profiles (e.g. [82]), is not taken under consideration so far. The comparison of the phase assemblage in experiments with natural samples may contribute to the understanding of alternative abiotic processes in nature.

Acknowledgments

Many thanks to Kai Tandon for performing experiments and measurements, especially on Au-loaded wastewaters and to Dr. Werner Ertel-Ingrish, Prof. Dr. Georg Amthauer (Univ. Salzburg) and Prof. Dr. Albert Gilg (TU Munich) for review and helpful discussions. We are grateful to Aladin Ulrich at the Univ. Augsburg for providing the SEM and TEM pictures and to Dr. Diemer, Bavarian Environment Agency, for measurement of gold in wastewater. The industrial partners are acknowledged, in particular Clariant AG for providing gold wastewater and WAFA Germany for copper wastewater. The main part of this work was financed by the Bavarian State Ministry of the Environment and Consumer Protection Grant BAF01SoFo-65214.

References

[1] Heinrich CA, Candela PA. Fluids and ore formation in the earth's crust. Chapter 13.1. In: Treatise on Geochemistry. Amsterdam, Netherlands: Elsevier; 2014; 13:1–28.
[2] Seward, TM, Williams-Jones AE, Migdisov AA. The chemistry of metal transport and deposition by ore-forming hydrothermal fluids. Chapter 13.2. In: Treatise on Geochemistry. Elsevier; 2014; 13:29–57.
[3] Arehard GB. Characteristics and origin of sediment-hosted disseminated gold deposits: a review. Ore Geol Rev 1996;11:383–403.
[4] Reich M, Vasconcelos PM. Geological and economic significance of supergene metal deposits. Elements 2015;11(5):305–10.
[5] Fu F, Wang Q. Removal of heavy metal ions from wastewaters: a review. J Environ Manage 2011;92(3):407–18.
[6] Karasyova ON, Ivanova LI, Lakshtanov LZ, Lövgren L, Sjöberg S. Complexation of gold (III)-chloride at the surface of hematite. Aquat Geochem 1998;4(2):215–31.
[7] Greffie C, Parron C, Benedetti M, Amouric M, Marion P, Colin F. Experimental study of gold precipitation with synthetic iron hydroxides: HRTM-AEM and Mössbauer spectroscopy investigations. Chem Geol 1993;107:297–300.
[8] Mann AW. Mobility of gold and silver in lateritic weathering profiles: some observations from Western Australia. Econ Geol 1984;79:38–49.
[9] Widler AM, Seward TM. The adsorption of gold(I) hydrosulphide complexes by iron sulphide surfaces. Geochim Cosmochim A 2002;66(3):383–402.
[10] Jean GE, Bancroft GM. Heavy metal adsorption by sulphide mineral surfaces. Geochim Cosmochim Acta 1985;50(7):1455–63.
[11] Knipe SW, Foster RP, Stanley CJ. Role of sulphide surfaces in sorption of precious metals from hydrothermal fluids. Trans Inst Mining Metall B 1992;101:83–8.
[12] Southam G, Lengke MF, Fairbrother L, Reith F. The biogeochemistry of gold. Elements 2009;5(5):303–7.
[13] Sillitoe R. Supergene silver enrichment reassessed. In: Titley SR, editor. Supergene environments, processes, and products. Society of Economic Geologists, special publication 2009;14:15–32.
[14] Rogers AF. Delafossite, a cupreous metaferrite from Bisbee, Arizona. Am J Sci 1913;207:290–4.
[15] Minakawa T, Dehara M. Delafossite from the oxidation zone of the besshi ore deposit. Jpn Mag Miner Petrol Sci 2005;37:91–5.

[16] Jelinek R. Nanoparticles. De Gruyter Textbook: Walter de Gruyter GmbH & Co KG; 2015.
[17] Choi DH, Moon SJ, Hong JS, An SY, Shim IB, Kim CS. Epitaxial growth of delafossite $CuFeO_2$ thin films by pulse laser deposition. Thin Solid Films 2009;517:3987–9.
[18] Sukeshini AM, Kobayashi H, Tabuchi M, Kageyama H. Lithium intercalation of delafossite, cuprous iron oxide. In: Intercalation compounds for battery materials. Proceedings of the International Symposium. Pennington, New Jersey: The Electrochemical Society; 2000:104.
[19] Galakhov VR, Poteryaev AI, Kurmaev EZ, Anisimov VI, Neumann M, Lu ZW, Zhao TR. Valence-band spectra and electronic structure of $CuFeO_2$. Phys Rev B 1997;56(8):4584.
[20] Chen HY, Wu JH. Transparent conductive $CuFeO_2$ thin films prepared by sol-gel processing. Appl Surf Sci 2012;258(11):4844–47.
[21] Wang W, Zhan Y, Wang X, Liu Y, Zheng C, Wang G. Synthesis and characterization of CuO nanowhiskers by a novel one-step, solid-state reaction in the presence of a nonionic surfactant. Mater Res Bull 2002;37(6):1093–1100.
[22] Bakhtiari F, Darezereshki E. One-step synthesis of tenorite (CuO) nanoparticles from $Cu_4(SO_4)(OH)_6$ by direct thermal-decomposition method. Mater Lett 2011;65(2):171–4.
[23] Armelao L, Barreca D, Bertapelle M, Bottaro G, Sada C, Tondello E. A sol-gel approach to nanophasic copper oxide thin films. Thin Solid Films 2003;442(1):48–52.
[24] Su YK, Shen CM, Yang HT, Li HL, Gao HJ. Controlled synthesis of highly ordered CuO nanowire arrays by template-based sol-gel route. Trans Nonferrous Metals Soc China 2007;17(4):783–6.
[25] Zhou K, Wang R, Xu B, Li Y. Synthesis, characterization and catalytic properties of CuO nanocrystals with various shapes. Nanotechnology 2006;17(15):3939.
[26] Moharam MM, Rashad MM, Elsayed EM, Abou-Shahba RM. A facile synthesis of delafossite $CuFeO_2$ powders. J Mater Sci Mater Electron 2014;25(4):1798–803.
[27] Qiu X, Liu M, Sunada K, Miyauchi M, Hashimoto K. A facile one-step hydrothermal synthesis of rhombohedral $CuFeO_2$ crystals with antivirus property. Chem Commun 2012;48.59:7365–67.
[28] Sheets WC, Mugnier E, Barnabé A, Marks TJ, Poeppelmeier KR. Hydrothermal synthesis of delafossite-type oxides. Chem Mater 2006;18.1:7–20.
[29] Ouyang S, Chen D, Wang D, Li Z, Ye J, Zou Z. From β-phase particle to α-phase hexagonal-platelet superstructure over $AgGaO_2$: phase transformation, formation mechanism of morphology, and photocatalytic properties. Crystal Growth Des 2010;10(7):2921–7.
[30] Yu M, Draskovic TI, Wu Y. Understanding the crystallization mechanism of delafossite $CuGaO_2$ for controlled hydrothermal synthesis of nanoparticles and nanoplates. Inorganic Chem 2014;53(11):5845–51.
[31] Xiong D, Zeng X, Zhang W, Wang H, Zhao X, Chen W, Cheng YB. Synthesis and characterization of $CuAlO_2$ and $AgAlO_2$ delafossite oxides through low-temperature hydrothermal methods. Inorganic Chem 2014; 53(8):4106–16.
[32] Nyirő-Kósa I, Rečnik A, Pósfai M. Novel methods for the synthesis of magnetite nanoparticles with special morphologies and textured assemblages. J Nanoparticle Res 2012;14(10):1150.
[33] O'Loughlin EJ, Shelly DK, Kemner KM, Csencsits R, Cook RE. Reduction of AgI, AuIII, CuII, and HgII by FeII/FeIII hydroxysulphate green rust. Chemosphere 2003;53:437–46.
[34] Kahani SA, Jafari M. A new method for preparation of magnetite from iron oxyhydroxide or iron oxide and ferrous salt in aqueous solution. J Magn Magn Mater 2009;321(13):1951–4.
[35] Heuss-Aßbichler S, John M, Klapper D, Bläß UW, Kochetov G. Recovery of Cu as zero-valent phase and/or Cu oxide nanoparticles from wastewater by ferritization. J Environ Manage 2016;181:1–7.
[36] John M, Heuss-Aßbichler S, Ullrich A. Conditions and mechanisms for the formation of nano-sized delafossite ($CuFeO_2$) at temperatures ≤90°C in aqueous solution. J Solid State Chem 2016;234:55–62.
[37] John M, Heuss-Aßbichler S, Ullrich A, Rettenwander D. Purification of heavy metal loaded wastewater from electroplating industry under synthesis of delafossite (ABO_2) by Lt-delafossite process. Water Res 2016;100:98–104.

[38] John M. Low temperature synthesis of nano crystalline zero-valent phases and (doped) metal oxides as $A_xB_{3-x}O_4$ (ferrite), ABO_2 (delafossite), A_2O and AO. A new process to treat industrial wastewaters? Dissertation, LMU, München; 2015.
[39] Hansen HCB. Composition, stabilization, and light adsorption of Fe(II) Fe(III) hydroxycarbonate (green rust) Clay minerals 1989;24(4):663–9.
[40] Tamaura Y. New ferrite-formation reactions in the suspensions of γ-FeOOH and "green rust II". Adv Ceram 1986;15:87–92.
[41] Géhin A, Ruby C, Abdelmoula M, Benali O, Ghanbaja J, Refait P, Génin JMR. Synthesis of the Fe (II-III) hydroxysulphate green rust by coprecipitation. Solid State Sci 2002;4(1):61–6.
[42] Ruby C, Géhin A, Abdelmoula M, Génin JMR, Jolivet JP. Coprecipitation of Fe(II) and Fe(III) cations in sulphated aqueous medium and formation of hydroxysulphate green rust. Solid State Sci 2003;5(7):1055–62.
[43] Bocher F, Géhin A, Ruby C, Ghanbaja J, Abdelmoula M, Génin JMR. Coprecipitation pf Fe(II-III) hydroxycarbonate green rust stabilised by phosphate adsorption. Solid State Sci 2004;6(1):117–24.
[44] Ahmed IAM, Benning LG, Kakonyi G, Sumoondur AD, Terrill NJ, Shaw S. Formation of green rust sulfate: a combined in situ time-resolved X-ray scattering and electrochemical study. Langmuir 2010;26(9):6593–603.
[45] Guilbaud R, White ML, Poulton SW. Surface charge and growth of sulphate and carbonate green rust in aqueous media. Geochim Cosmochim Acta 2013;108:141–53.
[46] Ruby C, Aïssa R, Géhin A, Cortot J, Abdelmoula M, Génin JM. Green rusts synthesis by coprecipitation of Fe II-Fe III ions and mass-balance diagram. CR Geosci 2006;338(6):420–32.
[47] Schwertmann U, Fechter A. Transformation to lepidocrocite. Clay Miner 1994;29:87–92.
[48] Refait P, Génin JMR. Mechanisms of oxidation of Ni (II)-Fe (II) hydroxides in chloride-containing aqueous media; role of the pyroaurite-type Ni-Fe hydroxychlorides. Clay Miner 1997;32(4):597–613.
[49] Refait P, Abdelmoula M, Génin JM. Mechanisms of formation and structure of green rust one in aqueous corrosion of iron in the presence of chloride ions. Corros Sci 1998;40(9):1547–60.
[50] Randall SR, Sherman DM, Ragnarsdottir KV. Sorption of As (V) on green rust ($Fe_4(II)Fe_2(III)(OH)_{12}SO_4·3H_2O$) and lepidocrocite (γ-FeOOH): surface complexes from EXAFS spectroscopy. Geochim Cosmochim Acta 2001;65(7):1015–23.
[51] Génin JM, Olowe AA, Refait P, Simon L. On the stoichiometry and Pourbaix diagram of Fe (II)-Fe (III) hydroxy-sulphate or sulphate-containing green rust 2: an electrochemical and Mössbauer spectroscopy study. Corros Sci 1996;38(10):1751–62.
[52] Misawa T, Asami K, Hashimoto K, Shimodaira S. The mechanism of atmospheric rusting and the protective amorphous rust on low alloy steel. Corros Sci 1974;14(4):279–89.
[53] Refait P, Benali O, Abdelmoula M, Génin JM. Formation of 'ferric green rust' and/or ferrihydrite by fast oxidation of iron (II-III) hydroxychloride green rust. Corros Sci 2003;45(11):2435–49.
[54] Ruby C, Abdelmoula M, Naille S, Renard A, Khare V, Ona-Nguema G, Génin JMR. Oxidation modes and thermodynamics of Fe II-III oxyhydroxycarbonate green rust: dissolution-precipitation versus in situ deprotonation. Geochim Cosmochim Acta 2010;74(3):953–66.
[55] John M, Heuss-Aßbichler S, Tandon K, Ullrich A. Recovery of Ag and Au from synthetic and industrial wastewater by 2-step ferritization and Lt-delafossite process via precipitation 2017 Submitted.
[56] Tamaura Y, Yoshida T, Katsura T. The synthesis of green rust II (Fe^{III}_1-Fe^{II}_2) and its spontaneous transformation into Fe_3O_4. Bull Chem Soc Jpn 1984;57(9):2411–6.
[57] Sumoondur A, Shaw S, Ahmed I, Benning LG. Green rust as a precursor for magnetite: an in situ synchrotron based study. Mineral Mag 2008;72(1):201–4.
[58] Ruby C, Usman M, Naille S, Hanna K, Carteret C, Mullet M, Abdelmoula M. Synthesis and transformation of iron-based layered double hydroxides. Appl Clay Sci 2010;48(1):195–202.

[59] Detournay PJ, Derie R, Ghodsi M. Etude de l'oxydation par aeration de Fe(OH)$_2$ en milieu chlorure. Zeitschrift für anorganische und allgemeine Chemie, 1976;427(3):265–73.

[60] Drissi SH, Refait P, Abdelmoula M, Génin JMR. The preparation and thermodynamic properties of Fe (II). Fe (III) hydroxide-carbonate (green rust 1); Pourbaix diagram of iron in carbonate-containing aqueous media. Corros Sci 1995;37(12):2025–41.

[61] Antony H, Legrand L, Chaussé A. Carbonate and sulphate green rusts – mechanisms of oxidation and reduction. Electrochim Acta 2008;53(24):7146–56.

[62] Lee W, Batchelor B. Abiotic reductive dechlorination of chlorinated ethylenes by iron-bearing soil minerals. 2. Green rust. Environ Sci Technol 2002;36(24):5348–54.

[63] O'Loughlin EJ, Kelly SD, Cook RE, Csencsits R, Kemner KM. Reduction of uranium (VI) by mixed iron (II)/iron (III) hydroxide (green rust): formation of UO$_2$ nanoparticles. Environ Sci Technol 2003;37(4):721–7.

[64] Benali O, Abdelmoula M, Refait P, Génin JMR. Effect of orthophosphate on the oxidation products of Fe (II)-Fe (III) hydroxycarbonate: the transformation of green rust to ferrihydrite. Geochim Cosmochim Acta 2001;65(11):1715–26.

[65] Génin JM, Ruby C, Géhin A, Refait P. Synthesis of green rusts by oxidation of Fe(OH)$_2$, their products of oxidation and reduction of ferric oxyhydroxides; Eh-pH Pourbaix diagrams. CR Geosci 2006;338(6):433–46.

[66] Inoue K, Shinoda K, Suzuki S, Waseda Y. Reduction of copper ions in green rust suspension and oxidation of green rust containing metallic copper. Mater Trans 2008;49(9):1941–6.

[67] Suzuki S, Shinoda K, Sato M, Fujimoto S, Yamashita M, Konishi H, Waseda Y. Changes in chemical state and local structure of green rust by addition of copper sulphate ions. Corros Sci 2008;50(6):1761–5.

[68] Tamaura Y. Ferrite formation from the Intermediate, Green Rust II, in the transformation reaction of γ-FeO (OH) in aqueous suspension. Inorganic Chem 1985;24(25):4363–6.

[69] Chaves LHG, Curry JE, Stone DA, Carducci MD, Chorover J. Nickel incorporation in Fe (II, III) hydroxysulfate green rust: effect on crystal lattice spacing and oxidation products. Rev Bras Ciência Solo 2009;33(5):1115–23.

[70] Kurniawan TA, Chan GY, Lo WH, Babel S. Physico-chemical treatment techniques for wastewater laden with heavy metals. Chem Eng J 2006;118(1):83–98.

[71] Jalali K, Baldwin SA. The role of sulphate reducing bacteria in copper removal from aqueous sulphate solutions. Water Res 2000;34(3):797–806.

[72] Katsura T, Tamaura Y, Terada H. 'Ferrite process' treatment of the wastewater from laboratories. Ind Water Jpn 1977:223.

[73] Okuda T, Sugano I, Tsuji T. Removal of heavy metals from waste-waters by ferrite co-precipitation. Filtrat Separ 1975;12.5:472–5.

[74] Barrado E, Prieto F, Vega M, Fernández-Polanco F. Optimization of the operational variables of a medium-scale reactor for metal-containing wastewater purification by ferrite formation. Water Res 1998;32.10:3055–61.

[75] Morgan BE, Loewenthal RE, Lahav O. Fundamental study of a one-step ambient temperature ferrite process for treatment of acid mine drainage waters. Water SA 2001:277–82.

[76] Chaiyaraksa C, Klaikeow C. Removal of heavy metals from electroplating wastewater by ferritisation. KMITL Sci Tech 2006;6.2:46–55.

[77] Huang YJ, Tu CH, Chien YC, Chen HT. Effect of [Fe^{2+}/total metal] on treatment of heavy metals from laboratory wasteliquid by ferrite process. Radiat Phys Chem 2006;75(11):1884–7.

[78] Pritosiwi G. Removal of metal ions from synthetic and galvanic wastewater by their incorporation into ferrites. TU Hamburg Harbur; Diss. 2012.

[79] Tu YJ, Chang CK, You CF, Wang SL. Treatment of complex heavy metal wastewater using a multi-staged ferrite process. J Hazard Mater 2012;209:379–84.

[80] Kochetov G, Zorya D, Grinenko J. Integrated treatment of rinsing copper-containing wastewater. Civil Environ Eng 2010;1:257–61.
[81] John M, Heuss-Aßbichler S, Ullrich A. Purification of wastewater form zinc plating industry. Environ Sci Technol 2016; 1735–72.
[82] Génin JMR, Refait P, Bourrié G, Abdelmoula M, Trolard F. Structure and stability of the Fe (II)-Fe (III) green rust "fougerite" mineral and its potential for reducing pollutants in soil solutions. Appl Geochem 2001;16(5):559–70.

Tsutomu Sato
8 Applied mineralogy for recovery from the accident of Fukushima Daiichi Nuclear Power Station

8.1 Introduction

Most of us use the products of modern technology without fully appreciating which minerals are required to make a cell phone, a modern internal combustion engine, an aluminum can, ceramics, and the concrete used in buildings [1]. For example, in Japan, all students in junior high school have to learn "what are minerals" and "what is the definition of minerals" with some examples from rock-forming minerals such as quartz, feldspar, mica, and so on, although the author does not know the situation of minerals in education at other countries. I suppose that the situation is not so different in different countries. However, as stated in the review in this book and in the special issue of *Elements* on "Social and economic impact of geochemistry", minerals are definitely central not only to our natural and technological environments but also to our social and economic environments [2]. Environmental mineralogy is a fast-growing multidisciplinary field, addressing major societal concerns about the impact of anthropogenic activities on the global ecosystem [3]. However, mineralogists are still not very good at communicating the social and economic impacts of mineralogy to the public. Of course, minerals may sometimes inspire us to design new materials for advanced technologies. Minerals and mineralogical processes such as adsorption, sorption, and precipitation may play an important role to solve problems in negative legacy such as pollution, health effect, and waste disposal.

In March, 2011, the fourth largest recorded earthquake occurred offshore from the Tohoku region of Japan and ruptured a 300-km-long by 200-km-wide portion of the subduction zone megathrust fault at the boundary of the Pacific plate [4]. The seafloor movement caused by the earthquake generated a huge tsunami that devastated communities along the entire northeastern coast of Japan. The 14- to 15-m-high tsunami waves hit the Fukushima Daiichi Nuclear Power Station (FDNPS) of the Tokyo Electrical Power Company (TEPCO). All electric power was lost in Units 1–4 of the FDNPS within 15 minutes after the first tsunami wave hit the station, and this led to a blackout of the station, followed by a series of hydrogen explosions in the FDNPS. These hydrogen explosions resulted in the displacement of approximately 470,000 citizens as well as negative legacies such as damaged nuclear power station, melted fuel debris, contaminated water, soils, vegetation, and debris, secondary wastes from on-site water treatment, and huge volumes of the Cs-contaminated soil from decontamination operation off-site. The author, as an environmental mineralogist, has been involved in the countermeasure activities against the negative legacies of the accident. In this chapter, the author shows how applied mineralogy has been and will be an important component in the recovery of areas surrounding the FDNPS and tries to promote the social and economic impacts of mineralogy based on the author's experiences after the Fukushima accident.

8.2 Mineralogical issues on- and off-site of the FDNPS

8.2.1 On-site

Immediately following the tsunami, all cooling and high-pressure makeup waters to the reactors were lost. The accident sequence of the FDNPS has been described in detail by other authors [5, 6]. Because of the long period without a core cooling system, substantial core damage occurred in the FDNPS. The fuel could no longer maintain a geometry capable of allowing cooling and temperatures soon exceeded the melting point (~2800°C) of the fuel. Therefore, to understand the state of the fuel debris and to develop debris-removal methods, mineralogists, and materials scientists should determine the phases of the fuel after meltdown and assess the potential reactions with construction materials such as the zirconium fuel cladding and concrete at the bottom of the building housing the reactor. Although mineralogy and chemistry of the fuel debris is important, the author cannot use space in this chapter to show these details. Please see the details in the relevant papers [7–10].

After the accident, to avoid self-sustaining nuclear chain reactions of the fuel, cooling water injection to the fuel had to be continued. The injected water that came into contact with the damaged fuel became highly contaminated with radioactive elements. Because coolant circulation was not possible, the discharged water had to be stored separately. Due to limited storage capacity, the cooling water should be used again after removal of the nuclides of concern, i.e., ^{134}Cs, ^{137}Cs, and ^{90}Sr. With the approach of the rainy season in June and July in 2011, there was urgent need for a secure space for this contaminated water and to establish coolant circulation systems with a decontamination capability.

The contaminated water contained high concentrations of non-radiogenic Na, K, Mg, Ca, Sr, and Cl, which originated from seawater, as well as radioactive Sr, Cs, and I, which were derived from the damaged fuel. The major element composition of the seawater was the main difference between the wastewater at Fukushima and the wastewater generated in the Three Mile Island accident in 1979, where there was no injection of seawater. The treatment system for the contaminated water had specific requirements to achieve sufficiently high decontamination factors, particularly for Cs, under high-salinity conditions. Therefore, data acquisition on the capacity and rate of Cs and Sr adsorption for various candidate materials under high-salinity conditions was urgently desired to construct the system. In this context, an academic team involving university professors, their students, ands nuclear chemistry and mineralogy researchers voluntarily conducted adsorption experiments and presented the results in the beginning of April 2011 (Fig. 8.1) [11].

8.2 Mineralogical issues on- and off-site of the FDNPS — 155

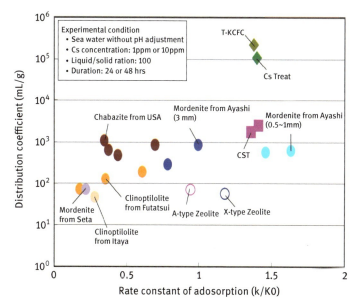

Fig. 8.1: Distribution coefficients of Cs between contaminated water and minerals and rate constant of adsorption for different candidate materials such as natural and synthetic zeolites, ferrocyanide, and silicotitanate compounds (T-KCFC and Cs Treat: commercial ferrocyanide compounds, CST: crystalline silicotitanate).

After data acquisition and test operation, TEPCO started running the treatment facility with a Cs/Sr removal system (Fig. 8.2) that includes synthetic zeolite (chabazite) and ferrocyanide ($Fe^{3+}_4[Fe^{2+}(CN)_6]_3 \cdot nH_2O$) for Cs and crystalline silicotitanate ($Na_2Ti_2O_3SiO_4 \cdot 2H_2O$) for Sr on June 17, 2011. This operation was important because the contaminated water in the trench was expected to overflow into Pacific Ocean without the treatment system. However, final disposal of the spent adsorbent containing high concentrations of Cs will be an important issue in near future. Also, this operation should be continued until the final removal of fuel debris from the damaged reactors because most of the radioactive Cs still remains in the reactor (Fig. 8.3) [12]. At present, the spent slurry and adsorbents have also been produced from other treatment facilities such as the multi-nuclide removal facility (Advanced Liquid Processing System: ALPS, Fig. 8.4 [13]). Because ALPS should reduce the radioactivity of the 62 nuclides [14] to below the limit specified by the Japanese reactor regulation, the system produces different kinds of spent slurry and adsorbents. Contributions from mineralogists and materials scientists are still desired in the selection and optimization of materials used in the above system for efficient and economical treatment of water, as well as the safe storage and disposal of spent adsorbents. In Section 8.3.1, the trials to find better materials for the efficient and economical storage and disposal of radioactive materials are discussed.

156 — 8 Applied mineralogy for recovery from the Fukushima accident

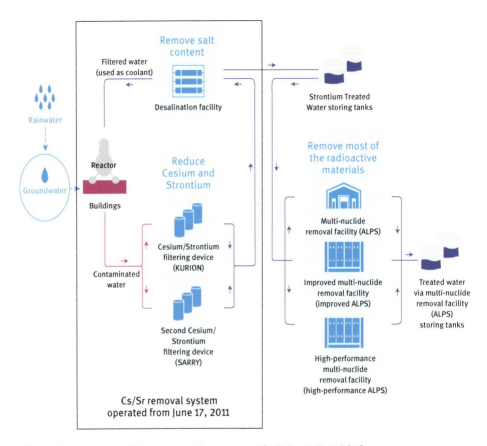

Fig. 8.2: Flow diagram of the contamination treatment facilities at FDNPS [12].

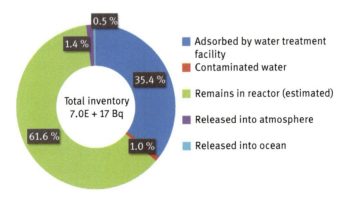

Fig. 8.3: Distribution of radioactive ^{137}Cs at FDNPS in 2014 after the accident [13].

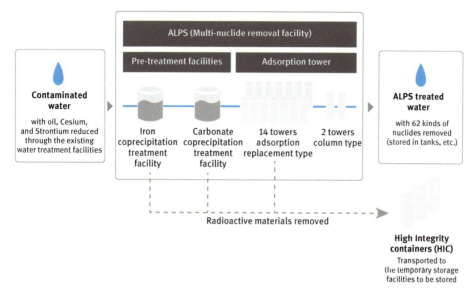

Fig. 8.4: Flow diagram of the ALPS multi-nuclide – removal facility [12]. Adsorbents include activated carbon for colloids, titanate for Sr, ferrocyanide compound for Cs, Ag-impregnated activated carbon for I, titanium oxide for Sb, chelate resin for Co, and resin adsorbent for Ru and negatively charged colloids.

8.2.2 Off-site

The accident at the FDNPS resulted in the release of huge amounts of volatile radionuclides such as ^{131}I, ^{134}Cs, and ^{137}Cs into the environment. To assess the health risks and impacts on food production in Fukushima prefecture and neighboring prefectures, mapping the spatial distribution of the radionuclides (Fig. 8.5) is of critical importance [15]. Many researchers, including the author and students with different research backgrounds, have been involved in the mapping team organized by the Ministry of Education, Culture, Sports, Science and Technology of Japan. The spatial and temporal distribution was attributed to emission rate, wind direction, deposition events, and other factors. Vertical profiles of nuclides are determined by the interaction between nuclides and components in soils after deposition and are extremely useful for present and future decontamination strategies as well as for forest and vegetation management. Since ^{131}I cannot be detected at the present time due to its short half-life (around 8 days), the most pertinent target radionuclides at the present time (after 5 years), therefore, are ^{134}Cs and ^{137}Cs. Examination of the vertical migration of the radionuclides in soils after deposition has shown that more than 90% of ^{134}Cs and ^{137}Cs deposited on the soil surface was retained in a surface layer no more than 5 cm thick and fixed into the structure of clay minerals [16, 17]. Decontamination of residential areas was then conducted by removing 3 cm (by hand) to 5 cm (by heavy machinery)

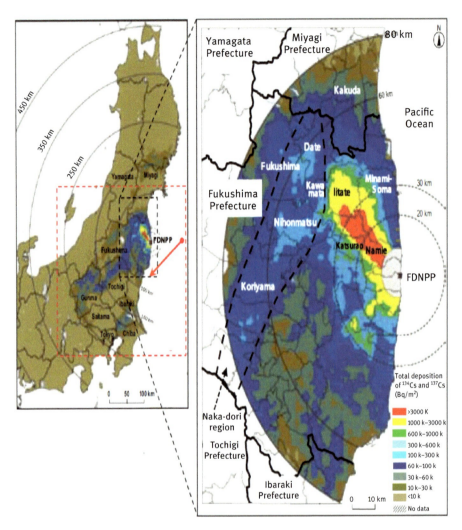

Fig. 8.5: Distribution and total deposition of ^{134}Cs + ^{137}Cs measured during airborne monitoring [15]. Distribution in the northern part of the main Island (left). Distribution within 80 km of the FDNPS (right). Decay corrected to July 2, 2011, and November 1, 2011. The airborne monitoring survey showed that the area in which deposition of ^{134}Cs + ^{137}Cs more than 3000 kBq/m² extended from the plant toward the northwest.

of topsoil. As a result of aggressive decontamination efforts, large volumes of the contaminated soil have been collected and placed under interim storage, as shown in Fig. 8.6. The estimated volume of contaminated soils is approximately 22 million m³. With this volume, it is difficult, both economically and logistically, to find appropriate disposal sites. Thus, the challenge is to reduce the volume of the contaminated soil destined for actual disposal and to develop technologies to separate highly radioactive

Fig. 8.6: Photograph of the interim storage site for the contaminated soils.

materials from bulk soils. In order to meet these challenges, both identification of the host minerals and understanding the relationship between Cs and host minerals are necessary. This scientific information would be also useful in the safety assessment of interim storage sites and the reuse of the separated soils with low radioactivity.

8.3 Case studies

8.3.1 Selection of adsorbents and processes for water treatment on site

As shown in Section 8.2.1, several kinds of facilities have been installed for water treatment. Several methods have been proposed for the separation and removal of radionuclides from the contaminated water such as evaporation, co-precipitation, solvent extraction, chemical treatment, ion exchange, micro-filtration, and membrane processes [18–23]. In such facilities, co-precipitation, ion exchange, and membrane processes have been employed and many kinds of minerals and crystalline materials have been installed to remove radioactive nuclides. Initially, three primary Cs removal technologies were used at FDNPS. One of these is the Areva (France) system, which involved co-precipitation with ferrocyanide. This system is currently in use for effluent treatment in several nuclear sites, including that of Areva. However, the use of this system in the FDNPS was discontinued after about 3 months due to the generation of large volumes of sludge that proved difficult to manage. The other two systems, Kurion (Kurion Inc., USA) and SARRY (Simplified Active Water Retrieve and Recovery System; Toshiba, Japan), employed the ion exchange process using zeolite. The Kurion system was also employed for water treatment in the Three Mile Island accident. At present, strontium-removal materials were added to both systems because ^{90}Sr is the most important nuclide contributing to the dose received by workers in the FDNPS after ^{134}Cs and ^{137}Cs removal. At the initial stage of installation, the main requirement for these materials is to have high selectivity for the target nuclides in seawater and saline waters. Eventually, in order to reduce the volume of the spent adsorbents, mineralogists and materials scientists should study hybrid adsorbents for Cs and Sr removal and the recycling of the adsorbent. A research group, led by Prof. Mimura in Tohoku University, has conducted research on hybrid materials that combine the high Cs selectivity of ferrocyanide with the high Sr selectivity of zeolite A [24]. Because this hybrid material can also trap Cs in gas and can be sintered, this material can safely encapsulate Cs in ceramic form. A research group led by Dr. Yaita of the Japan Atomic Energy Agency (JAEA) has tried to develop a recyclable hybrid adsorbent based on the novel macrocyclic ligand of o-benzo-p-xylyl-22-crown-6-ether (OBPX22C6) and successfully immobilized onto mesoporous silica for the preparation of hybrid adsorbent (Fig. 8.7) [25]. The potential and feasibility of the hybrid adsorbent to be Cs selective were evaluated in terms of sensitivity, selectivity, and reusability. Due to the high selectivity and reusability of these hybrid adsorbents, the volume of spent adsorbents can be significantly reduced.

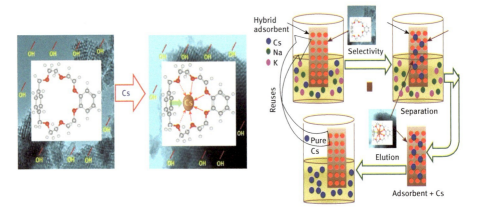

Fig. 8.7: Representation of Cs complexation with hybrid ligand for selective Cs removal in the presence of high amounts of potassium and sodium ions (left). Schematic design of the construction of a ligand-based hybrid adsorbent for Cs removal and recyclable of the adsorbent (right) [25].

The pretreatment facility of ALPS employs iron hydroxide and carbonate co-precipitation processes, producing huge volumes of slurry with high water content.

Handling and storage of the slurries are currently significant challenges due to the difficulty in the dewatering and volume-reduction processes for these slurries. However, these pretreatment facilities are necessary to remove ions such as Co, Mn, Mg, and Ca ions that may interfere in the adsorption column. Therefore, TEPCO wants to replace these facilities to reduce the waste. Currently, there are two different ALPS (improved and high performance) in the FDNPS, both of which are still under development. Thus, TEPCO needs scientific and engineering contributions from mineralogists and materials scientists to develop efficient treatment technologies that minimize the amount of wastes that are generated.

8.3.2 Solidification of the spent adsorbents for safe and economical storage and disposal

After using some materials as precipitates (slurry) and adsorbents, these materials must be solidified into a form that is safe and easy to handle. As mentioned above, there are many kinds of materials contaminated by radioactive nuclides. Many kinds of alternatives should be therefore prepared. Although the spent zeolite from the Three Mile Island accident was handled using vitrification, there is little or no actual experience for the solidification of the materials used in the FDNPS. In this context, some projects on solidification technology using cement, geopolymer, liquid glass, glass, melt, and sintered solidification were initiated in Japan with foreign collaboration in 2015. In terms of solidification, there are certain requirements for the process and the resulting solid: (1) technically feasible and can be manufactured economically, (2) physically and chemically stable for storage and/or disposal, (3) decreased hydrogen generation by radiation from the waste, and (4) minimal generation of secondary waste during production. Solidification by glass is possible even for materials with high radioactivity. However, the vitrified waste should be categorized as intermediate or high level waste and must be disposed in deeper geological media. On the contrary, cement and geopolymers for the slurry and adsorbents are relatively better due to the ease of operation, the reduction of secondary waste generation (i.e. no need for off-gas treatment as in most thermal processes), and the very high effectiveness of the matrices for Cs immobilization when the cement and geopolymer chemistry are designed appropriately. Ca-based Portland cement is not a highly effective matrix for Cs immobilization, so it cannot be selected as the primary immobilization matrix when Cs is the key radionuclide [26]. Recently, geopolymer technology has rapidly developed [27–29]. In geopolymer technology, we can tailor the waste form for different spent materials because the mineralogy in matrix and secondary phases of geopolymer can be selected using different alkali-activated materials (Fig. 8.8). However, not surprisingly, the long-term performance of geopolymer-conditioned spent material waste forms is not yet well understood, as there are no validated models that describe the transport, solubility and sorption processes that will control the long-term performance

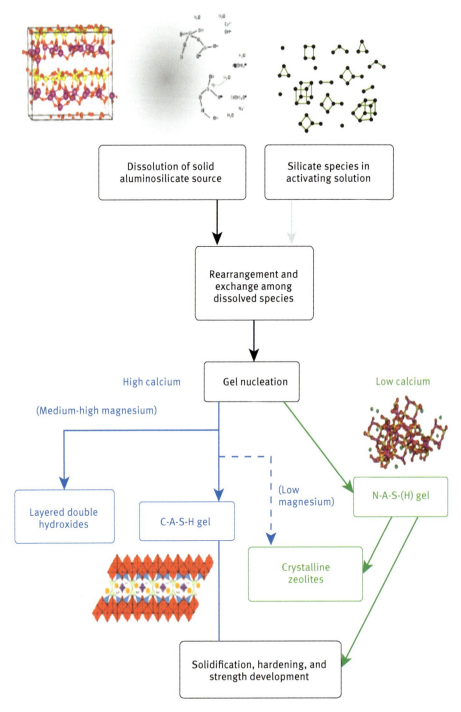

Fig. 8.8: Process and reaction of geopolymer by alkaline activation of a solid aluminosilicate precursor [28]. C-A-S-H and N-A-S-(H) gels are calcium aluminate silicate hydrate gel and sodium aluminate silicate (hydrate) gel, respectively.

of waste form. For this reason, and to support the safety case for construction of suitable disposal systems, it is necessary to urgently obtain key underpinning data on the leaching characteristics and long-term behavior of the geopolymer waste forms and the conditioned nuclides. Such data are relevant not only to recovery efforts in Fukushima, but also to waste conditioning efforts elsewhere.

8.3.3 Identification of the host mineral for Cs retention in off-site soils

Radioactive Cs contamination has also occurred around world due to aboveground nuclear tests and leaks from waste disposal sites and nuclear facilities. Contamination at the Hanford site in southeastern Washington, USA, and over a vast area of Europe and western Russia, following the Chernobyl accident in May of 1986 are famous examples of the contamination. These accidents triggered studies on the behavior of radioactive Cs. Many researchers have suggested, mainly based on laboratory experiments, that weathered biotite, illite, and vermiculite are important phases for sorption and retention of Cs in the ground or soil [30–35]. For instance, Cornel [33] summarized the potential sorption sites for Cs⁺ in the micaceous minerals as follows: (1) cation exchange sites on the surface, (2) layer edge sites, (3) frayed edge sites (FES), and (4) original interlayer sites inside the crystal. A number of studies have suggested that FES, which form around the edges of plate-like mica crystals during weathering, strongly and selectively sorb Cs [36–38] and is the final destination of radioactive Cs.

After the FDNPS accident, many mineralogists and materials scientists conducted research using actual contaminated soils to identify the host minerals for Cs retention in off-site soils. For example, a team led by Prof. Kogure of Tokyo University picked up particles that are highly contaminated by radioactive Cs by imaging plate (IP) autoradiography (Fig. 8.9) [39]. The particles were classified into three types based on

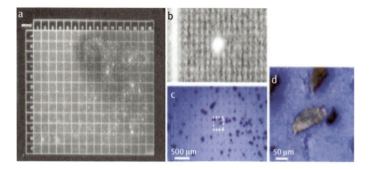

Fig. 8.9: (a) IP read-out image after 1 month of exposure to radiation from dispersed soil particles. (b) Magnified image of (a) showing bright spots formed by radiation. (c) Stereomicroscope image corresponding to the area of (b). (d) Magnified stereomicroscope image from the rectangle in (c) showing a radioactive soil particle [39].

their morphology and chemical composition, namely: (1) aggregates of clay minerals, (2) organic matter-containing clay mineral particulates, and (3) weathered biotite. From the chemical treatment and precise observation by transmission electron microscopy (TEM) after focused ion beam treatment, they concluded that the radionuclides were associated with the clay minerals, especially weathered biotite. They also checked the reduction of the radiation intensity before and after trimming of the plate edges using FIB to examine if radioactive Cs is sorbed at FES. The radiation intensity was attenuated in proportion to the volume decrease by the edge trimming, implying that radioactive Cs was sorbed uniformly within the porous weathered biotite (Fig. 8.10) [39].

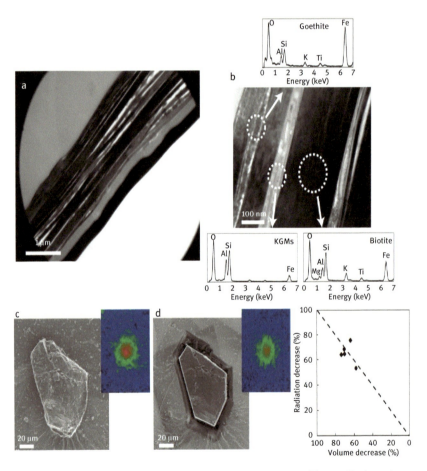

Fig. 8.10: (a) Cross-sectional TEM image of the weathered biotite. (b) Magnified TEM image and EDX spectra from each portion of the particle. Kaolin group minerals (KGMs) and goethite are formed inside the cleavage spaces of the biotite. (c, d) An example of the edge-removal treatment. SEM images of the particle and read-out IP images (upper right) (c) before and (d) after the edge-removal. (e) The results of the edge-removal treatment for five radioactive biotite grains. The volume and radiation decrease by the edge-removal are almost proportional, indicating that radionuclides are not concentrated around the edges but distributed rather homogeneously in the grain [39].

Kogure's group also conducted ^{137}Cs (10^{-11}-10^{-9} mol/L) adsorption experiments using clay minerals from Fukushima Prefecture [40]. By characterizing the solids using autoradiography, they observed that the weathered biotite selectively sorbed ^{137}Cs much more strongly than the other clay minerals present in the same substrate (Fig. 8.11). When the weathered biotite is removed from the substrate in the adsorption experiments, the radioactivity of the other clay minerals, particularly ferruginous smectite, increased considerably. This implied that the presence/absence of weathered biotite is a key factor affecting the dynamics and fate of radioactive Cs in Fukushima. In Fukushima, most of radioactive Cs was fixed into the weathered biotite because host rock distributed at Fukushima is mainly granite.

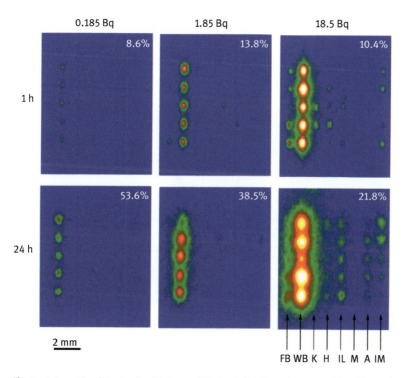

Fig. 8.11: A matrix of the read-out images of IPs covering the substrates with various mineral particles (five particles for each species) sorbed radiocesium from the solutions. The radioactivity input to the solution and reaction time are at the top and left, respectively. The figure at the top right of each image is the percentage of radioactivity (or ^{137}Cs) sorbed to the whole mineral particles, estimated from the IP signal. The abbreviations at the bottom-right mean FB: fresh biotite, WB: weathered biotite, K: kaolinite, H: halloysite. IL: illite, M: montmorillonite, A: allophane, IM: imogolite [40].

In the accident at FDNPS, other particles with radioactive Cs were also observed by electron microscopy: Adachi et al. [41] discovered spherical microparticles with diameters of 2.0–2.6 µm and radioactivities ranging from 0.7 to 3 Bq of ^{137}Cs in a single particle ("Cs ball"). Such Cs-bearing microparticles remained in the filter after acid digestion, indicating that the microparticles were insoluble. Kogure's group also characterized the mineralogy of these "Cs balls" (Fig. 8.12) [42, 43]. They found that the matrix of the microparticles is composed dominantly of silicate glass with Cl, K, Fe, Zn, Rb, Sn, and Cs as constituent elements. Although the dissolution rate of these particles is very slow under the soil pH conditions at Fukushima, we have to consider carefully the leaching of radioactive Cs from these Cs balls when we use low-radioactivity-contaminated soils as construction materials mixed with cement and steel slag materials. Therefore, the stability and fate of radioactive Cs in weathered biotites under neutral and alkaline condition should be studied to assess the safety of their secondary usage.

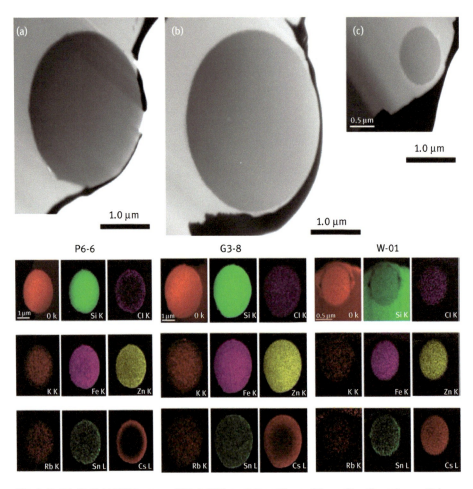

Fig. 8.12: Bright-field TEM images of "Cs ball" from thin sections of the radioactive microparticles (top) and their elemental maps from the three microparticles (bottom) [43]. Cs ball is microparticle of radioactive cesium (Cs)-bearing silicate glass emitted from FDNPS. Fe and Zn were relatively homogeneously distributed, whereas the concentration of alkali ions varied radially. Generally, Cs was rich and K and Rb were poor outward of the particles but the degree of such radial dependence was considerably different among the particles.

8.4 Concluding remark

The disaster at FDNPS has proven to be technically, politically, economically, and emotionally costly. At the end of 2016, the Japanese government estimated the overall cost of cleaning up the Fukushima nuclear disaster at more than 20 trillion Japanese yen (0.18 trillion US$) [44, 45]. However, the estimated cost does not include the final disposal of the wastes generated by on- and off-site treatment efforts. Therefore, final disposal will account for additional cost. In addition, if mineralogists and engineers cannot solve the mineralogical issues in on- and off-site treatment described above, the cost may significantly increase in the future. However, from the view of commonality, recovery from the accident at FDNPS needs to accelerate to alleviate lingering personal and economic hardships for the people affected by the disaster as well as for the people involved in recovery efforts. However, reduction of cost is indispensable to our posterity by the developed technology. Mineralogists must also understand that treatment efforts must not only be conducted in scientifically sound ways but also in ways that are acceptable to the public. With such a mindset, mineralogists would be able to effectively communicate the social and economic impacts of using mineralogy to address pervasive environmental problems. This could ensure greater public understanding and support for the work of mineralogists, allowing them to thrive and continue their research. In this context, the projects contributing to the recovery from the accident at FDNPS will serve as tests of mineralogists' scientific and social skills.

Acknowledgment

The author thanks Dr. Paul Clarence Francisco of the Japan Atomic Energy Agency for his valuable discussion and for checking the manuscript.

References

[1] Brown G. Bounties from the Earth. Elements 2015;11:227–8.
[2] Ludden J, Albarède Coleman M. The impact of geochemistry. Elements 2015;11:239–40.
[3] Calas G, Paul F, McMillan PF, Bernier-Latmani R. Environmental mineralogy: new challenges, new materials. Elements 2015;11:247–52.
[4] Ritsema J, Lay T, Kanamori H. The 2011 Tohoku earthquake. Elements 2012;8:182–8.
[5] Blandford ED, Ahn J. Examining the nuclear accident at Fukushima Daiichi. Elements 2012;8:189–94.
[6] Acton JM, Hibbs M. Why Fukushima was preventable. Washington (DC): Carnegie Endowment for International Peace; 2012.
[7] Nishihara K, Yamaguchi I, Yashuda K, Ishimori K, Tanaka K, Kuno T. Radionuclide release to stagnant water in the Fukushima-1 nuclear power plant. J Nucl Sci Technol 2015;52:152–61.
[8] Sasaki T, Takeno Y, Kirishima A, Sato N. Leaching test of gamma-emitting Cs, Ru, Zr, and U from neutron-irradiated UO_2/ZrO_2 solid solutions in non-filtered surface seawater. J Nucl Sci Technol 2015;52:147–51.

[9] Kirishima A, Sasaki T, Sato N. Solution chemistry study of radioactive Sr on Fukushima Daiichi NPS site. J Nucl Sci Technol 2015;52:301–7.

[10] Sasaki T, Takeno Y, Kobayashi T, Kirishima A, Sato N. Leaching behavior of gamma-emitting fission products and Np from neutron-irradiated UO_2-ZrO_2 solid solutions in non-filtered surface seawater. J Nucl Sci Technol 2016;53:303–11.

[11] Division of Nuclear Fuel Cycle and Environment (NUCE), Atomic Energy Society of Japan (AESJ). Contaminated liquid water treatment for Fukushima Daiichi NPS. Available at: http://nuce.aesj.or.jp/clwt:start. Accessed 1 Dec 2016.

[12] TEPCO. Contaminated water treatment. Available at: http://www.tepco.co.jp/en/decommision/planaction/alps/index-e.html. Accessed 1 Dec 2016.

[13] IRID. IRID Annual Symposium 2014. Available at: http://irid.or.jp/wp-content/uploads/2014/07/Sympo_Miyamoto_E.pdf. Accessed 1 Dec 2016.

[14] TEPCO. Target radioactive materials to be removed. Available at: http://www.tepco.co.jp/en/decommision/planaction/images/150517.pdf. Accessed 1 Dec 2016.

[15] Yoshida N, Takahashi Y. Land-surface contamination by radionuclides from the Fukushima Daiichi nuclear power plant accident. Elements 2012;8:201–6.

[16] Koarashi J, Atarashi-Andoh M, Matsunaga T, Sato T, Nagao S, Nagai H. Factors affecting vertical distribution of Fukushima accident-derived radiocesium in soil under different land-use conditions. Sci Total Environ 2012;431:392–401.

[17] Matsunaga T, Koarashi J, Atarashi-Andoh M, Nagao S, Sato T, Nagai H. Comparison of the vertical distributions of Fukushima nuclear accident radiocesium in soil before and after the first rainy season, with physicochemical and mineralogical interpretations. Sci Total Environ 2013;447:301–14.

[18] Han F, Zhang GH, Gu P. Removal of Cs^+ from simulated waste with countercurrent two-stage adsorption followed by microfilteration. J Hazard Mater 2012;225:107–13.

[19] Tsai SC, Wang TH, Li MH, Wei YY, Teng SP. Cesium adsorption and distribution onto crushed granite under different physicochemical conditions. J Hazard Mater 2009;161:854–61.

[20] Mouri G, Golosov V, Shiiba M, Hori T. Assessment of the caesium-137 flux adsorbed to suspended sediment in a reservoir in the contaminated Fukushima region in Japan. Environ Pollut 2014;187:31–41.

[21] Rogers H, Bowers J, Anderson DG. An isotope dilution-precipitation process for removing radioactive cesium from wastewater. J Hazard Mater 2012;243:124–9.

[22] Lin Y, Fryxell GE, Wu H, Engelhard M. Selective sorption of cesium using self-assembled monolayers on mesoporous supports. Environ Sci Technol 2001;35:3962–66.

[23] Chen R, Chen R, Tanaka H, Kawamoto T, Asai M, Fukushima C, Na H, Kurihara M, Watanabe M, Arisaka M, Nankawa T. Selective removal of cesium ions from wastewater using copper hexacyanoferrate nanofilms in an electrochemical system. Electrochim Acta 2013;87:119–25.

[24] Yin XB, Wu Y, Mimura H., Niibori Y, Wei YZ. Selective adsorption and stable solidification of radioactive cesium ions by porous silica gels loaded with insoluble ferrocyanides. Sci China Chem 2014;57:1470–6.

[25] Awual MR, Yaita T, Miyazaki Y, Matsumura D, Shiwaku H, Taguchi T. A reliable hybrid adsorbent for efficient radioactive cesium accumulation from contaminated wastewater. Sci Rep 2016;6:19937.

[26] McCulloch CE, Angus MJ, Crawford RW. Rahman AA, Glasser FP. Cement in radioactive waste disposal: some mineralogical consideration. Mineral Mag 1985;49:211–21.

[27] Provis JL. Geopolymers and other alkali activated materials: why, how, and what? Mater Struct 2014;47:11–25.

[28] Provis JL, Bernal S. Geopolymers and related alkali-activated materials. Annu Rev Mater Res 2014;44:299–327.

[29] Provis JL, Palomo A, Shi C. Advances in understanding alkali-activated materials. Cem Concr Res 2015;78:110–25.
[30] Francis CW, Brinkley FS. Preferential adsorption of Cs-137 to micaceous minerals in contaminated freshwater sediment. Nature 1976;260:511–3.
[31] Evans DW, Alberts JJ, Clark RA. Reversible ion-exchange fixation of cesium-137 leading to mobilization from reservoir sediments. Geochim Cosmochim Acta 1983;47:1041–9.
[32] Comans RNJ, Haller M, Depreter P. Sorption of cesium on illite – non-equilibrium behavior and reversibility. Geochim Cosmochim Acta 1991;55:433–40.
[33] Cornell RM. Adsorption of cesium on minerals – a review. J Radioanal Nucl Chem 1993;171:483–500.
[34] Poinssot C, Baeyens B, Bradbury MH. Experimental and modelling studies of caesium sorption on Illite. Geochim Cosmochim Acta 1999;63:3217–27.
[35] Zachara JM, Smith SC, Liu CX, McKinley JP, Serne RJ, Gassman PL. Sorption of Cs^+ to micaceous subsurface sediments from the Hanford site, U.S.A. Geochim Cosmochim Acta 2002;66:193–211.
[36] Brouwer E, Baeyens B, Maes A, Cremers A. Cesium and rubidium ion equilibria in Illite Clay. J Phys Chem 1983;87:1213–9.
[37] McKinley JP. Zachara JM, Heald SM, Dohnalkova A, Newville MG, Sutton SR. Microscale distribution of cesium sorbed to biotite and muscovite. Environ Sci Technol 2004;38:1017–23.
[38] Nakao A, Thiry Y, Funakawa S, Kosaki T. Characterization of the frayed edge site of micaceous minerals in soil clays influenced by different pedogenetic conditions in Japan and northern Thailand. Soil Sci Plant Nutr 2008;54:479–89.
[39] Mukai H, Hatta T, Kitazawa H, Yamada H, Yaita T, Kogure T. Speciation of radioactive soil particles in the Fukushima contaminated area by IP autoradiography and microanalyses. Environ Sci Technol 2014;48:13053–13059.
[40] Mukai H, Hirose A, Motai S, Kikuchi Y, Tanoi K, Nakanishi TM, Yaita T, Kogure T. Cesium adsorption/desorption behavior of clay minerals considering actual contamination conditions in Fukushima. Sci Rep 2016;6:21543.
[41] Adachi K, Kajino M, Zaizen Y, Igarashi Y. Emission of spherical cesium-bearing particles from an early stage of the Fukushima nuclear accident. Sci Rep 2013;3:2554.
[42] Yamaguchi N, Mitome M, Kotone A-H, Asano M, Adachi K, Kogure T. Internal structure of cesium-bearing radioactive microparticles released from Fukushima nuclear power plant Sci Rep 2016;6:20548.
[43] Kogure T, Yamaguchi N, Segawa H, Mukai H, Motai S, Akiyama-Hasegawa K, Mitome M, Hara T, Yaita T. Constituent elements and their distribution in the radioactive Cs-bearing silicate glass microparticles released from Fukushima nuclear plant. Microscopy 2016;65:451–9.
[44] Cost of Fukushima disaster expected to soar to ¥20 trillion. Available at: http://www.japantimes.co.jp/news/2016/11/28/national/cost-fukushima-disaster-expected-soar-%C2%A520-trillion/#.WFycMFwuTEY. Accessed 25 Dec 2016.
[45] Fukushima nuclear plant decommission, compensation costs to almost double. Available at: https://www.japantoday.com/category/national/view/fukushima-nuclear-decommission-compensation-costs-to-almost-double). Accessed 25 Dec 2016.

Hartmut Schlenz, Stefan Neumeier, Antje Hirsch,
Lars Peters and Georg Roth

9 Phosphates as safe containers for radionuclides

9.1 Introduction

Vitrification is currently the most widely used technology for the treatment of high-level radioactive waste (HLW) in most countries using nuclear power [1]. Alkali borosilicate and aluminophosphate glass are applied for the immobilization of actinides and fission products depending on the waste streams that arise, respectively. Future waste generation is primarily driven by interest in sources of clean energy and this has led to an increased interest in advanced nuclear power production in many countries, with few exceptions such as Germany, where nuclear power generally is no longer an option for future electric power generation strategies. Therefore, advanced nuclear waste forms are being designed for properly thought-out future nuclear waste management strategies. Potential advanced waste forms can be single-phase or poly-phase crystalline ceramic (mineral) waste forms that chemically incorporate radionuclides and hazardous species atomically in their crystal structures. In this context, phosphates play an important role because many of them show outstanding and desired properties like a high radiation resistance and a high chemical durability.

In this chapter, we will focus on phosphates that are currently discussed as potential single-phase ceramic waste forms for the safe disposal of the actinides U, Pu, and the minor actinides Np, Cm, and Am, and in some cases, even for the disposal of fission products like ^{90}Sr. The mineral monazite and monazite-type synthetic ceramic phases will be described in more detail; simply because such phases are especially well suited for the permanent incorporation of actinides. Additionally, natural analogues can be used as references that enable a secured estimation of the long-term properties of phases designed in scientific laboratories.

9.2 Nuclear waste management

The safe disposal of radioactive waste is an urgent scientific and social challenge that must be tackled over the coming decades. The peaceful use of nuclear power results in low and intermediate radioactive waste with predominantly short-lived radionuclides and highly radioactive waste. In the Federal Republic of Germany, e.g. the produced low- and medium-radioactive waste will be introduced into the Konrad repository. Konrad is located near the city of Salzgitter in the northern part of Germany, and it is a former iron ore mine (host rock is an oolitic limestone formation), which is intended to be a repository for low-level waste (LLW) and intermediate-level waste (ILW), respectively [2, 3]. The highly radioactive waste shall also be placed in a repository, which must be ensured for very long periods (about 1 million years). Currently, there is no repository for highly radioactive waste in operation worldwide.

For long-term safety of a repository, some long-lived fission products (e.g. ^{99}Tc, ^{129}I) and the actinides (U, Th, Pu, Am, Cm, Np) are a major challenge due to their long half-lives and high radiotoxicity. Therefore, innovative disposal strategies with regard to final disposal and transmutation are required. Most HLWs are in either one of two forms as (1) used nuclear fuel that is destined for direct disposal or (2) waste from the reprocessing of commercially generated spent nuclear fuel (SNF), or in some countries, from the reprocessing of fuel used to generate ^{239}Pu for weapons. The SNF retains a high inventory of transuranium elements (about 1 atom%) in its uranium matrix, and the waste from reprocessing is depleted in actinides, mainly ^{235}U and ^{239}Pu (about 99% removed), recovered during chemical processing.

Liquid HLW streams are stored either as neutralized nitric acid streams in mild steel tanks or as nitric acid streams in stainless steel tanks. Although borosilicate glass has become the preferred waste form for the immobilization of HLW solutions in the majority of the nuclear nations, the chemical variability of the wastes from the different reactor and reprocessing flowsheets coupled with the additional variability imposed by neutralization vs. direct storage or processing of acidic wastes has led to a diverse HLW chemistry [4]. HLW contains about 75% of the elements in the periodic table [1].

There have been many comprehensive reviews on waste forms (glass, glass-ceramic, crystalline ceramic [mineral], cementitious, geopolymer, bitumen, and other encapsulant waste forms) and their properties (e.g. some recent ones are [5–11]), and this will not be elaborated on in this chapter. However, conditioning and the immobilization of radionuclides are essential issues of recent nuclear waste management strategies. In this context, the radionuclides are structurally incorporated into tailor-made high-specific materials [12] and end-deposited. Regarding the immobilization of the actinides U, Th, Pu, Am, Cm, and Np, these are mainly ceramic host phases. Such ceramics are distinguished by their excellent resistance, which considerably reduces the risk of release and thus a radiotoxic hazard to the geosphere and biosphere. The advantages and disadvantages of different possible phosphate ceramic waste forms will be discussed in more detail within the next section, and we will justify why we consider phosphates to be a good choice.

9.3 Why single-phase phosphate ceramics?

The HLW generated by reprocessing of spent nuclear fuel contains higher amounts of phosphates, i.e. usually up to 15 wt% P_2O_5 [13, 14]. It is a real challenge to develop ceramic waste forms that can be adapted to different waste streams, because possible amounts of other metal oxides also have to be taken into account, e.g. up to 30 wt% UO_2.

9.3 Why single-phase phosphate ceramics?

Actinide waste forms can be divided into four main groups [10, 11]: Simple oxides, complex oxides, silicates, and phosphates. Simple oxides under consideration are zirconia (ZrO_2), uraninite (UO_2), and thorianite (ThO_2). More complex oxides and titanates are zirconolite ($CaZrTi_2O_7$), perovskite ($CaTiO_3$), pyrochlore ($(Na,Ca,U)_2(Nb,Ti,Ta)_2O_6$) and murataite ($(Na,Y)_4(Zn,Fe)_3(Ti,Nb)_6O_{18}(F,OH)$). Well-known silicate minerals are zircon ($ZrSiO_4$), thorite ($ThSiO_4$), garnet ($(Ca,Mg,Fe^{2+})_3(Al,Fe^{3+},Cr^{3+})_2SiO_4$), britholite ($(Ca,Ce)_5(SiO_4)_3(OH,F)$) and titanite ($CaTiSiO_5$). Finally, relatively widespread phosphate minerals are monazite ($LnPO_4$; Ln = La to Gd), xenotime ($LnPO_4$; Ln = Tb to Lu, and Y), and apatite ($Ca_{4-x}Ln_{6+x}(PO_4)_y(O,F)_2$). In principle, ceramic waste forms are distinguished in poly-phase and single-phase waste forms [7]. Phosphates are mainly intended as single phases and Tab. 9.1 (see also [14]) shows the most prominent candidate materials that are currently under investigation (minerals and synthetic phases).

Tab. 9.1: Potential ceramic phosphate waste forms (Ln = La to Gd).

Structure type	Composition	Space group
Monazite	$LnPO_4$	$P2_1/n$
Apatite	$Ca_{10-y}Ln_y(SiO_4)_y(PO_4)_{6-y}(F,OH,O)_2$	$P6_3/m$
Xenotime	YPO_4	$I4_1/amd$
Kosnarite (NZP)	$KZr_2(PO_4)_3$	$R\bar{3}c$
Florencite	$LnAl_3(PO_4)_2(OH)_6$	$R\bar{3}m$
Th-phosphate-diphosphate (β-TPD)	$\beta\text{-}Th_4(PO_4)_4P_2O_7$	$Pcam$
Th-U-phosphate-diphosphate (β-TUPD)	$\beta\text{-}Th_{4-x}U^{IV}_x(PO_4)_4P_2O_7$	$Pcam$
Th-Pu-phosphate-diphosphate (β-TPPD)	$\beta\text{-}Th_{4-x}Pu^{IV}_x(PO_4)_4P_2O_7$	$Pcam$
Th-Np-phosphate-diphosphate (β-TNPD)	$\beta\text{-}Th_{4-x}Np^{IV}_x(PO_4)_4P_2O_7$	$Pcam$
Monazite-cheralite solid solutions	$LnPO_4\text{-}CaTh(PO_4)_2$	
Monazite-huttonite solid solutions	$LnPO_4\text{-}ThSiO_4$	
Monazite-brabantite solid solutions	$LnPO_4\text{-}Ca_{0.5}Th_{0.5-x}U_xPO_4$	
Monazite-β-TUPD composite matrices	$LnPO_4\text{-}\beta\text{-}Th_{4-x}U_x(PO_4)_4P_2O_7$	

The importance of advanced waste forms can be elucidated by the following list of basic requirements for a reliable actinide waste form: a high waste loading (about 20–35%), easy processing (established methods being preferred), high radiation stability (the waste form should show a high tolerance to radiation effects, particularly those that yield from α-decay), chemical flexibility, durability, natural analogues (as references for the judgment of the long-term performance), and finally criticality issues. The latter can be accomplished by reducing the concentrations of the incorporated actinides, which in turn reduces the maximum waste loading. Another choice can be the incorporation of neutron absorbers like Hf and Gd into the crystal structure.

The best way to accomplish all properties listed above is the usage of single-phase ceramics that can be processed even on larger scales quite easily, and whose properties can be controlled more reliably than it is possible for poly-phases, with unknown interactions of the components under extreme conditions, over very long time periods.

9.4 Crystal structures and properties

The symmetry of the crystal structures of the phosphates listed in Tab. 9.1 varies significantly, from the relatively low monoclinic symmetry of the monazite-type structure, up to the highly symmetric hexagonal structure of apatite. All these structures exhibit specific atomic sites that can be used for the permanent incorporation of lanthanide or actinide cations. The coordination number of these M-cations ranges from 6 (kosnarite) to 12 (florencite). However, all mentioned crystal structures are, e.g. suitable for the incorporation of the most prominent actinides U and Pu. The most important properties relevant for nuclear waste management, like radiation resistance and aqueous durability, possibly vary between these phases (e.g. normalized dissolution rates can even vary several orders of magnitude for monazite-type phases; [6, 14]), but none of them appears to be unsuitable for the disposal of actinides, or in some cases even for the disposal of long-lived fission products. In the following subsections, some more details about selected crystal structures will be given to illustrate the structural variability in phosphates considered as potential waste matrices. For the special case of monazite, some recently investigated specific structural features are discussed separately at the end of this subsection.

9.4.1 Monazite type

Monazite is a naturally occurring rare-earth element (REE) orthophosphate ($M^{3+}PO_4$). Typically, M^{3+} are the light (and big) REE from La to Gd. It is found as an accessory mineral in metamorphic and plutonic rocks [15–17]. The Greek origin μοναçειν (to be solitary) refers to natural monazite crystals usually occurring solely due to their extreme geological stability with ages up to 2 billion years [18–20]. Natural monazites contain up to 15 wt% UO_2 and up to 32 wt% ThO_2, respectively [11]. Although these elements are known for their α-decay, monazite is usually found in a crystalline state due to its low temperature of recrystallization [16, 21].

Because of this high resistance to radiation damage and corrosion [16] combined with the incorporated amount of U, monazite is used in geochronology and discussed as a potential host for nuclear waste [9, 22, 23].

Monazite has a monoclinic structure (space group $P2_1/n$, non-standard setting; [17]). All atoms are located on general Wyckoff positions 4e (x, y, z). The lattice parameters for $LaPO_4$ are approximately $a = 6.8$ Å, $b = 7.1$ Å, $c = 6.5$ Å, and $\beta = 103.3°$. A schematic drawing of the structure is given in Fig. 9.1.

M^{3+} cations are nine-fold coordinated by oxygen. These [$M^{3+}O_9$]-polyhedra share edges with the isolated [PO_4] tetrahedra forming alternating chains along **c** [17, 24]. Within the **ab**-plane, [$M^{3+}O_9$] are connected via corners [7]. The nine-fold coordination allows the structural incorporation of various cations without severe constraints on symmetry, size or charge of the cation [25].

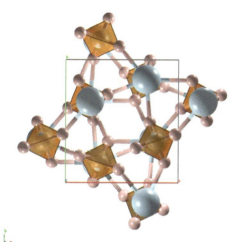

Fig. 9.1: Crystal structure of monazite, e.g. LaPO$_4$, viewed along the crystallographic c-axis. [PO$_4$]-tetrahedra are orange, M^{3+} are light blue spheres, oxygen small pink spheres.

In addition to trivalent ions, monovalent, divalent, and tetravalent cations can be introduced and charge-compensated for by vacancies or coupled and/or double substitution on both cation sites [14]. This feature is pronouncedly interesting for the incorporation of minor actinides and Pu (see [26], and references therein).

9.4.2 Apatite type

As one of the most abundant phosphate minerals, apatite can be described as $M_{10}(TO_4)_6X_2$. The two crystallographically distinct M-sites are usually occupied by Ca^{2+} but can accommodate REE as well. T represents P and/or Si, while X can be F, OH, Cl, or O. The mineral name is derived from the Greek word for "misleading" or "deceive" (απατείν) because different varieties of apatite were regularly mistaken for other minerals.

It is common in placers and as an accessory mineral in igneous rocks [27–29]. U-bearing apatite from the Oklo natural reactor in Gabon was found to be 2 billion years old, indicating a high long-term stability [30, 31].

A detailed study of the structure of apatite was presented by Hughes and Rakocvan [27]. Having the hexagonal space group $P6_3/m$, the approximate lattice parameters of apatite are around a = 9.4 Å and c = 6.8 Å. The occupied Wyckoff positions are given in Tab. 9.2.

Tab. 9.2: Occupied Wyckoff positions in the apatite-type structure in space group $P6_3/m$.

Ion	Wyckoff site	x	y	z
M^1	4f	1/3	2/3	z
M^2	6h	x	y	1/4
T	6h	x	y	1/4
O^1	6h	x	y	1/4
O^2	6h	x	y	1/4
O^3	12i	x	y	z
X	2a	0	0	1/4

Apatite is composed of rigid $[TO_4]$-tetrahedra. T is usually occupied by P, but phosphorus can be replaced by Si (then named britholite). These tetrahedra form a honeycomb-like structural network. Since the M-ions occupy two distinct crystallographic sites, two different coordination environments are observed: Ions on the M^1 position are nine-fold coordinated by oxygen forming a tricapped trigonal prism. The M^2 position is seven-fold coordinated by six oxygen atoms and one X. The latter is located in the center of the hexagonal tunnels along **c** which are formed by the $[TO_4]$-tetrahedra [27, 28, 32]. By coupled substitutions, not only divalent ions can be incorporated at the two distinct M sites, but also trivalent and tetravalent ions such as REE, U and Th [33].

Because of its significant chemical and structural flexibility, and its easy "self-healing", apatite is considered as a nuclear waste container material [13, 34]. However, Chartier et al. [32] pointed out that the possible migration/mobility of the M-ions in the structure might indicate significant disadvantages for the use of apatite-type materials for the immobilization of radionuclides.

9.4.3 Xenotime type

The name xenotime is derived from the greek κενοσ (vain, apparent, foreign) and τιμη (honor, value). This *vainglory* stands for the original claim of a newly discovered element which was later found to be the already known yttrium. Over time, the name changed to its contemporary form [20]. Xenotime was first described by Berzelius in 1824 [35](cited after [36]) and is a rather common accessory mineral found in granitoids, granitic pegmatites, and low- to high-grade metamorphic rocks [37].

Several structural investigations were carried out, e.g. by Milligan et al. [38] and Ni et al. [17]. Like monazite, xenotime is an orthophosphate of general formula $M^{3+}PO_4$. Here, M^{3+} typically are the heavy (and small) REE from Tb to Lu, and Sc and Y. The structure is the tetragonal zircon ($ZrSiO_4$)-type structure with space group $I4_1/amd$ [7, 17]. For YPO_4, the lattice parameters are around $a = 6.9$ Å and $c = 6.0$ Å. The occupied

Wyckoff positions in the xenotime (zircon)-type structure are given in Tab. 9.3. A schematic drawing of the crystal structure is shown in Fig. 9.2.

Tab. 9.3: Occupied Wyckoff positions in the xenotime-type structure in space group $I4_1/amd$.

Ion	Wyckoff site	x	y	z
M^{3+}	4a	0	3/4	1/8
P	4b	0	1/4	3/8
O	16h	0	y	z

Because of the smaller size of the heavy REE, in xenotime (unlike in monazite), M^{3+} are eight-fold coordinated by oxygen.

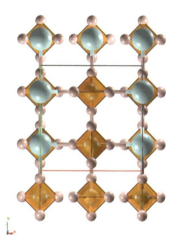

Fig. 9.2: Crystal structure of xenotime, e.g. YPO$_4$, viewed along the crystallographic c-axis. [PO$_4$]-tetrahedra are orange, M^{3+} are light blue spheres, oxygen small pink spheres.

Milligan et al. [38] described the [MO$_8$]-polyhedron to be a dodecahedron. This shape results from the specific point-group-consistent bisphenoidal distortion of a cube (coordination number: 8). Comparable to monazite, the [MO$_8$]-polyhedra and the [PO$_4$]-tetrahedra share common edges to form alternating chains along [001] while being interlinked in the (100) plane. In general, xenotime is 10% less dense than monazite [7] and is not as stable as monazite in terms of radiation damages [13].

9.4.4 Kosnarite type

The chemical composition of the name-giving mineral kosnarite (for Richard A. Kosnar [1946–2007], see [39]) is KZr$_2$(PO$_4$)$_3$. A general formula for kosnarite-type compounds can be given as $M^{1+}M_2^{4+}(PO_4)_3$. Natural $KZr_2(PO_4)_3$ occurs as a late-stage secondary phosphate mineral in pegmatites [39]. Kosnarite is trigonal (rhombohedral) with space group $R\bar{3}c$ and unit cell dimensions of about $a = 8.7$ Å and $c = 23.9$ Å in the hexagonal setting (see [40]). The occupied Wyckoff positions in kosnarite are given in Tab. 9.4. A schematic drawing of the crystal structure is shown in Fig. 9.3.

Tab. 9.4: Occupied Wyckoff positions in the kosnarite-type structure in space group $R\bar{3}c$ (hexagonal setting).

Ion	Wyckoff site	x	y	z
M^{1+}	6b	0	0	0
M^{4+}	12c	0	0	z
P	18e	x	0	¼
O^1	36f	x	y	z
O^2	36f	x	y	z

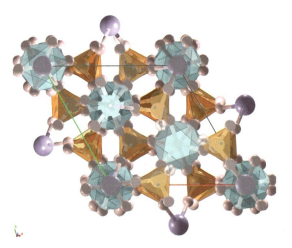

Fig. 9.3: Schematic drawing of the kosnarite-type crystal structure, $M^{1+}M^{4+}_2(PO_4)_3$ in a projection along the crystallographic c-axis. [PO$_4$]-tetrahedra are orange, [$M^{4+}O_6$]-octahedra are light-blue, M^{1+} cations are light violet, and oxygen atoms pink spheres.

Kosnarite is an orthophosphate; hence, all [PO$_4$]-tetrahedra are isolated. All oxygen atoms are linked to both phosphorus and zirconium atoms. Zirconium is coordinated by six oxygen atoms in the shape of an octahedron. Two [ZrO$_6$]-octahedra then are linked in c-direction (along the $\bar{3}$ roto-inversion axis) via a "bridging" [PO$_4$]-tetrahedron, but also in the **ab**-plane, thus forming a 3D-network of [Zr$_2$(PO$_4$)$_3$]-groupings [40]. Along the c-direction, space left open by this framework is filled by K atoms.

Aside from eulytine ($Bi_8(SiO_4)_6$), kosnarite is one of the two known phases which can accommodate tetravalent actinides on an octahedrally coordinated site (see e.g. the chapter by Locock in Krivovichev et al. [41]). Ewing and Wang [13] stressed the immense chemical flexibility of the kosnarite-type structure, including the materials *NASICON*, $Na_{1+x}Zr_2Si_xP_{3-x}O_{12}$, and *NZP*, $NaZr_2(PO_4)_3$. Both are well known for their ionic conductivity properties (e.g. [42]). By coupled substitutions, "nearly two-thirds of the periodic table" (p. 687 in [13]) can be accommodated in this structure type.

In summary, although kosnarite-type materials could qualify as a nuclear waste matrix owing to their chemical flexibility (see, [13], and references therein), their ion-exchange properties might turn out as severe disadvantage in the attempt to immobilize radioactive nuclides.

9.4.5 Florencite type

Florencite is a REE mineral of isostructural hydrous Al-phosphates ($M^{3+}Al_3(PO_4)_2(OH)_6$). In florencite, M^{3+} is mostly Ce, occasionally La and Nd. Other minerals in this group contain divalent Ca (crandallite), Sr (goyazite), Ba (gorceixite), and Pb (plumbogummite) and an additional proton. Instead of P, variations are found with V or As. Therefore, it belongs to the alunite-crandallite group [43–45].

Florencite is named after the Brazilian mineralogist W. Florence. It is found worldwide, often in association with monazite, xenotime, goyazite and/or gorceixite in mica schists or diamond-bearing sands, in granitic pegmatites, carbonatites, hydrothermal deposits, placers, shales, and weathered zones [44, 46–48].

The crystal structure of florencite was first described by Kato [49] to be analogous to goyazite [50]. In the hexagonal setting of trigonal (rhombohedral) space group $R\bar{3}m$, approximate lattice parameters are $a = 6.9$ Å and $c = 16.2$ Å. In Tab. 9.5, the Wyckoff positions of the atoms in the florencite-type structure are given. Fig. 9.4 shows a schematic drawing of the crystal structure.

Tab. 9.5: Occupied Wyckoff positions in the florencite structure type in space group $R\bar{3}m$ (hexagonal setting) after Kato [49].

Ion	Wyckoff site	x	y	z
M^{3+}	3a	0	0	0
Al	9d	½	0	½
P	6c	0	0	z
O^1	6c	0	0	z
O^2	18h	x	-x	z
O(H)	18h	x	-x	z
H	18h	x	-x	z

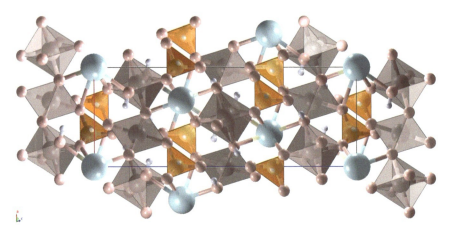

Fig. 9.4: Crystal structure of florencite $M^{3+}Al_2(PO_4)_2(OH)_6$, viewed along the crystallographic **b**-axis. [PO_4]-tetrahedra are orange, [AlO_6]-octahedra are light-violet, M^{3+} cations are blue, and oxygen atoms pink spheres. Small blue spheres are hydrogen atoms.

Al is coordinated by two oxygen atoms and four (OH) groups. These slightly distorted Al-centered octahedra are connected via common oxygen atoms, forming alternating layers perpendicular to **c** with "hexagonal" and "triangular" voids. The [PO_4]-tetrahedra are isolated and share three corners with the octahedral layer. Due to this arrangement, the fourth oxygen points away from the triangular hole.

Through the hexagonal ring, two [PO_4]-tetrahedra point towards one another. Between the octahedral layers, the large cations (*M*) are surrounded by six hydroxide groups and six oxygen atoms [49, 51]. The coordination polyhedron of these *M* cations is an icosahedron (see Locock in [41]).

9.4.6 Th-phosphate-diphosphate (β-TPD) type

In contrast to all cases discussed above, $β-Th_4(PO_4)_4P_2O_7$ – to the best of our knowledge – has not yet been found as a naturally occurring mineral, but has only been prepared synthetically (see [52]). The same holds for the U-, Np- and Pu-substituted analogues. In general terms, the formula hence reads $β-M^{4+}_4(PO_4)_4P_2O_7$.

The crystal structure of β-TPD was first described by Bénard et al. [53]. In a later work, Dacheux et al. [54] as well as Dacheux et al. [55] showed the isotypes of the solid solutions of β-TPD containing U, Np, and Pu. The maximum substitution of Th_{4-x} hereby was $x = 3.00$, 2.09, and 1.63 for U, Np, and Pu, respectively [56]. All these phases have the orthorhombic space group *Pcam* (non-standard setting of *Pbcm*) with approximate lattice parameters $a = 12.9$ Å, $b = 10.4$ Å, and $c = 7.1$ Å for $β-Th_4(PO_4)_4P_2O_7$. The Wyckoff positions of the atoms in the β-TPD–type structure are given in Tab. 9.6.

Tab. 9.6: Occupied Wyckoff positions in the β-TPD–type structure in space group *Pcam* after Bénard et al. [53].

Ion	Wyckoff site	x	y	z
$M^{4+,1}$	4d	x	y	1/4
$M^{4+,2}$	4d	x	y	1/4
P^1	4c	1/4	y	0
P^2	4d	x	y	1/4
P^3	4d	x	y	1/4
O^{11}	4c	1/4	y	0
O^{12}	8e	x	y	z
O^{13} (occ = 1/2)	4d	x	y	1/4
O^{21}	4d	x	y	1/4
O^{22}	8e	x	y	z
O^{23}	4d	x	y	1/4
O^{31}	4d	x	y	1/4
O^{32}	4d	x	y	1/4
O^{33}	8e	x	y	z

In their original work, Bénard et al. [53] described the β-TPD type as made up of layers of [PO$_4$]-tetrahedra and [P$_2$O$_7$]-pyrophosphate-groups parallel to (010). These phosphate layers then alternate with parallel planes containing all M^{4+}-ions. The pyrophosphate groups show positional disorder, which can also be read from the central oxygen (O^{13}) having an occupancy of only 0.5. This does not facilitate the graphic display of the structure. Both symmetrically non-equivalent M^{4+}-positions are in eight-fold coordination by oxygen, with reasonable M^{4+}–O distances. No easy description of the coordination-polyhedra geometry was presented.

9.4.7 Specific structural features of monazite

Some recent systematic and detailed investigations of various monazite solid solutions gave a rather deep insight into specific features of this structure type. Such investigations generally are important, since the formation of solid solutions strongly influences structural, physical, and (thermo-)chemical properties of materials. Aside from global, long-range information by (synchrotron and laboratory) X-ray diffraction experiments, local, short-range information was obtained from infrared (IR)- and Raman-spectroscopic investigations (see [57–62]).

Figure 9.5 shows combined lattice parameter data for a considerable number of monazite-type binary solid solutions as a function of the average effective ionic radius of M^{3+} (see [63]). Corresponding references for the data are given in Tab. 9.7.

Additionally, the unit cell volume of some solid solution series is shown in Fig. 9.6 as a function of the cube of the ionic radius.

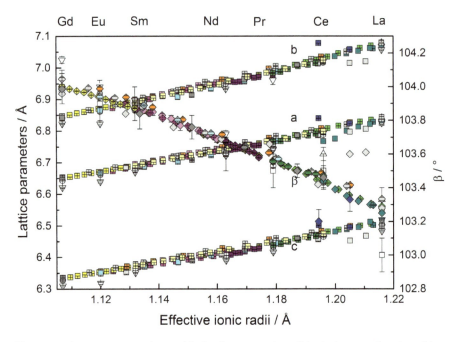

Fig. 9.5: Lattice parameters a, b, c and β of various monazite solid solutions as a function of the average effective ionic radius of M^{3+}. References for the data are given in Tab. 9.7.

Tab. 9.7: Monazite solid solutions and references used for Fig. 9.5.

Composition	Symbol	Reference
(La,Pr)PO$_4$	■	[61]
(Pr,Sm)PO$_4$	■	[64]
(Nd,Sm)PO$_4$	■	[65]
(Pr,Nd)PO$_4$	■	pers. commun., C. Claßen, unpublished.
(Nd,Eu)PO$_4$	■	[66]
(Sm,Gd)PO$_4$	■	[67]
(La,Ce)PO$_4$	■	[68, 69]
(La,Eu)PO$_4$	■	[70]
(La,Gd)PO$_4$	■	[70, 71]
(Ce,Pr)PO$_4$	■	[72]
(Ce,Sm)PO$_4$	■	[73]
(Ce,Eu)PO$_4$	■	[74]
(Ce,Gd)PO$_4$	■	[75]
(La,Pu)PO$_4$	■	[76]
(La,Gd)PO$_4$	■	[77]
End-members	▽; ○;	[17, 24, 78, 79]
	□; △;	[71, 80]

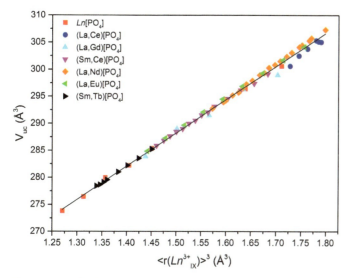

Fig 9.6: Linear fit of the unit cell volume of selected monazite-type end members and solid solutions as a function of the cube of the corresponding average effective ionic radius. The data for $LnPO_4$, (La,Ce)PO_4, (La,Gd)PO_4 and (Sm,Ce)PO_4, respectively, are identical to that used in Schlenz et al. [14]. The data for (La,Nd) PO_4 are from Schlenz et al. [62]. (La,Eu)PO_4 is from Arinicheva et al. [81] and (Sm,Tb)PO_4 from Heuser [73].

These two figures clearly demonstrate a systematic structural behavior of monazite, which will probably even allow for predictions as a function of the average ionic radius of any M^{3+}:

i. All unit cell lengths (a, b, c) increase linearly with an increase in effective ionic radius of M^{3+}.
ii. The monoclinic angle β also decreases linearly with an increase in effective ionic radius of M^{3+}.
iii. Both (i) and (ii) combined lead to an almost linear dependence of the unit cell volume on the M^{3+}-cation volume, here represented as the cube of the ionic radius.

In conclusion, for all solid solution series shown here, it is most probably the well-known lanthanides' contraction that translates directly (and rather simply) into the change of the size and shape of the unit cell.

Some more details of the structural behavior of a monazite solid solution were, e.g. recently presented for $La_{1-x}Pr_xPO_4$ [61]. The authors showed that for all solid solution members studied by X-ray diffraction in that work, the [PO_4]-tetrahedron, while slightly distorted, could be considered more or less as a rigid body [82].

To gain further insight into this issue, the authors used the complementary method of (infrared and Raman) spectroscopy. Exemplary IR and Raman spectra for the characteristic modes of the [PO_4]-tetrahedron in monazites are shown in Figs. 9.7 and 9.8.

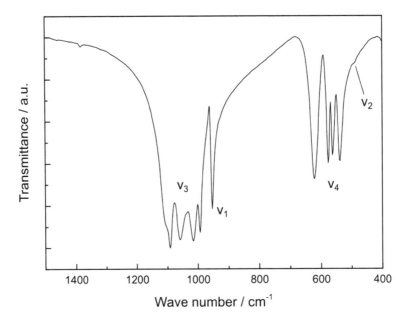

Fig. 9.7: IR spectrum for the characteristic modes of the [PO$_4$]-tetrahedron in monazites for LaPO$_4$: v_1 and v_3 are the symmetric and anti-symmetric stretching modes, respectively, while v_2 and v_4 are the symmetric and anti-symmetric bending modes.

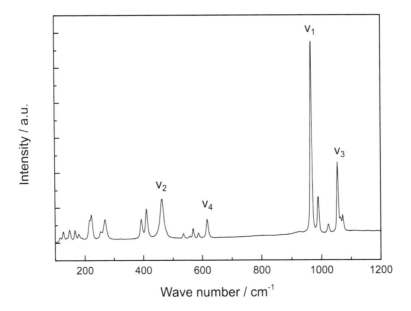

Fig. 9.8: Raman spectrum for the characteristic modes of the [PO$_4$]-tetrahedron in monazites for LaPO$_4$: v_1 and v_3 are the symmetric and anti-symmetric stretching modes, respectively, whereas v_2 and v_4 are the symmetric and anti-symmetric bending modes.

Hirsch et al. [61] stated that, within resolution, the IR spectra for the characteristic modes of the [PO$_4$]-tetrahedron in monazites were not visibly dependent on composition. This was different for the Raman modes. An updated and amended example for this behavior can be seen in Fig. 9.9 for three solid solution series for the most composition-sensitive [61] symmetrical stretching mode v_1 of the [PO$_4$] tetrahedron.

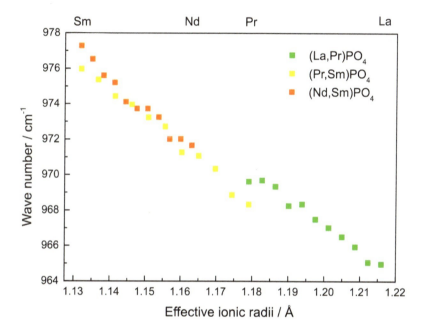

Fig. 9.9: Raman wave numbers of the v_1 mode.

The observed blue shift of the characteristic Raman modes with decreasing effective ionic radius of M^{3+} was attributed to the corresponding M^{3+}–O bond length contraction, resulting in a "stiffening" of the affected P–O bonds [61].

In conclusion, this type of extensive structural characterization of various solid solution series would be most welcome for all structure types considered to be used as waste matrices. The aim of these studies should be the ability to predict the structural properties of any composition of any phase used to immobilize radionuclides.

9.5 Chemical thermodynamics of single phases and solid solutions

Any application in general and the application of phosphate materials as nuclear waste forms in particular require a model based on reliable thermodynamic data that allows for a description and prediction of the waste form stability under repository-relevant conditions over a very long time scale. A refined understanding of the thermochemistry of solid solutions and their constituents provides the information for a reliable description of the phosphates' stability, their tendency to destabilization due to immiscibility and phase separation.

Experimentally obtained thermodynamic data are accessible from calorimetric measurements. New developments in these methods significantly expanded the abilities to obtain thermodynamic data up to very high temperature (2000°C) in different environments (e.g. oxygen, argon) using milligram amounts of material. Three main classes of calorimetry can be distinguished [83]. The first group includes drop solution calorimeters to investigate directly the enthalpy of chemical reactions by dissolving the samples at high temperature. Among them, the Calvet-type twin calorimeter e.g. for oxide melt solution calorimetry is the most advanced system [84]. The second group includes the widely spread differential scanning calorimeters to collect data about phase transitions and heat capacities with a very limited capability to investigate chemical reactions. Another option to obtain thermodynamic data of phase transitions and heat capacities offers the third class including the conventional drop calorimetry [85, 86].

All methods were used to investigate phosphate materials. Most studies were performed on pure $REEPO_4$ materials [80, 87, 88] as well as on $REEPO_4$ solid solutions [61, 62, 70, 89] and to smaller extent on diphosphate solid solutions [90, 91]. The enthalpies of formation of pure $REEPO_4$ were found to follow roughly a linear trend consistent with the decrease of ionic radii of lanthanides and the ionic binding in these materials [80].

For an application as a waste form, the data from investigations on solid solutions are very important because they provide insight into the thermochemical behavior of materials which display more realistically the excess properties of a waste form after the incorporation of e.g. actinides into the crystal structure of the host matrix. Excess properties such as the heat contents were measured using high-temperature calorimetry of phosphate solid solutions ($La_{1-x}Ln_xPO_4$; Ln = Nd, Eu, Gd [76] and $Ca_xTh_xLn_{1-2x}PO_4$, Ln = La, Ce [91]). The phosphate compositions were found to be regular solid solutions and the non-ideal behavior is a result of the mismatch of the size of substituting ions. These data represent an integral of excess heat capacity of the regular solid solutions from ambient to high temperature, but they do not give directly the excess enthalpy of mixing. Such data are accessible most accurately by drop solution calorimetry, because the dissolution enthalpy can be measured directly while the sample is dissolved at precisely adjusted isothermal conditions. Very recently, the excess enthalpy of mixing of $La_{1-x}Ln_xPO_4$ (Ln = Pr, Nd, Eu, Gd) solid solutions were measured by drop solution calorimetry [61, 62, 70]. The data clearly demonstrate that the mismatch of ionic radii contributes as well as lattice strain due to bond length distribution in the solid solutions, to the excess enthalpy of mixing of these solid solutions. The appearance of lattice strain as a consequence of bond length distribution is in good agreement with detailed structural investigations and modeling studies of phosphate solid solutions [92, 93]. Moreover, *ab initio* methods were applied to compute the Margules interaction parameters and resulting excess enthalpies of mixing for series of monazite solid solutions from measured or computed volumes of end-members and Young's moduli [94, 95].

Only very few data are available on actinide containing solid solutions because these experiments can exclusively be performed in restricted areas due to the radioactivity of such samples and are therefore extremely costly. Th–bearing $Ca_xTh_xLn_{1-2x}PO_4$ (Ln = La, Ce) solid solution series measured by Konings et al. [91] showed thermodynamic behavior comparable to the lanthanide mixed solid solutions. Popa et al. [77] measured the low-temperature heat capacity on a $La_{0.9}Pu_{0.1}PO_4$ solid solution and reported an antiferromagnetic effect of Pu^{3+} in that system. Reliable data on Pu or minor actinides (Np, Am, Cm) containing solid solution series do not exist so far and can only be realized by a combination of very few and well selected experiments and atomistic modeling.

9.6 Computer simulations as a valuable supplement to experimental results

In recent years, substantial progress could be made concerning computer simulations of materials that contain 4*f*- and 5*f*-elements [96]. Because lanthanides and certainly actinides play an important role regarding phosphates as ceramic nuclear waste forms, such progress can be very valuable in securing experimental results through computer simulations. Additionally, in some cases, where experimental data are desirable but for some reason not achievable, advanced simulation techniques can be the only choice. Currently, computer simulations based on the density functional theory (DFT) and molecular dynamics (MD) are developing particularly strongly, and the results that yield from such simulations provide new insights into the properties of phosphate ceramics. During the past 5 years, especially several studies on monazite- and xenotime-type materials were published, as well as some papers on solid solutions of these phases.

Now it is not only possible to calculate cell constants and bond distances with a significantly higher precision and accuracy as they were accessible via simulations in former years [97]. It is also possible to calculate e.g. heat capacities [98], elastic moduli and thermal conductivities [95, 99], enthalpies of formation [70], enthalpies of mixing of solid solutions [94], the radiation resistance of a monazite- or xenotime-type ceramic [100, 101], and finally DFT calculations were also shown to be helpful in the assessment of chemical durability [102].

The DFT is currently the method of choice for the accurate prediction of the electronic properties of solid materials at low computational expenses ([42, 103, 104]). On the one hand, DFT enables first-principles studies of materials properties, but on the other hand, DFT suffers from approximations that can result in inaccurate descriptions of electronic band structures, especially for the case of actinides [96, 105].

An advanced extension of DFT helps to solve latter problems, the so-called DFT + U method using a Hubbard U parameter, that explicitly accounts for the correlations between f-electrons (for more detailed information about the theoretical background of the DFT + U method, the interested reader is kindly referred to the attached references). Blanca-Romero et al. [97] demonstrated that this method results in very good predictions, e.g. of the Ln-O distances (Ln = La to Dy) in the crystal structures of monazite and of the formation enthalpies of such orthophosphates. With similar success, Beridze et al. [96] extended this study to xenotime-type materials.

Another good example is the reliable computation of heat capacities (C) of lanthanide and actinide monazite-type ceramics [98]. Different materials and different phases of the same material have different heat capacities that vary as a function of temperature. Additionally, heat capacity values of all materials are different for heating at constant volume (C_V) or at constant pressure (C_P). The heat capacity of a material is generally due to the uptake of kinetic energy by atoms and molecules, and the heat capacity of a solid (C_P) can be written as follows [106]:

$$C_P = C_{lat} + (C_P - C_V) + C_E + C_M + C_\lambda + C_{Sch} \tag{9.1}$$

The lattice heat capacity (C_{lat}) is generally the largest term, followed by the thermal expansion term ($C_P - C_V$), that is the next most important contribution for most nonmetals. The electronic heat capacity (C_E) generally does not become significant until high temperatures (except for metals). Other contributions are magnetic transitions (C_M), order/disorder effects in crystal structures (C_λ), and transitions between electronic energy levels with d and f electrons, the latter being present in lanthanides and actinides (C_{Sch}, Schottky transitions). The first law of thermodynamics results in a relationship between heat capacity and enthalpy, and the second law of thermodynamics likewise creates a link with the entropy. Heat capacities can be determined experimentally using calorimetric methods, or they can be calculated, e.g. using DFT + U methods.

The importance of the heat capacity as a central materials property is emphasized by the aforementioned, and the reader easily gets the impression, that modeling heat capacities precisely and accurately might not be a trivial task. Kowalski et al. [98] computed heat capacities and the standard entropies for monazite-type phases, and they achieved a good agreement between available experimental data and the results of their DFT simulations for lanthanides ($LnPO_4$; Ln = La to Gd). Therefore, they suggested that in general missing thermodynamic data on actinide monazites could be computed in a similar manner.

One of the most important properties of a nuclear waste form is its radiation resistance. During the past three decades, several experimental irradiation studies had been performed (see [11; 107–111], for some recent work and reviews). However, irradiation experiments on actinide-containing materials require a lot of effort and are generally not easy to perform. Therefore, computational tools that are able to do the same job a lot easier are desirable. Grechanovsky et al. [100] used MD simulations in order to calculate the radiation resistance of $LaPO_4$ (monazite) and $YbPO_4$ (xenotime), respectively. Ji et al. [101] also simulated the radiation resistance of $LaPO_4$ and additionally that of $(La_{0.2}Gd_{0.8})PO_4$, using atomistic modeling techniques. Experimental results could be confirmed, and the determined threshold displacement energies (E_d) will help to select proper setup parameters for future irradiation experiments [101]. Through detailed analysis of radiation-induced damages that originate from α-decay, and especially by the simulation of Frenckel pairs being produced by such irradiation processes, Grechanovsky et al. [100] confirmed experimental results that the radiation resistance of $LnPO_4$ (Ln = La to Gd) is generally higher than that of zircon-type phosphate structures. Again, this demonstrates the value of computer simulations as a supplement to laboratory experiments. The results of Grechanovsky et al. [100] also explain the structural changes due to irradiation, and this increases our knowledge about radiation damage and possible reconstruction, respectively.

For the future, it can be expected that the increased use of supercomputers will enable more impressive results that emerge from advanced computer simulations being performed on radioactive materials in general, and nuclear waste forms in particular.

9.7 Summary

Suitable nuclear waste forms require several properties, and only some of them could be mentioned shortly within this chapter (physical properties and chemical durability are also important issues).

Our aim was to emphasize the significance of phosphates for conditioning within the framework of advanced nuclear waste management strategies in general, and the importance of monazite-type phases in particular. Therefore, we restricted our discussion to four prominent topics with regard to the assessment of the suitability of such phases for conditioning: Crystal structures, thermodynamic properties, relevant analytical methods for the experimental determination of necessary data, and finally complementary and progressive computer simulations using supercomputers, in order to fill knowledge gaps. Collecting all this information about phases under investigation, one gets a quite complete picture related to the qualification of phosphates as ceramic nuclear waste forms.

However, many of the listed references and the experimental studies behind these papers used lanthanides as surrogates for actinides, because often experiments with active materials are very elaborate or simply not feasible, and therefore, in comparison, the number of references that practically incorporated actinides is significantly lower [112, 113]. Future investments in modern laboratory equipment should enable more experimental studies using radioactive samples (e.g. even for irradiation experiments), and further developed simulation strategies will be of more importance in order to gain a comprehensive knowledge of such phases, which shall be stable for very long time periods.

Furthermore, we would like to emphasize that the task of finding suitable matrices for long-lived radionuclides and fission products will be a challenge even for generations to come, and therefore it has to be stressed that also sufficient investment in education and training, regarding radiochemistry and handling of radionuclides, remains essential.

References

[1] Jantzen CM, Lee WE, Ojovan MI. Radioactive waste conditioning, immobilization, and encapsulation processes and technologies: overview and advances. Chapter 6. In: radioactive waste management and contaminated site clean-up: processes, technologies and international experience; 2013:171–272. DOI: 10.1533/9780857097446.1.171.

[2] Nieder-Westermann GH, Bollingerfehr W. Germany: experience of radioactive waste (RAW) management and contaminated site clean-up. Chapter 14. In: Radioactive waste management and contaminated site clean-up: processes, technologies and international experience; 2013:462–88. DOI: 10.1533/9780857097446.2.462.

[3] Printz R, Valencia L, Merx H. Research locations preparing for commissioning of the Konrad repository. ATW-Int J Nucl Power 2009;54:307.

[4] Stefanovsky SV, Yudintsev SV, Vinokurov SE, Myasoedov BF. Chemical-technological and mineralogical-geochemical aspects of the radioactive waste management. Geochem Int 2016;54:1136–55.

[5] Dacheux N, Clavier N, Robisson A-C, Terra O, Audubert F, Lartigue J-E, Guy C. Immobilisation of actinides in phosphate matrices. CR Chim 2004;7:1141–52.

[6] Dacheux N, Clavier N, Podor R. Monazite as promising long-term radioactive waste matrix: benefits of high-structural flexibility and chemical durability. Am Mineral 2013;98:833–47.

[7] Lumpkin GR. Ceramic waste forms for actinides. Elements 2006;2(6):365–72.

[8] Terra O, Dacheux N, Audubert F, Podor R. Immobilization of tetravalent actinides in phosphate ceramics. J Nucl Mater 2006;352:224–32.

[9] Ewing RC. Ceramic matrices for plutonium disposition. Prog Nucl Energy 2007;49:635–43.

[10] Weber WJ, Navrotsky A, Stefanovsky S, Vance ER, Vernaz E. Materials Science of High-Level Nuclear Waste Immobilization. Mater Res Soc Bull 2009;34:46–53.

[11] Ewing RC, Weber WJ. Actinide waste forms and radiation effects. Chapter 35. In: Morss LR, Edelstein NM, Fuger J, editors. The chemistry of the actinide and transactinide elements. 4th ed. Springer, Dordrecht; 2010:3813–87.

[12] Brégiroux D, Audubert F, Charpentier T, Sakellariou D, Bernache-Assollant D. Solid-state synthesis of monazite-type compounds $LnPO_4$ (Ln = La to Gd). Solid State Sci 2007;9:432–9.

[13] Ewing RC, Wang LM. Phosphates as nuclear waste forms. In: Phosphates: geochemical, geobiological, and materials importance. Vol. 48. Mineralogical Society of America; 2002:673–99.

[14] Schlenz H, Heuser J, Neumann A, Schmitz S, Bosbach D. Monazite as a suitable waste form. Z Kristallogr 2013;228:113–23.

[15] Gramaccioli C, Segalstad T. A uranium- and thorium-rich monazite from a southalpine pegmatite at Piona, Italy. Am Mineral 1978;63:757–61.
[16] Boatner L, Sales B. Monazite. In: Lutze W, Ewing RC, editors. Radioactive waste forms for the future. New York: Elsevier; 1988; 495–564.
[17] Ni Y, Hughes JM, Mariano AN. Crystal chemistry of the monazite and xenotime structures. Am Mineral 1995;80:21–6.
[18] Schärer U, Deutsch A. Isotope systematics and shock-wave metamorphism: II. U-Pb and Rb-Sr in naturally shocked rocks, the Haughton Impact Structure, Canada. Geochim Cosmochim Acta 1990;54:3435–47.
[19] Haaker RF, Ewing RC. Naturally occurring crystalline phases: analogues for radioactive waste forms. Technical Report. Richland (WA): Battelle Pacific Northwest Labs, 1981.
[20] Boatner LA. Synthesis, structure and properties of monazite, pretulite, and xenotime. In: Phosphates: geochemical, geobiological, and materials importance. Vol. 48. Mineralogical Society of America; 2002:87–121.
[21] Ewing RC, Haaker R. The metamict state: implications for radiation damage in crystalline waste forms. Nucl Chem Waste Manage 1980;1:51–7.
[22] Boatner L, Beall G, Abraham M, Finch C, Huray P, Rappaz M. Monazite and other lanthanide orthophosphates as alternate actinide waste forms. In: Scientific basis for nuclear waste management. Springer, Boston; 1980:289–96.
[23] Weber WJ, Ewing RC, Catlow CRA, de la Rubia TD, Hobbs LW, Kinoshita C, Matzke H, Motta AT, Nastasi M, Salje EKH, Vance ER, Zinkle SJ. Radiation effects in crystalline ceramics for the immobilization of high-level nuclear waste and plutonium. J Mater Res 1998;13:1434–84.
[24] Mullica D, Milligan W, Grossie DA, Beall G, Boatner L. Ninefold coordination LaPO$_4$: pentagonal interpenetrating tetrahedral polyhedron. Inorg Chim Acta 1984;95:231–6.
[25] Beall GW, Boatner LA, Mullica DF, Milligan WO. The structure of cerium ortho-phosphate, a synthetic analog of monazite. J Inorg Nucl Chemi 1981;43:101–5.
[26] Clavier N, Podor R, Dacheux N. Crystal chemistry of the monazite structure. J Eur Ceram Soc 2011;31:941–76.
[27] Hughes JM, Rakovan J. The crystal structure of apatite, $Ca_5(PO_4)_3(F,OH,Cl)$. Rev Mineral Geochem 2002;1:1–12.
[28] Montel J-M. The importance of radioactivity in geosciences. Actual Chim 2011:22–6.
[29] Brégiroux D, Audubert F, Champion E, Bernache-Assollant D. Mechanical and thermal properties of hot pressed neodymium-substituted britholite $Ca_9Nd(PO_4)_5(SiO_4)F_2$. Mater Lett 2003;57:3526–31.
[30] Horie K, Hidaka H, Gauthier-Lafaye F. Isotopic evidence for trapped fissiogenic REE and nucleogenic Pu in apatite and Pb evolution at the Oklo natural reactor. Geochim Cosmochim Acta 2004;68:115–25.
[31] Oelkers E, Montel J. Phosphates and nuclear waste storage. Elements 2008;4:113–16.
[32] Chartier A, Meis C, Gale JD. Computational study of Cs immobilization in the apatites $Ca_{10}(PO_4)_6F_2$, $Ca_4La_6(SiO_4)_6F_2$ and $Ca_2La_8(SiO_4)_6O_2$. Phys Rev B 2001;64:085110.
[33] Pan Y, Fleet M. Compositions of the apatite-group minerals: substitution mechanisms and controlling factors. Rev Mineral Geochem 2002;48:13–50.
[34] Ouchani S, Dran J-C, Chaumont J. Evidence of ionization annealing upon helium-ion irradiation of pre-damaged fluorapatite. Nucl Instr Methods Phys Res B 1997;132:447–51.
[35] Berzelius J. Undersökning af några Mineralier. 1. Phosphorsyrad Ytterjord. K Svenska Vet-Akad Handl 1824;2:334–8.
[36] Dana JD. Dana's system of mineralogy. 7th ed. Vol. II. Rewritten by Palache C, Berman H, Frondel C. New York: John Wiley & Sons; 1951.

[37] Förster HJ. The chemical composition of REE-Y-Th-U-rich accessory minerals in peraluminous granites of the Erzgebirge-Fichtelgebirge region, Germany. Part II: xenotime. Am Mineral 1998;82:1302–15.

[38] Milligan WO, Mullica DF, Beall GW, Boatner LA. Structural investigations of YPO_4, $ScPO_4$, and $LuPO_4$. Inorg Chim Acta 1982;60:39–43.

[39] Brownfield ME, Foord EE, Sutley SJ, Botinelly T. Kosnarite, $KZr_2(PO_4)_3$, a new mineral from Mount Mica and Black Mountain, Oxford County, Maine. Am Mineral 1993;78:653–6.

[40] Šljukić M, Matković B, Prodić B, Anderson D. The crystal structure of $KZr_2(PO_4)_3$. Z Kristallogr 1969;130:148–61.

[41] Krivovichev SV, Burns P, Tananaev I, editors. Structural chemistry of inorganic actinide compounds. Elsevier, Amsterdam; 2007.

[42] Clearfield A. Inorganic-ion exchangers with layered structures. Annual Review of Materials Science 1984; 14: 205–29.

[43] Janeczek J, Ewing RC. Florencite-(La) with fissiogenic REEs from a natural fission reactor at Bangombe, Gabon. Am Mineral 1996;81:1263–69.

[44] Lefebvre J-J and Gasparrini C. Florencite an occurrence in the Zairian copperbelt. Can Mineral 1980;18:301–11.

[45] Strunz H, Nickel EH. Strunz mineralogical tables. 9th ed. E. Stuttgart: Schweizerbart'sche Verlagsbuchhandlung (Nägele u. Obermiller); 2001.

[46] Sawka WN, Banfield JF, Chappell BW. A weathering-related origin of widespread monazite in S-type granites. Geochim Cosmochim Acta 1986;50:171–5.

[47] Schwab RG, Herold H, da Costa ML, de Oliveira NP, Balasubramanian KS, Evangelov VP. The formation of aluminous phosphates through lateritic weathering of rocks. Weathering 1989;2:369–86.

[48] Pouliot G, Hofmann HJ. Florencite: a first occurence in Canada. Can Mineral 1981;19:535–40.

[49] Kato T. The crystal structure of florencite. Neu Jb Mineral Mh 1990;5:227–31.

[50] Kato T. The crystal structure of goyazite and woodhouseite. Neu Jb Mineral Mh 1971:241–7.

[51] Schwab RG, Herold H, Gotz C, de Oliveira NP. Compounds of the crandallite type: synthesis and properties of pure rare earth element-phosphates. Neu Jb Mineral Mh 1990;6:241–54.

[52] Goffé B, Janots E, Brunet F, Poinssot C. Déstabilisation du phosphate–diphosphate de thorium (PDT), $Th_4(PO_4)_2P_2O_7$, entre 320 et 350 °C, porté à 50 MPa en milieu calcique, ou pourquoi le PDT n'a pas d'équivalent naturel. CR Geosci 2002;334:1047–52.

[53] Bénard P, Brandel V, Dacheux N, Jaulmes S, Launay S, Lindecker C, Genet M, Louër D, Quarton M. $Th_4(PO_4)_4P_2O_7$, a new thorium phosphate: synthesis, characterization, and structure determination. Chem Mater 1996;8:181–8.

[54] Dacheux N, Thomas AC, Brandel V, Genet M. Investigation of the system ThO_2-NpO_2-P_2O_5. Solid solutions of thorium-neptunium (IV) phosphate-diphsophate. J Nucl Mater 1998a;257:108–17.

[55] Dacheux N, Podor R, Chassigneux B, Brandel V, Genet M. Actinides immobilization in new matrices based on solid solutions: $Th_{4-x}M^{IV}_x(PO_4)_4P_2O_7$, ($M^{IV}$=^{238}U, ^{239}Pu). J Alloys Compounds 1998b;271–273:236–9.

[56] Genet M, Dacheux N, Thomas AK, Chassigneux B, Pichot E Brandel V. Thorium phosphate-diphosphate as a ceramic for the immobilization of tetravalent uranium, neptunium and plutonium. Waste Management '99 Conf. Proc., Laser Options Inc., Tuscon (AZ); 1999.

[57] Begun GM, Beall GW, Boatner LA, Gregor WJ. Raman spectra of the rare earth orthophosphates. J Raman Spectrosc 1981;11:273–8.

[58] Podor R. Raman spectra of the actinide-bearing monazites. Eur J Mineral 1995;7:1353–60.

[59] Silva EN, Ayala AP, Guedes I, Paschoal CWA, Moreira RL, Loong C-K, Boatner LA. Vibrational spectra of monazite-type rare-earth orthophosphates. Opt Mater 2006;29:224–30.

[60] Ruschel K, Nasdala L, Kronz A, Hanchar JM, Többens DM, Škoda R, Finger F, Möller A. A Raman spectroscopic study on the structural disorder of monazite-(Ce). Mineral Petrol 2012;105:41–55.

[61] Hirsch A, Kegler P, Alencar I, Ruiz-Fuertes J, Shelyug A, Peters L, Schreinemachers C, Neumann A, Neumeier S, Liermann H-P, Navrotsky A, Roth G. Structural, vibrational, and thermochemical properties of the monazite-type solid solution $La_{1-x}Pr_xPO_4$. J Solid State Chem 2017;245:82–8.

[62] Schlenz H, Dellen J, Kegler P, Gatzen C, Schreinemachers C, Shelyug A, Klinkenberg M, Navrotsky A, Bosbach D. Structural and mixing properties of $(La_{1-x}Nd_x)PO_4$ monazite-type solid solutions. Acta Materialia 2017; submitted.

[63] Shannon RD. Revised effective ionic radii and systematic studies of interatomic distances in halides and chalcogenides. Acta Crystallogr A 1976;32:751–67.

[64] Bigdeli D. Synthese und Charakterisierung von Monazit-Mischkristallen im System Pr_2O_3-Sm_2O_3-P_2O_5. MSc thesis, RWTH Aachen University; 2016.

[65] Kuleci Z. Synthese und Charakterisierung von Monazit-Typ Mischkristallen im System Nd_2O_3-Sm_2O_3-P_2O_5. MSc thesis. RWTH Aachen University; 2015.

[66] Schumacher A. Synthese und Charakterisierung der Monazit-Typ-Orthophosphate $(Nd, Eu)PO_4$. BSc thesis. RWTH Aachen University; 2016.

[67] Ladenthin N. Synthese und Charakterisierung von Monazit-Typ-Mischkristallen im System $(Sm, Gd)PO_4$. BSc thesis. RWTH Aachen University; 2017.

[68] de Biasi RS, Fernandes AAR, Oliveira JCS. Cell volumes of $LaPO_4$-$CePO_4$ solid-solutions. J Appl Crystallogr 1987;20:319–20.

[69] Hou ZY, Zhang ML, Wang LL, Lian HZ, Chai RT, Zhang CM, Cheng ZY, Lin J. Preparation and luminescence properties of Ce^{3+} and/or Tb^{3+} doped $LaPO_4$ nanofibers and microbelts by electrospinning. J Solid State Chem 2009;182:2332–2.

[70] Neumeier S, Kegler P, Arinicheva Y, Shelyug A, Kowalski PM, Schreinemachers C, Navrotsky A, Bosbach D. Thermochemistry of $La_{1-x}Ln_xPO_4$-monazites (Ln = Gd, Eu). J Chem Thermodyn 2017;105:396–403.

[71] Terra O, Clavier N, Dacheux N, Podor R. Preparation and characterization of lanthanum-gadolinium monazites as ceramics for radioactive waste storage. New J Chem 2003;27:957–67.

[72] Zeng P, Teng Y, Huang Y, Wu L, Wang X. Synthesis, phase structure and microstructure of monazite-type $Ce_{1-x}Pr_xPO_4$ solid solutions for immobilization of minor actinide neptunium. J Nucl Mater 2014;452:407–13.

[73] Heuser J. Keramiken des Monazit-Typs zur Immobilisierung von minoren Actinoiden und Plutonium. PhD thesis. RWTH Aachen University; 2015.

[74] Wang X, Teng Y, Huang Y, Wu L, Zeng P. Synthesis and structure of $Ce_{1-x}Eu_xPO_4$ solid solutions for minor actinides immobilization,. J Nucl Mater 2014;451:147–52.

[75] Yang H, Teng Y, Ren X, Wu L, Liu H, Wang S, Xu L. Synthesis and crystalline phase of monazite-type $Ce_{1-x}Gd_xPO_4$ solid solutions for immobilization of minor actinide curium. J Nucl Mater 2014;444:39–42.

[76] Popa K, Konings RJM, Geisler T. High-temperature calorimetry of $(La_{1-x}Ln_x)PO_4$ solid solutions. J Chem Thermodyn 2007a;39:236–9.

[77] Popa K, Colineau E, Wastin F, Konings RJM. The low-temperature heat capacity of $(Pu_{0.1}La_{0.9})PO_4$. Solid State Commun 2007b;144:74–7.

[78] Mullica DF, Grossie DA, Boatner LA. Structural refinements of praseodymium and neodymium orthophosphate. J Solid State Chem 1985a;58:71–7.

[79] Mullica DF, Grossie DA, Boatner LA. Coordination geometry and structural determinations of $SmPO_4$, $EuPO_4$ and $GdPO_4$. Inorg Chim Acta 1985b;109:105–10.

[80] Ushakov SV, Helean KB, Navrotsky A. Thermochemistry of rare-earth orthophosphates. J Mater Res 2001;16:2623–33.

[81] Arinicheva Y, Bukaemskiy A, Neumeier S, Modolo G, Bosbach D. Studies on thermal and mechanical properties of monazite-type ceramics for the conditioning of minor actinides. Prog Nucl Energy 2014;72:144–8.

[82] Popović L, de Waal D, Boeyens JCA. Correlation between Raman wavenumbers and P-O bond lengths in crystalline inorganic phosphates. J Raman Spectrosc 2005;36:2–11.

[83] Sarge SM, Höhne GWH, Hemminger WF. Calorimetry: fundamentals, Instrumentation and applications. Weinheim: Wiley-VCH; 2014.

[84] Navrotsky A. High temperature oxide melt calorimetry of oxides and nitrides. In: Huffman Lecture at ICCT 2000. J Chem Thermodyn 2001;33:859–71.

[85] Navrotsky A. Progress and new directions in calorimetry: a 2014 perspective. J Am Ceram Soc 2014;97:3349–59.

[86] Navrotsky A, Lee W, Mielewczyk-Gryn A, Ushakov SV, Anderko A, Wu H, Riman RE. Thermodynamics of solid phases containing rare earth oxides. J Chem Thermodyn 2015;88:126–41.

[87] Popa K, Konings RJM. High-temperature heat capacities of $EuPO_4$ and $SmPO_4$ synthetic monazites. Thermochim Acta 2006;445:49–52.

[88] Popa K, Sedmidubský D, Beneš O, Thiriet C, Konings RJM. The high-temperature heat capacity of $LnPO_4$ (Ln = La, Ce, Gd) by drop calorimetry. J Chem Thermodyn 2006a;38:825–9.

[89] Popa K, Konings RJM, Geisler T. High-temperature calorimetry of $(La_{1-x}Ln_x)PO_4$ solid solutions. J Chem Thermodyn 2006b;39:236–9.

[90] Ushakov SV, Navrotsky A, Farmer JM, Boatner LA. Thermochemistry of the alkali rare-earth double phosphates, $A_3RE(PO_4)_2$. J Mater Res 2004;19:2165–75.

[91] Konings RJM, Walter M, Popa K. Excess properties of the $(Ln_{2-2x}Ca_xTh_x)(PO_4)_2$ (Ln = La, Ce) solid solutions. J Chem Thermodyn 2008;40:1305–8.

[92] Geisler T, Popa K, Konings RJM. Evidence for lattice strain and non-ideal behavior in the $(La_{1-x}Eu_x)PO_4$ solid solution from X-ray diffraction and vibrational spectroscopy. Front Earth Sci 2016;4:article 64.

[93] Huittinen N, Arinicheva Y, Kowalski PM, Vinograd VL, Neumeier S, Bosbach D. Probing structural homogeneity of $La_{1-x}Gd_xPO_4$ monazite-type solid solutions by combined spectroscopic and computational studies. J Nucl Mater 2017;486:148–57.

[94] Li Y, Kowalski PM, Blanca-Romero A, Vinograd V, Bosbach D. Ab initio calculation of excess properties of $La_{1-x}(Ln, An)_xPO_4$ solid solutions. J Solid State Chem 2014;220:137–41.

[95] Kowalski PM, Li Y. Relationship between the thermodynamic excess properties of mixing and the elastic moduli in the monazite-type ceramics. J Eur Ceram Soc 2016;36:2093–6.

[96] Beridze G, Birnie A, Koniski S, Ji Y, Kowalski PM. DFT+U as a reliable method for efficient ab initio calculations of nuclear materials. Prog Nucl Energy 2016;92:142–6.

[97] Blanca-Romero A, Kowalski PM, Beridze G, Schlenz H, Bosbach D. Performance of the DFT+U method for prediction of structural and thermodynamic parameters of monazite-type ceramics. J Comput Chem 2014;35:1339–46.

[98] Kowalski PM, Beridze G, Vinograd VL, Bosbach D. Heat capacities of lanthanide and actinide monazite-type ceramics. J Nucl Mater 2015;464:147–54.

[99] Feng J, Xiao B, Zhou R, Pan W. Anisotropy in elasticity and thermal conductivity of monazite-type $REPO_4$ (RE = La, Ce, Nd, Sm, Eu, and Gd) from first-principles calculations. Acta Mater 2013;61:7364–83.

[100] Grechanovsky AE, Eremin NN, Urusov VS. Radiation resistance of $LaPO_4$ (monazite structure) and $YbPO_4$ (zircon structure) from data of computer simulation. Phys Solid State 2013;55:1929–35.

[101] Ji Y, Kowalski PM, Neumeier S, Deissmann G, Kulriya PK, Gale JD. Atomistic modeling and experimental studies of radiation damage in monazite-type $LaPO_4$ ceramics. Nucl Instrum Methods Phys Res B 2017; 393: 54–8.

[102] Veilly E, du Fou de Kerdaniel E, Roques J, Dacheux N, Clavier N. Comparative behavior of britholites and monazite/brabantite solid solutions during leaching tests: a combined experimental and DFT approach. Inorg Chem 2008;47:10971–79.

[103] Rustad JR. Density functional calculations of the enthalpies of formation of rare-earth orthophosphates. Am Mineral 2012;97:791–9.

[104] Agapito LA, Curtarolo S, Buongiorno Nardelli M. Reformulation of DFT + U as a pseudohybrid Hubbard density functional for accelerated materials discovery. Phys Rev X 2015;5:011006-1-16.

[105] Beridze G, Kowalski PM. Benchmarking the DFT+U method for thermochemical calculations of uranium molecular compounds and solids. J Phys Chem A 2014;118:11797–810.

[106] Fegley Jr, B. Practical chemical thermodynamics for geoscientists. Elsevier, Amsterdam; 2013.

[107] Ewing RC, Meldrum A, Wang L, Wang S. Radiation-induced amorphization. Chap. 12. In: Redfern SAT, Carpenter MA, editors. Transformation processes in minerals. Rev Mineral Geochem 2000;39:319–61.

[108] Tamain C, Dacheux N, Garrido F, Thomé L. Irradiation effects on thorium phosphate diphosphate: chemical durability and thermodynamic study. J Nucl Mater 2007;362:459–65.

[109] Tamain C, Garrido F, Thomé L, Dacheux N, Özgümüs A. Structural behavior of β-thorium phosphate diphosphate (β-TPD) irradiated with ion beams. J Nucl Mater 2008;373:378–86.

[110] Nasdala L, Grötzschel R, Probst S, Bleisteiner B. Irradiation damage in monazite-(Ce): an example to establish the limits of Raman confocality and depth resolution. Can Mineral 2010;48:351–9.

[111] Rafiuddin MR, Grosvenor AP. Probing the effect of radiation damage on the structure of rare-earth phosphates. J Alloys Compounds 2015;653:279–89.

[112] Popa K, Raison PE, Martel L, Martin PM, Prieur D, Solari PL, Bouëxière Konings RJM, Somers J. Structural investigations of Pu^{III} phosphate by X-ray diffraction, MAS-NMR and XANES spectroscopy. J Solid State Chem 2015;230:169–74.

[113] Dellen J, Labs S, Schlenz H, Bosbach D. Syntheses and structural characterization of $Sm_x(Ca,Th)_{1-x}PO_4$ and $Sm_x(Ca,U)_{1-x}PO_4$ solid solutions. In: MRS Meeting Montpellier, Scientific Basis for Nuclear Waste Management XXXIX, Book of Abstracts 2015;116.

[114] Wen X-D, Martin RL, Henderson TM, Scuseria GE. Density Functional Theory of the Electronic Structure of Solid State Actinide Oxides. Chem Rev 2013;113:1063–96.

S. Stöber and H. Pöllmann
10 Immobilization of high-level waste calcine (radwaste) in perovskites

10.1 Introduction

In the field of applied mineralogy, the technical applications of perovskites are of great interest due to their extraordinary ferroelectric [1–3], dielectric [4, 5], pyroelectric [6, 7], and piezoelectric [8] properties. Therefore, they are applied for different purposes like automobile exhaust purification [9], fuel cells [10, 11], N_2O decomposition [12], solar cells (photovoltaics) [13], and in many other fields of technology. The broad fields of application are based on the variety of perovskite structures, due to their structural modularity [14] and the great chemical variability. Those skills are presented in the outstanding compilation of the perovskite family, *Perovskites: Modern and Ancient*, by Mitchell [15].

Usually, perovskites with the general formula ABO_3 are well known, and it is assumed that perovskites should have cubic symmetry sg. *Pm3m* (aristotype), which is true for many perovskite phases, but dependent on temperature, pressure, and bulk chemistry; perovskites exhibit complicated tilt systems, which are responsible for the stabilization of hettotypes related to the aristotype by group-subgroup relations [16–21]. Different types and degrees of disorder for the coordination polyhedra of A- or B-site are a result of the decrease in symmetry.

Aside from ABX_3-perovskites with a cubic close packing, a large group of layered perovskites with a hexagonal close packing also exists. Their unit cells can be transformed easily and hexagonal perovskites exhibit polytypes with large stacking layer sequences [22]. We have to deal with ordered perovskites, the so-called double perovskites with 1:1, 1:2, and 1:3 ordering either on A- or B-site. Ion ordering on both A- and B-sites is also possible. Quadruple-layered defect perovskites belong to this species. They are related closely to layered superconducting perovskites of the 1212C-type, where most of them are derived from Ruddlesdon-Popper [23, 24] and Arivillius compounds [25]. Layered phases of the Dion-Jacobson series [26] and $A_nB_nX_{3n+2}$ compounds [27] also belong to the large perovskite family.

Another important property of perovskite phases is the high degree of A- or X-site deficiency comprising bronzes like ReO_3 or the framework defect perovskites, such as the brownmillerite group $A_2B_2X_5$, whose members of the solid solution system $Ca_2Fe_2O_5$–$Ca_2Al_{1.33}Fe_{0.66}O_5$ are important in the field of cement chemistry.

Not only artificial perovskites are of interest for mineralogists, because up to 93% of the lower mantle consists of silicate perovskites [28]. At high pressures and temperatures, comparable to the conditions at Earth's core-mantle boundary, $MgSiO_3$ perovskite transforms to a new high-pressure form, the so-called post perovskite [29]. The majority of titanates and niobates in the earth crust and certain halides and hydroxide (söhngenite, schoenfliesite, and stottite groups) are minerals with a perovskite-type structure [15].

In this work, the application of $CaTiO_3$, a member of the synroc (synthetic rock) phases, should be proofed to act as a host for the fixation of radwaste (radioactive waste) ions. Of special interest is the crystal chemistry of very different perovskite phases and their solid solutions series, which are formed under the heat treatment of synroc ceramics together with different types of radwastes at non-ambient temperatures. The close relation between natural and technical phases should be demonstrated.

10.2 Immobilization of radioactive waste in ceramics

Different multiphase types of titanite ceramics were developed for the incorporation/immobilization for example high-level waste (HLW) streams including Barnwell and Rockwell-Hanford wastes of thermal oxide reprocessing plants and unprocessed fuel rods [30]. It is possible to prepare ceramics of varying complexity and with sufficient resistance in every instance. In all cases, TiO_2 is predominantly used. Together with lower oxide quantities, typically, a combination of CaO, ZrO_2, Al_2O_3, and BaO were added to encourage the crystallization of specific radwaste-bearing phases. In order to obtain a tailored ceramic, hazardous elements are fixed in the crystal structure of the relevant phases. The immobilization of specific radionuclides is often enhanced. Higher waste loadings can be achieved. Quite often, it is possible to predict the long-term and radiation stability of waste ceramics, as they consist of minerals, whose degradation and alteration by low-level irradiation have been studied in epithermal environments [30].

10.2.1 Supercalcine ceramic

The first ideas about the immobilization of radwaste in an artificial stone were presented by Hatch [31] and were proofed to be feasible in the Materials Science Laboratories of Pennsylvania State University [32–37]. This type of phase assemblages in the artificial stone would be much closer to thermodynamic equilibrium than in amorphous matter, so that the risks, arising from devitrification of glasses, could be prevented. Furthermore, some of the applied phases for radwaste immobilization are stable in a geological environment for an extremely long period ($>10^6$ years) [34]. The advantages of supercalcine ceramics versus glassy matter were summarized by Roy [34]. Perovskites and phases with perovskite-type structures were also considered as a leach-resistant phase, where CaO and TiO_2 should be substituted partially or completely by SrO and RuO_2 [38, 39].

10.2.2 Applications and phase assemblages of synroc ceramics

Another type of natural similar ceramics is the synroc material, which differs by its chemical and mineralogical composition from the supergene material and possesses superior advantages for the immobilization of radwaste ions [40]. Different types of synrocs were proposed and their usability investigated.

Synroc A, produced by Ringwood [41], is composed predominantly by titanates (perovskite, zirconolite, "hollandite") and contain lower concentrations of feldspar components $BaAl_2Si_2O_8$, kalsilite $KAlSiO_4$, and leucite $KAlSi_2O_6$. Synroc A was produced by melting 90 wt% inert additives (13 wt% SiO_2, 33 wt% TiO_2, 10 wt% ZrO_2, 16 wt% Al_2O_3, 6 wt% CaO, and 17 wt% BaO) with 10 wt% radwaste at 1300°C. After cooling at a rate of 2 K/min to 1150°C, a complete crystalline mineral assemblage was yielded [40]. Predominantly lower concentrations of Sr, REE, and actinides were fixed chemically in the perovskite phase. The stability and leaching tests showed that Synroc A, exposed to pure water at 400°C and 1000 bars of pressure for 24 hours, is indeed unaffected, but the application of 10 wt% NaCl solution under the same conditions forced the substitution of Cs^+ by Na^+ in kalsilite and leucite [40].

Synroc B is an improved mineral assemblage, without silicates, which was synthesized applying the following oxide concentrations 90 wt% inert additives (60.4 wt% TiO_2, 9.9 wt% ZrO_2, 11.0 wt% Al_2O_3, 13.9 wt% CaO, 4.2 wt% BaO, and 0.6 wt% NiO) [40]. Synroc B was produced by a hot-pressing technique at about 1300°C. A typical phase assemblage of Synroc B exhibit just three phases, $BaAl_2TiO_{16}$ with a hollandite-like structure (35 wt%), zirconolite $CaZrTi_2O_7$ (31 wt%), and perovskite $CaTiO_3$ (22 wt%), together with additional amounts of TiO_2 (7 wt%) and Al_2O_3 (5 wt%) [35, 40, 42–45].

The incorporation of relevant possible radwaste elements in the perovskite lattice for Ca were proposed to be Sr^{2+}, Ba^{2+}, Na^+, all rare earth elements $(REEs)^{3+}$, Y^{3+}, Cd^{2+}, Cm^{3+}, Am^{3+}, Pu^{3+}, and Ti^{4+} might be substituted by Nb^{5+} Zr^{4+}, Mo^{4+}, Pu^{4+}, Th^{4+}, U^{4+}, Sn^{4+}, Ru^{4+}, Fe^{3+}, Cr^{3+}, Al^{3+} [40]. Synroc B was homogenized with an evaporated and calcined residue of spent fuel from a nuclear power plant. It contained predominantly fission products (REE 26.4 wt%, Zr 13.2 wt%, Mo 12.2 wt%, Ru 7.6 wt%, Cs 7.0 wt%, Pd 4.1 wt%, Sr 3.5 wt%, Ba 3.5 wt%, Rb 1.3 wt%), actinides (U + Th 1.4 wt%, Am + Cm + Pu + Np 0.2 wt%), processing contaminants (Fe 6.4 wt%, (PO_4) 3.2 wt%, Na 1.0 wt%) and others 9.0 wt% [40]. After the heat treatment, the perovskite-phase $CaTiO_3$ contained in Synroc B predominantly Sr (7.5 wt%), HREE Gd (7 wt%) and Y (7 wt%), LREE La (5 wt%), Mo (2.3–1.3 wt%), Th^{4+}, U^{4+}, Cm^{4+}, Pu^{4+}, Am^{3+}, Na^+ (approximately 1–2 wt%). It was considered to change the oxide formulation of the inert additives. Only hollandite and zirconolite should be present in Synroc B, because Sr can be fixed readily in the hollandite structure while the remaining radwaste elements occurring in perovskite also can enter "hollandite" and zirconolite. As ^{90}Sr, a particular hazardous radwaste element [40], has a high affinity with perovskite, the three phase assemblages retained. Further results concerning leaching tests and radiation damage were published [40, 46, 47].

Synroc C was intended mainly for the immobilization of liquid high-level civilian reactor wastes (HLW) arising from the reprocessing of light water reactor fuel. Usually, the sodium concentration must not be considered, but the calcine may contain as much as 17% Na_2O in order to neutralize the acid waste with NaOH [48]. Synroc C was tested for the immobilization of Na^+ especially in perovskite $CaTiO_3$ [48]. The concentration of sodium in HLW is quite low at normal operation, but higher sodium concentrations are present in intermediate-level wastes (ILWs). At the Allied General Nuclear Services (AGNS) plant in Barnwell, SC, the admixture of ILW could result in wastes, which contain about 10% Na_2O after calcination [49]. In order to investigate the distribution of sodium in Synroc C, natural analogues of sodium perovskites, zirconolites, and pyrochlores were evaluated for their application performing melting experiments. Natural zirconolites contain little Na_2O concentrations (<0.5%). Minerals of the pyrochlore group ($NaCaNb_2O_6F$) may contain several percentages of Na_2O. However, natural REE-perovskites like loparite $NaCeTi_2O_6$, one of the end-members in the $CaTiO_3$–$NaCeTi_2O_6$–$Na_2Nb_2O_6$ system are important host minerals for the long-term sodium fixation [30]. The fixation of sodium in Synroc C was investigated performing laboratory experiments. Two specific phases $NaCeTi_2O_6$, and $NaYTi_2O_6$ were homogenized with Synroc C [48]. Cerium represents the typical behavior of light lanthanides and yttrium behaves like an excellent crystal chemical analogue for the heavy rare earths. At low $NaCeTi_2O_6$ or $NaYTi_2O_6$ concentrations (5 wt% + Synroc C), solid solutions were detected between $NaCeTi_2O_6$–$CaTiO_3$ and $NaYTi_2O_6$–$CaTiO_3$. Na and Ca diffused into zirconolite. The admixtures Synroc + 20 wt% $NaYTi_2O_6$ or + 20 wt% $NaYTi_2O_6$ exhibited the formation of increased amounts of $NaYTi_2O_6$–$CaTiO_3$ or $NaYTi_2O_6$–$CaTiO_3$ solid solutions but sodium diffused predominantly into the perovskite phase after hot pressing.

Titanite ceramics, close to the chemical composition of Synroc C (perovskite, hollandite, zirconolite and Magneli phases) were destined for the incorporation of light water reactor (LWR) fuel Purex (United States) and JW-A waste (Japan). The hot-pressing process between 1100°C and 1150°C was performed at specific oxygen fugacities to provide a ceramics with minimal phase assemblages. Concerning the crystal chemistry of perovskite phases, different substitution mechanisms $Ca_{Ca}^x\ Ti_{Ti}^x\ Ti_{Ti}^x \Leftrightarrow U_{Ca}^{..}(Ti,Al)_{Ti}'\ (Ti,Al)_{Ti}'$ (1), $Ca_{Ca}^x\ Ti_{Ti}^x \Leftrightarrow REE_{Ca}^{.}(Ti,Al)_{Ti}'$ (2), $Ca_{Ca}^x\ Ca_{Ca}^x \Leftrightarrow REE_{Ca}^{.}\ Na_{Ca}'$ (3), $Ca_{Ca}^x\ Ca_{Ca}^x \Leftrightarrow Na_{Ca}'\ Ti_{Ca}^{.}$ (4) were proposed for the incorporation of Na^+, REE^{3+}, and U^{4+} in perovskites [30].

Sixteen Synroc samples containing simulated Purex (PW-4b-D) HLW at loadings of 10, 15, 19, and 23 wt% were hot-pressed, and the partitioning of uranium and REEs was investigated [50]. It was concluded that the distribution of uranium and REEs in synroc at low sodium concentration was influenced by the ionic radii of U^{4+}, Y^{3+}, and Gd^{3+} ions. They were fixed preferentially into zirconolite and Nd^{3+} and Ce^{3+} ions with larger ionic radii-preferred $CaTiO_3$. Increased sodium concentrations of 3 wt% destabilized the zirconolite end-members. Uranium, REE together with sodium, entered $CaTiO_3$, forming perovskite like $(Na,REE)(Ti,U)_2O_6$ [50].

The previous results, incorporating sodium and REE in synroc were applied using a mixture of high-level and intermediate-level wastes PW-7a high in Na_2O and P_2O_5 plus additional Gd_2O_3 from the AGNS plant at Barnwell, SC [51]. Compared to the PW-4b material [49], PW-7a contains lower U and Pu concentrations. It can be concluded that the phosphorus-, sodium-, and gadolinium-rich PW-7a waste can be successfully incorporated in Synroc C. Sodium, gadolinium, and different REE concentrations were predominantly fixed in $CaTiO_3$, yielding a complex solid solution. The phosphate concentration of PW-7a waste was fixed in apatite. However, some minor sodium concentrations were detected in apatite $CaNaPO_4$ solid solutions, which crystallized during cooling. If the hot-pressing process was carried out by adding titanium under slightly reducing conditions the partition behavior of sodium was affected significantly. Sodium entered the crystal structure of apatite and REE diffused into perovskite and zirconolite. The coupled substitution of Ca + Ti by Na^+ REE was restrained and REE dominated perovskite phases crystallized.

Synroc D is a ceramic waste-form developed for the specific purpose of immobilizing USA's defense HLW sludges [51] and to evaluate the capacity of synroc to immobilize PW-7a waste [52]. Moreover, the feasibility of sintering different blends of Fe- and Al-rich wastes were also evaluated [53]. If waste material was sintered without additives, the sinter cake of USA's defense HLW sludges contained corundum, magnetite (Fe_3O_4), magnetoplumbite ($(Ca,Sr)Al_{12}O_{19}$), a hauyne-type sodalite ($Ca_2Na_6Al_6Si_6O_{24}(SO_4)_2$), uraninite ($UO_2$), and a calcium-sodium uranate [54]. Except for the hauyne and uranate phases, the aluminum and iron phases were very insoluble and provided the basis for a high-waste–loaded ceramic [54]. The addition of silica together with sodium formed at lower temperatures (1040°C) nepheline, and the addition of titania, zirconia, and calcia were responsible for the crystallization of the stable host minerals perovskite and zirconolite. Synroc D is composed of zirconolite (~16 wt%), perovskite (~11 wt%), nepheline (~18 wt%), spinels (hercynite $FeAl_2O_4$ and ulvo-type Fe_2TiO_4) (~55 wt%), vitreous phases, and iron alloy. The ceramic was designed to immobilize about 60% of calcined defense HLW [55].

The advantages of those titanates is their very low solubility when compared to vitreous phases [56]. They are capable of incorporating radionuclides in their crystal structures at low concentrations. Uranium replaces Zr in Zirconolite, while the actinides and REEs are partitioned between zirconolite and perovskite. Nepheline ($NaAlSiO_4$) is a silicate phase, which crystallized due to the presence of sodium and silicon in the defense waste. Caesium present in the radwaste partitions strongly into the nepheline phase.

The chemical compositions of perovskite phases are comparable to perovskite solid solutions formed in Synroc C in presence of sodium and niobium. Strontium is substituted preferentially for calcium. The addition of further rawly refined niobium from bastnäsite is a cheap option to increase the percentage of perovskite compared to nepheline in Synroc D [56]. Perovskites contained the radionuclides Pu^{3+}, Am^{3+}, Cm^{3+} Np^{3+}, which were also immobilized in zirconolite [56]. USA's defense nuclear wastes used for those experiments contained additional contents of process contaminants Al-, Fe-, Mn-, Ni-oxides and hydroxides, together with 0.5 to 5% of fission products and actinides. After the hot-pressing sinter process, ferric and ferrous iron, manganese, and aluminum were detected in the spinel phase [56].

Finally, it was decided not to apply Synroc D for the immobilization of USA's defense HLW sludge. Instead of the ceramic matter, borosilicate glass was applied, because it was cheaper and it was supposed that its production technology could be developed more rapidly [55].

Further mechanical-thermal physical properties and leach test results concerning final phase assemblages of Synroc D ceramics were performed and discussed [57–59].

Synroc E is a new TiO_2-rich formulation designed to immobilize ≤7% HLW. It consists essentially of the same minerals as Synroc C but differs due to the large excess of rutile (≈80 wt%). The phase assemblage consists of hollandite, zirconolite, perovskite, pyrochlore, and alloy (totaling ~20%). In opposite to Synroc C, the phases were encapsulated in a continuous matrix of fully densified rutile and utilized the principle of "synthetic rutile microencapsulation" in its microstructure [60]. Therefore, the ceramic was dense and composed of fine grains in the range 0.05 to 1.5 µm, where perovskites appeared as the largest crystals. Pseudo-symmetric twinning in the large perovskite crystals yielded domains related to each other by either a 180° or a 90° rotation about an axis perpendicular to {101} [61].

Synroc F was developed for the disposal of unprocessed spent fuel from light water [62] and Candu reactors [63]. It was intended for an incorporation of 50 wt% spent nuclear fuel. It was fabricated using the same technology currently utilized for Synroc C [64]. Synroc F is an assemblage of three crystalline phases: 90 wt% pyrochlore-structured $CaUTi_2O_7$, 5 wt% Ba-hollandite-type phase, $Ba_{1.14}Al_{2.27}Ti_{5.71}O_{16}$, and 5 wt% rutile TiO_2. Different samples, which where sintered between 1250 and 1400°C for 3 to 8 hours under reducing atmospheres, contained uranium-bearing perovskites with the chemical composition $(Ca,U)TiO_3$. Based on EPMA, SEM/EDX, and TEM/EDX experiments, $(Ca,U)TiO_3$ crystallized due to the substitution $2 Ca^{2+} \Leftrightarrow U^{4+}$ + vacancy or $Ca^{2+} + 2Ti^{4+} \Leftrightarrow U^{4+} + 2Ti^{3+}$ [63]. Alternative formulations for Synroc F were 85 wt% of pyrochlore-structured $CaUTi_2O_7$ + 5 wt% each of perovskite, hollandite, and rutile. The perovskite content is quite often reduced to 0 wt%, although 3 wt% of fission products needs the presence of perovskite for the fixation of strontium. Strontium also can be incorporated in the Ba-hollandite-type phase, $Ba_{1.14}Al_{2.27}Ti_{5.71}O_{16}$, and at higher Sr concentrations, minor $SrTiO_3$ contents (Fig. 2.1) can crystallize with available TiO_2.

10.3 Crystal chemistry of perovskite-type structures suitable for the fixation of radwaste

10.3.1 Perovskite minerals and their technical analogues

The investigations of such perovskites, crystallized during the hot-pressing process of Synroc-radwaste blends, yielded modified ternary perovskites due to the incorporation of fission products Sr, Na, REE, and actinides in the crystal structure of $CaTiO_3$. Aside from those technical perovskites, perovskite minerals with close chemical compositions exist and were investigated comprehensively. Now, their properties can be used to understand the crystallization paths, element distribution patterns, crystallography, and structural variations (polymorphs and polytypes) of "synroc perovskites". In case of inconsistent data or doubtful results, the synthesis and consequent analyses of pure phases under standardized and reproducible laboratory conditions are another tool to transfer the knowledge gained to analyze and investigate perovskite minerals and "synroc perovskites". In the following sections, important perovskite minerals in connection with the synroc process together with synthesized analogues are presented in order to give a brief overview on the complex crystal chemistry of those different perovskites phases.

10.3.1.1 Perovskite $CaTiO_3$ (perovskite-lueshite group 4.CC.30) [65]

The mineral $CaTiO_3$ was named originally as perovskite according to the mineralogist Lew Alexejewitsch Perowski (1792–1856). The mineralogist Gustav Rose discovered the mineral in a rock drift from Achmatovsk (Ural) [66], describing its cleavage, hardness, and crystal system.

Since its discovery, many scientific investigations concerning symmetry and lattice parameters were determined [67–69]. Kay and Baily [70] mentioned the close relation between artificial and natural $CaTiO_3$, possessing the same orthorhombic crystal system and space group $Pcmn$, which was determined by single crystal structure analysis. Further structural analyses were performed using neutron and single crystal data [71, 72]. Electron difference density measurements were performed 9 years later in order to determine the structural parameters of $CaTiO_3$ [73]. The first high-temperature powder X-ray diffraction (PXRD) analysis was performed in the range of 298–1373 K. They revealed a possible phase transition close to 1600 K [74]. The first structure analysis using natural single crystals with poor twinning from the Benitoite Gem Mine (San Benito) were conducted by Beran et al. [75]. Reversible phase transitions and octahedral tilting of $CaTiO_3$ were published by Yashima et al. [76]. The initial phase transition $Pbnm$-$I4/mcm$ with $a^-a^-c^+$ to $a^0a^0c^-$ occurred at 1512±13 K and the second phase transition $I4/mcm$-$Pm\bar{3}m$ was observed at 1635±2 K. The tilt system changed from $a^0a^0c^-$ to $a^0a^0a^0$.

Meanwhile, the term "perovskite" is used generally for compounds, which belong to the family of perovskites, for instance, for technical important halides [77], sulfides, and complex oxides and hydrides [78]. A very detailed overview of the structural variability of perovskites was provided by Mitchell in his book *Perovskites: Modern and Ancient* [15]. He summarizes the structure-related properties and peculiarities of natural and artificial perovskites. Recently, an up-to-date compilation of perovskites, titled *Perovskites: Structure-Property Relationships*, by R. J. D. Tilley, was published [79]. Following the idea of a so-called perovskite space [80], natural perovskites cover the ratio oxygen/A-cation = 3 (ABO_3) and are plotted along the joint ABO_3–$ABO_{3.5}$ until the oxygen/A-cation ratio = 3.5 ($ABO_{3.5} \equiv A_2B_2O_7$) is reached. For instance, the complete substitution of Nb_2O_5 for TiO_2 yields the orthorhombic compound $Ca_2Nb_2O_7$ with perovskite-type slabs [81]. The substitution Fe_2O_3 for $2TiO_2$ in opposite direction stabilizes oxygen-frustrated perovskites (oxygen/A-cation ratio = 3–2.5). The complete substitution of $2TiO_2$ by Fe_2O_3 yields the end-member of the brownmillerite solid solution series $Ca_2Fe_2O_5$ (srebrodolskite) [82]. Generally, natural oxide perovskites show specific chemical compositions. The A-site is usually occupied by K^+, Na^+, Ca^{2+}, Sr^{2+}, REE^{3+}, Pb^{2+}, and Ba^{2+}, and the B-site is occupied by Ti^{4+}, Nb^{5+}, $Fe^{2+/3+}$, Ta^{5+}, Th^{4+}, and Zr^{4+}. The structural variety, together with the huge number of A/B-site cation combinations, enables the formation of complex solid solutions and end-members. Most of the solid solution series of natural perovskites can be expressed by seven end-members, which put up an improved perovskite space system. The chemical compositions of the majority of natural perovskites are plotted in the ternary system perovskite ($CaTiO_3$)-lueshite ($NaNbO_3$)-loparite ($NaCeTi_2O_6$)-tausonite ($SrTiO_3$). Three different ternary systems and one quaternary system are applied sufficiently for the classification of perovskites in alkaline rocks [15]:

1. Nb-poor, Ca–, Sr–, and REE-rich perovskites: $CaTiO_3$–$NaCeTi_2O_6$–$SrTiO_3$
2. Sr-poor, Nb–, Ca–, and REE-rich perovskites: $CaTiO_3$–$NaCeTi_2O_6$–$NaNbO_3$
3. Ca-poor, Na–, REE–, Nb–, and Sr-rich perovskites: $NaNbO_3$–$NaCeTi_2O_6$–$SrTiO_3$
4. Sr-poor, Na–, Ca–, and Nb-rich perovskites: $NaNbO_3$–Ca_2FeNbO_6–$CaTiO_3$–$Ca_2Nb_2O_7$

10.3.1.2 Tausonite $SrTiO_3$ (loparite-macedonite group 4.CC.35) [65]

The mineral tausonite [65] was dedicated to L. V. Tauson (1917–1989). It was first described in the Little Murun potassic alkaline complex [15, 83, 84]. Tausonite shows an extensive compositional variation with Ce_2O_3 (0.1–9.8 wt%), CaO (1.5–5.5 wt%), Nb_2O_5 (0–1.28 wt%), ThO_2 (0–2.7 wt%), and BaO (0.3–1.2 wt%) in opposition to the general formula $SrTiO_3$ [83]. Furthermore, zoning and resorption features were determined for tausonite crystals from the Little Murun complex [83].

Tausonite was also described at Sarambi and Chiriguelo, Paraguay, and at Salitre I (Brazil), as an accessory mineral in rheomorphic sanidine-aegirine-nepheline dikes [85]. The specimen from Sarambi is enriched in SrO, REE, Na_2O, and TiO_2. The specimen from Salitre I contains lower SrO contents but higher $Na_2O + Nb_2O_5$ concentrations, in comparison with the specimen from Sarambi. Approximately 50% of the solid solution series between tausonite and loparite is present [85]. Tausonite (Fig. 10.1) exhibits cubic symmetry with sg. $Pm\bar{3}m$ and $a ≈ 3.8$ Å. Tausonite is the only cubic natural perovskite, whose crystal system was predicted with the help of the so-called tolerance factor (tausonite t ≈ 1.00) [86].

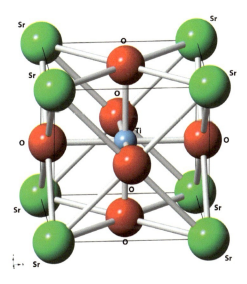

Fig. 10.1: Crystal structure of cubic $SrTiO_3$ with space group $Pm\bar{3}m$ (221) [116] ICSD code: 91899 [86].

10.3.1.3 Lueshite NaNbO₃ (perovskite-lueshite group 4.CC.30) [87]

Safiannikoff [88] described the mineral lueshite for the first time. It was named according to the Lueshe carbonatite complex, North Kivu (Kongo). At the same time, a mineral with a comparable chemical composition was detected in the Ilimaussaq complex SW (Greenland, Denmark) [89]. They called it "igdloite", but this term was changed in favor of lueshite [90]. Furthermore, lueshite was identified from several carbonatite complexes in the Kola Peninsula. The results published in Russian literature were compiled by Chakhmouradian and Mitchell [91, 92]. Further localities, where lueshite was detected, are agpaitic rocks from Mont Saint-Hilaire (Quebec, Kanada) [93] and Lovozero (Russia) [94].

Between the 1950 and 1960, different attempts were made to obtain PXRD data in order to determine the lattice parameters $a = 5.51$ Å, $b = 5.53$, and $c = 15.50$ Å and the space group $P222_1$ of lueshite [83, 95]. Further PXRD studies at room temperature and also at non-ambient conditions were performed in 1961. Solovev et al. [96] proposed at room temperature a monoclinic cell for an artificial phase with $a_{pc} = 3.9092$ Å, $b_{pc} = 3.8713$ Å, $c_{pc} = 3.9092$ Å, $β = 90.75°$, and $P2/m$ (pc = pseudo-cubic). In the range of 450–470°C a so-called pseudo-cubic cell was observed, between 530–630°C a tetragonal cell and higher than 650°C cubic $NaNbO_3$ was detected. According to

Sakowski-Cowley et al. [97] artificial distorted perovskite $NaNbO_3$ is orthorhombic with lattice parameters a = 5.566 Å, b = 15.520 Å, c = 5.506 Å, and space group *Pnma*. Niobium occupies the center position of the octahedron, and it is displaced from the geometrical center by 0.13 Å in a direction lying almost exactly in the plane of the O(3), O(4) square. The Nb-octahedron is quite regular. The edge-sharing octahedra are rotated 8° about [010] and tilted 9.5° about [100]. Sodium is placed on two different positions, Na(1) and Na(2), with a 9- and 8-fold coordination, respectively. A comprehensive study concerning the complex sequence of phase transitions of $NaNbO_3$ was subsequently carried out. Six phases were observed: phase P *Pnma* [97] tilt systems $a^-b^+c^-$, $a^-b^-c^-$ (room temperature to 373°C), phase R tilt systems $a^-b^+c^+$; $a^-b^0c^+$ (373–480°C), phase S [98] tilt system $a^-b^+c^+$ *Pnmm* (480–520°C), phase T_1 *Ccmm* [98] tilt system $a^-b^0c^+$ (520–575°C), phase T_2 *P4/mbm* [99] (Fig. 10.2) tilt system $a^0a^0c^+$ (575–641°C) [99], and U cubic $a^0a^0a^0$ [16, 17] (>641°C). Between 753 and 793 K, new $NaNbO_3$ structures were detected, which have not been reported in literature so far [100]. In order to determine the crystallography in the range of 12 to 350 K, neutron diffraction studies were conducted. Rietveld analysis of the diffraction data yielded that the antiferroelectric to ferroelectric phase transition occurs on cooling at 73 K and the reversible process starts at 245 K. The phase transition was not complete until 12 K was achieved. Before 12 K, a ferroelectric phase with *R3c* coexisted with an antiferroelectric phase (*Pbcm*) [101, 102].

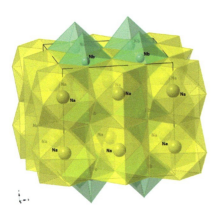

Fig. 10.2: Crystal structure of tetragonal lueshite with space group *P4/mbm* (127) [98] ICSD code: 23563 [86].

The high-temperature phases of $NaNbO_3$ were re-examined using the high-resolution powder diffractometer HRPD at the ISIS neutron spallation source. A new study demonstrated the diversity of nonpolar antiferrodistortive to ferroelectric and antiferroelectric phase transitions. Furthermore, the phase transition from orthorhombic P to R phase is sole first order. All others are second-order phase transitions. A new phase diagram of $NaNbO_3$ with adapted phase transition temperatures was presented. Between 633 and 680 K, two modifications exist and the space groups for phase T1 were changed to *Cmcm*, phase S to *Pbnm*, and phase R to *Pbnm*, too.

Furthermore, trigonal NaNbO$_3$ with space group $R3c$ is present exclusively below 10 K [103]. Further combined PXRD and neutron diffraction studies together with a symmetry mode analysis on phases R and S of NaNbO$_3$ should help complete the phase system of NaNbO$_3$. Phase S stable in the range of 480–510°C has a 2 × 2 × 4 super lattice of the cubic structure with space group $Pmmn$. Phase R comprises two possible structural models in the temperature range of 370–470°C. Both models have a 2 × 2 × 6 super lattice and can be distinguished by their complex tilt system along the "long" axis [104].

After solving satisfactorily the phase relations for artificial NaNbO$_3$, four different lueshite minerals from different occurrences plus an artificial NaNbO$_3$ were investigated [105]. This paper demonstrates impressively the differences concerning structure types and phase transitions. PXRD and neutron diffraction experiments yielded the orthorhombic structure with $Pbnm$, opposite to the artificial NaNbO$_3$, which adopted space group $Pbcm$ at room temperature. However, TOF-neutron data could not be refined satisfactorily applying $Pbnm$. Another orthorhombic/monoclinic space group or the existence of domains of differing crystal structures can be responsible for this condition. The evaluation of high-temperature TOF neutron diffraction studies yielded the sequence of phase transition $Cmcm \Leftrightarrow P4/mbm \Leftrightarrow Pm\bar{3}m$ above 500°C, which is identical to NaTaO$_3$ [106]. Because NaTaO$_3$ has an orthorhombic cell with $Pbnm$ at room temperature, it is assumed that lueshite behaves similarly [106].

10.3.1.4 Loparite (Na,Ce,Ca)$_2$(Ti,Nb)$_2$O$_6$ (loparite-macedonite group 4.CC.35) [65]

The name "lopar" is related with the native people (laps) at the Kola peninsula [107]. Ramsay and Hackman [108] described the mineral loparite for the first time in the Lovozero peralkaline nepheline syenite complex (Kola Peninsula, Russia). The first full description of loparite from the Khibina alkaline complex has been provided by Kutsnezov [109]. Despite the early discovery of loparite, its structure was only inadequately determined, since the mineral is twinned complexly or intergrown with other minerals [110, 111].

Different structural models were proposed for the complex perovskite with a strong variability of cations in the A- and B-site. The first crystal structure determination of loparite was performed in 1930, applying a single crystal with the chemical composition (Na$_{0.5}$Ce$_{0.3}$Ca$_{0.2}$)(Ti$_{0.8}$Nb$_{0.2}$)O$_3$, a = 3.854(18) Å, and space group $Pm\bar{3}m$. It adopts the aristotype of perovskites owing the tilt system a°a°a° [112, 113]. The chemical compositions and crystal structures of three different loparite-(Ce) samples, niobian calcian loparite-(Ce) Khibina complex (Russia) named "loparite K", calcian niobian loparite-(Ce) from the Lovozero complex (Russia) named "loparite L", and strontian calcian loparite-(Ce) from the Bearpaw Mountains (Montana, USA) named "loparite BM", were examined [114].

Another structure determination of loparite-(Ce) using a single crystal from the Lovozero alkaline massif was carried out with an extremely complicated chemical composition $(Na_{0.475}Ca_{0.118}Ce_{0.19}La_{0.08}Nd_{0.04}Th_{0.01}Sr_{0.05}Pr_{0.01}Eu_{0.0075}Sm_{0.01})(Ti_{0.865}Nb_{0.135})O_3$ [115]. The structural proposal seemed to be implausible, due to different reasons as well as the choice of unusual space groups, which did not obey group-subgroup relations [116].

"Loparite K" $(Na_{0.682}Ce_{0.318})(Ti_{0.834}Nb_{0.166})O_3$ was found to be orthorhombic (*Pbnm*, 62), exhibiting tilt system $a^-b^+b^+$ [16–18], and adopting the $GdFeO_3$ perovskite-type structure (Fig. 10.3) [116]. Lattice parameters are $a = 5.5108(14)$ Å, $b = 5.5084(14)$ Å, and $c = 7.7964(20)$ Å. "Loparite K" is composed of tilted and distorted $(Ti,Nb)O_6$-polyhedra. Na^+ and Ca^{2+} cations are placed in distorted 12-fold-coordinated sites $((Na,Ce,Ca)O_{12}$-polyhedra).

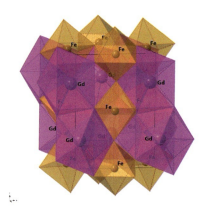

Fig. 10.3: Crystal structure of orthorhombic $GdFeO_3$ with space group *Pbnm* (62) [115] ICSD code: 27278 [86].

"Loparite L" $(Na_{0.632}Ce_{0.368})(Ti_{0.912}Nb_{0.088})O_3$ and "loparite BM" $(Na_{0.536}Ce_{0.209}Sr_{0.255})(Ti_{0.944}Nb_{0.056})O_3$ are both tetragonal with space group *I4/mcm* and the tilt scheme $a^0a^0c^-$, which allows just a rotation with positive sense around [001]. The high strontium concentration in "loparite BM" is responsible for the lower degree of rotation and distortion, quite close to ideal conditions of the cubic aristotype. Therefore, bond lengths of the B-site octahedron are minor distorted (0.46%) and the A-site (AO_{12}-polyhedron) also is less distorted than in "loparite K". The polyhedra volume ratio V_A/V_B, ideally $V_A/V_B = 5$ for cubic perovskite like tausonite, was calculated from structural data as 4.97. Mitchell et al. [113] have drawn the conclusion that loparite is a mineral species that needs to be redefined because of different space groups driven by the chemical compositions of the solid solution members. The influence of variable high strontium contents in loparite was investigated for the tausonite-loparite series $(Sr_{1-2x}Na_xLa_x)TiO_3$ [117].

The Rietveld analyses of PXRD data derived from $(Sr_{1-2x}Na_xLa_x)TiO_3$ synthetic powders yielded in the range $0 \le x \le 0.1$ non-distorted cubic phases with $Pm\bar{3}m$ and $a = 3.905–3.902$ Å. A phase in the range $0.15 \le x \le 0.35$ crystallized tetragonal with space group *I4/mcm*, $a \sim \sqrt{2}a_{pc}$, $c \sim 2a_{pc}$. The progressive substitution of $Sr_{Sr}^x \Leftrightarrow Na_{Sr}' + La_{Sr}^{\cdot}$ in the range $x = 0.35$ and 0.40 is responsible for the stabilization of a trigonal perovskite phase with $R\bar{3}c$ ($a \sim \sqrt{2}a_{pc}$, $c \sim 2\sqrt{3}a_{pc}$). The formation of an intermediate orthorhombic phase with the space group $Imma$ was not detected [118, 119]. Strontian-loparite was also found at Sarambi and Chiriguelo, Paraguay, and at Salitre I, Brazil, as an accessory mineral in rheomorphic sanidine-aegirine-nepheline dikes [85].

Not only the substitutions presented in the three different ternary systems for loparite were detected; a Th-rich loparite from the Khibina alkaline complex, Kola Peninsula, was found and analyzed [120]. Thorian loparite belongs undoubtedly to the loparite (NaLREETi$_2$O$_6$)-lueshite (NaNbO$_3$)–ThTi$_2$O$_6$–ThNb$_4$O$_{12}$ quaternary system. The structure of a Th-rich loparite was solved ($P4_2/mnm$, $a = 4.5936(6)$ Å, $c = 2.9582(6)$ Å). Na, La, and Th occupy site $4b$ $(0,1/2,1/4)$, which is coordinated by 12 oxygen atoms O1 and O2. Ti is placed solely in the B-site $4c$ $(0,0,0)$ [121].

For the scheduled investigations in the system (NaNbO$_3$)$_{1-x}$(NaLnTi$_2$O$_6$)$_x$ ($0 < x < 1$), the crystal structures of the intermediate members (NaNbO$_3$)$_{0.5}$(NaLnTi$_2$O$_6$)$_{0.5}$ with Ln = La, Pr, Nd, Sm, Eu, Gd, Th, Dy, Ho, Er, Tm were synthesized and their crystal structures were solved in advance, applying Rietveld analyses. The orthorhombic compounds Na$_{0.75}$Ln$_{0.25}$Ti$_{0.5}$Nb$_{0.5}$O$_3$ with Ln = Pr, Nd, Sm, Eu, Gd, Th, Dy, Ho, Er, Tm adopt the GdFeO$_3$ perovskite-type structure with space group $Pbnm$ (Fig. 10.3) and the tilt sequence $a^-b^+b^+$. These finding are close to the characteristics of "loparite K" (Na$_{0.682}$Ce$_{0.318}$)(Ti$_{0.834}$Nb$_{0.166}$)O$_3$. The unit cell parameters are expressed in terms of the pseudo-cubic lattice parameter a_{pc} $a(\sim\sqrt{2}a_{pc}) < b(\sim\sqrt{2}a_{pc}) < c(\sim 2a_{pc})$ and $a < c/\sqrt{2} < b$. The cell volumes of the compounds decrease regularly with increasing atomic number of the Ln cations.

Na and Ln occupy exclusively the eight fold coordinated A-site of the structure, so that the mean A-cation ion radii (R_ALn + R_ANa) decreases from 1.175 to 1.134 Å. The eight fold coordination must be expressed as a four-fold anti-prism with eight short A-O(1, 2) bonds in the first (I) coordination sphere of the A-cation. Four longer A-O(2) bonds give rise to the second (II) coordination sphere. The occupation of the six fold coordinated B-site is 0.5 Ti + 0.5 Nb in all Na$_{0.75}$Ln$_{0.25}$Ti$_{0.5}$Nb$_{0.5}$O$_3$ compounds, so that (R_BTi + R_BNb) has a constant value of 0.623 Å. Both A- and B-site polyhedra show a certain distortion indicated by their bond lengths and bond angle variances.

Na$_{0.75}$La$_{0.25}$Ti$_{0.5}$Nb$_{0.5}$O$_3$ also crystallized in the orthorhombic system, but adopted the SrZrO$_3$ type-structure with space group $Cmcm$. The lattice parameters are $a(\sim 2a_{pc}) = 7.7827(2)$ Å, $b(\sim 2a_{pc}) = 7.78641(2)$, $c(\sim 2a_{pc}) = 7.8039(1)$ Å. The structure contains corner shared (Ti,Nb)O$_6$ octahedra, which are tilted about cubic [010] and [001] according to the notation $a^0b^-c^+$ and the A-site polyhedra fill the empty space between the BO$_6$ octahedra. In the Na$_{0.75}$La$_{0.25}$Ti$_{0.5}$Nb$_{0.5}$O$_3$ structure, two A-sites, A1 and A2, occupied by sodium and lanthanum are 12-fold coordinated. Both polyhedra AO$_{12}$ and BO$_6$ show a slight distortion.

The authors of this article have the opinion that the compounds are stable at ambient temperatures and could be used for the fixation of actinide fission products such as ^{147}Pm, ^{151}Sm, ^{154}Eu of radioactive waste. Even Pu^{3+} might be fixed in the crystal structures of Na$_{0.75}$Ln$_{0.25}$Ti$_{0.5}$Nb$_{0.5}$O$_3$ phases [122].

10.3.1.5 Isolueshite (Na,La,Ce)(Nb,Ti)O$_3$ (loparite-macedonite group 4.CC.35) [65]

According to Chakhmouradian et al. [123] isolueshite crystallizes in hydrothermally altered pegmatite veins, found at the Khibina alkaline complex. Those phases can be described as intermediate solid solutions in the binary system NaNbO$_3$–NaLaTi$_2$O$_6$. Crystals are typically zoned with regard to Nb$_2$O$_5$, ThO$_2$ and REE. The La/Ce ratio and the Nb$_2$O$_5$ concentration decrease from the core to the edge and becomes enriched in TiO$_2$ and LREE.

Different type-structures of isolueshite were described by Krivovichev et al. [124]. The structure of the natural specimen was refined on the basis of an ideal perovskite lattice with cubic symmetry. In the cubic structure with space group $Pm\bar{3}m$ and a = 3.909(1) Å, Na, La, and Ca occupy the A-site $1a$, the B-site is occupied by Nb and Ti ($1b$), and O was placed on the Wyckoff position $3c$. The large thermal parameter indicates either a lower symmetry or dislocation of O from its position. The problem was solved by placing O in position 12h (x,0,1/2); thus, O could be moved along the x coordinate. The calculated structural formula (Na$_{0.75}$La$_{0.19}$Ca$_{0.06}$)(Nb$_{0.5}$Ti$_{0.5}$)O$_3$ almost matches the empirical formula of isolueshite. Moreover, the complexity of isolueshite is not only due to the complex cationic substitutions at the A- and B-sites, iso-lueshite contains certain OH-concentrations, which were detected by IR-spectroscopy. The stability of the cubic lattice for isolueshite was supposed to be a result of the fixation of LREE and Ti in the A- and B-sites. For this reason, a phase with the chemical composition (Na$_{0.75}$La$_{0.25}$)(Nb$_{0.5}$Ti$_{0.5}$)O$_3$ was synthesized, applying oxides La$_2$O$_3$, Nb$_2$O$_5$, and TiO$_2$ plus Na$_2$CO$_3$. Stoichiometric concentrations were mixed, ground, and sintered initially for 24 hours at 1000°C and after regrinding for 48 hours at 1200°C in air. The sample was rapidly cooled in air to room temperature. The powder was analyzed by PXRD, and the structure was refined by applying the Rietveld method. In opposition to the cubic symmetry of the natural isolueshite, the synthetic compound is orthorhombic. Several structural models of monoclinic symmetry ($I2/m$), orthorhombic symmetry ($Cmcm$ [Fig. 10.4] and $Imma$), and tetragonal symmetry (sg. $I4/mcm$) were tried. The fit of the powder diffraction pattern with the various structural models gives reasonable results, but on the basis of the R values and the isotropic thermal parameters of the oxygen atoms, the structural model with space group $Cmcm$ was chosen.

A structural model with the most abundant space group $Pnma$ for orthorhombic perovskite used for the Rietveld refining process did not converge. The lattice parameters determined for (Na$_{0.75}$La$_{0.25}$)(Nb$_{0.5}$Ti$_{0.5}$)O$_3$ are a = 7.7842(6) Å, b = 7.8033(2) Å, and c = 7.7828(6). The lattice parameters a, b, and c are approximately twice the value of the pseudo-cubic cell a_{pc}. Na and La occupy statistically both A-sites A1 $4c$ (0.002,1/4,3/4) and A2 $4c$ (0.499,1/4,3/4) in which Na and La are coordinated by

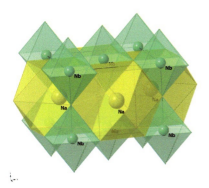

Fig. 10.4: Crystal structure of orthorhombic synthetic isolueshite with space group *Cmcm* (63) [123] ICSD code: 90041 [86].

10 oxygen O1, O2 and O3. Nb and Ti are placed on the B-site $8d$ (1/4,1/4,0), the center position of a distorted octahedra with bond lengths $2 \times$ B-O1 = 1.970 Å, $2 \times$ B-O2 = 1.952 Å, and $2 \times$ B-O3 = 1.984 Å. Oxygen ions occupy three different special Wyckoff positions, $8e$, $8f$, and $8g$.

10.3.1.6 Latrappite (Ca,Na)(Nb,Ti,Fe)O$_3$ (perovskite-lueshite group 4.CC.30) [65]

Latrappite is a member of the quarternary solid solution system lueshite (NaNbO$_3$)-latrappite (Ca,Na)(Nb,Ti,Fe)O$_3$-perovskite (CaTiO$_3$)-Ca$_2$Nb$_2$O$_7$. It is named according to the La Trappe monastery (La Trappe, Quebec) half mile from the deposit where the mineral was discovered [125]. Pseudo-cubic latrappite crystals were also found at the Kaiserstuhl (Upper Rhine Valley, Germany) [126]. For the investigated compound [(Ca$_{0.55}$Na$_{0.34}$Fe$_{0.01}$(REE,Sr,K,Mn)$_{0.02}$] [Nb$_{0.59}$Ti$_{0.29}$Fe$_{0.1}$Zr$_{0.01}$(Ta,Si,Al,V)$_{0.01}$]O$_3$(OH)$_{0.02}$, an orthorhombic cell and lattice parameters a = 5.46(1) Å, b = 15.6(15) Å, and c = 5.49(1) Å were determined. According to Nickel [125], page 122 latrappite is defined as follows: "A mineral with the perovskite structure and a composition corresponding to the general formula ABO$_3$, in which the 'A' and 'B' cation sites are occupied predominantly (in excess of 50 mol%) by calcium and niobium, respectively. Therefore, he suggested that the mineral names 'knopite' and 'dysanalyte' are unnecessary and should be named as latrappite (accept by International Mineralogical Association (IMA) nomenclature committee). The name niobian perovskite is to be retained for those niobium-bearing perovskites in which titanium predominates". The chemical composition is quite complicated (Ca$_{0.75}$Na$_{0.21}$REE$_{0.02}$)(Nb$_{0.54}$Ti$_{0.21}$Fe$_{0.16}$Mg$_{0.08}$Mn$_{0.02}$)$_{1.01}$O$_3$ but resembles the chemical composition of the specimen from the Kaiserstuhl. The unit cell dimensions are a = 5.448 Å, b = 7.777 Å, and c = 5.553 Å. For the reinvestigation of latrappite in 1998, different samples from the type localities at Oka (Quebec), Badloch Quarry, Kaiserstuhl complex (Germany), and Magnet Cove carbonatite complexes [127] were used. Electron microprobe analyses (EMPA) showed that latrappite from Oka predominantly contains CaO, Fe$_2$O$_3$, TiO$_2$, and Nb$_2$O$_5$, and most of the specimens investigated are true latrappite, as defined by Nickel [125]. The valence of iron was determined by ^{57}Fe Mössbauer spectroscopy, which yielded that iron is exclusively trivalent and placed in the coordinated octahedrally B-site [127, 128]. The crystal structure of latrappite from Oka, determined by Rietveld analysis, corresponds to the GdFeO$_3$ perovskite-type structure with sg. *Pbnm* (Fig. 10.3) and the lattice parameters $a = 5.4479(3)$ Å, $b = 5.5259(3)$ Å, and $c = 7.7575(5)$ Å. Compared to the perovskite structure of CaTiO$_3$, differences occur in longer Me-O bond and greater distortion angles due to the fixation of the Nb^{5+} and Fe^{3+} ions in the crystal structure of latrappite.

10.3.2 Artificial ternary actinide perovskites

10.3.2.1 Ternary A^{+II}B^{+IV}O$_3$ perovskites with A = Ba, Sr, and B = Pu, Np, Am, Cm, U

BaPuO$_3$ was synthesized by a ball-milling process of stoichiometric concentrations PuO$_2$ and BaCO$_3$ for 44 hours with subsequent sintering at 1470 K in an argon atmosphere for 24 hours. The dark brown powder was analyzed through neutron and x-ray radiation. The neutron scattering experiments were performed at Los Alamos Neutron Powder Diffraction Center using a neutron powder diffractometer installed at a pulsed neutron source.

The structure of BaPuO$_3$ [129] with lattice parameters $a = 6.219\,(1)$ Å, $b = 6.19\,(1)$ Å, and $c = 8.744(1)$ Å; V = 336.8 Å^3; Z = 4 is isotypical to the distorted GdFeO$_3$-type structure (Fig. 10.3) [130] with space group *Pbnm*. The properties of the structure are comparable with the one of BaPrO$_3$ [131], which were used as initial parameters for the structure refinement.

BaPrO$_3$ structure is a build-up of corner-sharing octahedra with Pu as the central atom and Ba ions are coordinated by 12 oxygens. The Ba polyhedra share edges with the Pu octahedra. The Pu octahedra are only distorted slightly with bond lengths of 2 × 2.2306(5), 2 × 2.2295(12), and 2 × 2.2230(12) Å, and O-Pu-O angles at about 88–89°. The Ba-O distances are in the range of 2.695–3.532 Å, with a mean value of 3.113 Å. According to the tilt systems of coordination polyhedra in perovskite structures [17–21, 132], the notation a$^+$b$^-$b$^-$ is used for ternary ABX$_3$-perovskites with space group *Pnma* (62). The notation a$^+$b$^-$b$^-$ describes the tilting of the octahedron regarding the lattice parameters of the unit cell a, b, c.

Aside from the distorted orthorhombic phase, cubic BaPuO$_3$ was proposed and its crystal structure solved, too. At temperatures of 900°C, cubic BaPuO$_3$ with $a = 4.322(3)$ Å, and space group $Pm\bar{3}m$ was identified [132]. In a series of Pu containing phases, cubic BaPuO$_3$ with $a = 4.39$ Å was synthesized with some impurities of BaCO$_3$ and PuO$_2$ [133]. The same researchers tried to synthesize SrPuO$_3$ and CaPuO$_3$. They synthesized successfully pure SrPuO$_3$, but different approaches for the synthesis of CaPuO$_3$ failed.

SrPuO$_3$ was synthesized in order to determine its vaporization behavior, which is important concerning the stability of the so-called gray phase. The composition of this phase largely depends on the temperature gradient in the fuel, burnup and on the oxygen potential [134]. The "gray phase" contains different complex perovskites with the general composition (Ba, Sr)(U, Pu, Zr, Ln, Mo)O$_3$ in irradiated mixed oxide fuels (MOX) [135]. Different structures with orthorhombic and cubic [136] symmetry comparable with the findings for SrPuO$_3$ were determined.

The fixation of tetravalent species of Np, Am, Cm, and U on the B-site of ABX$_3$ perovskites with A = Sr or Ba were investigated and revealed different compounds with cubic or orthorhombic symmetry (Tab. 10.1).

Tab. 10.1: Lattice parameters of A^{+II}B^{+IV}O$_3$ perovskites.

	a (Å)	b (Å)	c (Å)	V (Å^3)	Space group	Reference
BaNpO$_3$	4.357			82.71	$Pm\bar{3}m$	[135]
SrNpO$_3$						
BaAmO$_3$	4.35(1)			82.31	$Pm\bar{3}m$	[135]
SrAmO$_3$						
BaCmO$_3$						
SrCmO$_3$						
BaUO$_3$	4.40					[135, 136]
SrUO$_3$	6.01	8.60	6.17			[135, 136]

10.3.2.2 Ternary A^{+III}AlO$_3$-Perovskites with A = Pu, Np, Am, Cm

The transmutation of minor actinides (MA: Np, Am, Cm) in fast reactors or accelerator driven systems is a possibility to reduce their radioactivity [138]. For a MA-matrices MgAl$_2$O$_4$ [139] was chosen as a host material, however, at sinter temperatures in the region of 1900 K, a new Am-bearing phase crystallized and a perovskite-like phase AmAlO$_3$ appeared during the sinter process in porous ceramic samples with 11% ^{241}Am [140].

AmAlO$_3$ was synthesized purely by sintering a coprecipitate of Am- and Al-hydroxides at 1250°C two times for 8 hours. Pale rose AmAlO$_3$ was obtained [141]. The phase crystallizes trigonal with space group $R\bar{3}m$ (Fig. 10.5).

In order to understand the behavior of Pu, generated by neutron capture reactions in UO$_2$ fuel different computational methods were applied. For PuAlO$_3$ the GdFeO$_3$ perovskite-type structure was predicted, however refined lattice parameters yielded a trigonal symmetry (Tab. 10.2) [142, 143] and similar ternary oxides of the rare earths LaAlO$_3$ [144], PrAlO$_3$ [145], and NdAlO$_3$ [145] also are trigonal. Further geometrical properties concerning trigonal ternary oxides of the rare earths [145] were discussed and investigated intensively [113, 146–149]. The predominant space groups in rhombohedral perovskites are $R3m$, $R3c$, and $R\bar{3}c$, respectively, with PrAlO$_3$ crystallizing in space group 166 ($R\bar{3}m$). The slight deviation of the rhombohedral alpha angle of 60° shows the strong similarity of the rhombohedral cell with a pseudo-cell derived from the cubic F cell (Tab. 10.2).

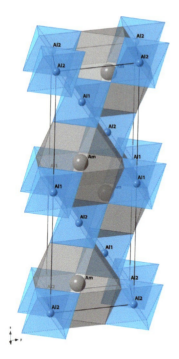

Fig. 10.5: Crystal structure of trigonal synthetic AmAlO$_3$ with space group $R\bar{3}m$ (160) [140].

Tab. 10.2: Lattice parameters of A^{+III}B^{+III}O$_3$ perovskites.

	Hexagonal cell		Rhombohedral cell		Cubic/pseudo-cubic cell		Reference
	a (Å)	c (Å)	a (Å)	α (°)	a·√2 (Å)	α (°)	
PuAlO$_3$	5.367	13.43	5.33	56.07°	3.78	90.4°	[140, 141]
AmAlO$_3$	5.336(5)	13.12(2)	5.36	60.1°	3.79	90.08°	[140, 141]
CmAlO$_3$ >500°C					3.790	90	[143]
CmAlO$_3$ <500°C					3.779	90.13°	[143]
CmAlO$_3$ >900°C 4% H$_2$ + 96% He					3.756	90	[143]

10.4 Discussion and summary

Synroc ceramics (perovskite + zirconolite + pyrochlore) were homogenized with variable concentrations and different calcined radwastes. During the hot-pressing process, new phases crystallized. Aside from the alteration of zirconolite and hollandite, the chemical composition of the original phase in synroc CaTiO$_3$ was affected strongly. ^{90}Sr, a key radwaste element, is substituted for calcium in synroc CaTiO$_3$. Natural analogues are tausonite-perovskite solid solutions. The incorporation of ^{90}Sr, REE

(Gd^{3+}, Y^{3+}), actinides, and Na^+ yielded different perovskite phases, which can be compared with tausonite-lopparite solid solutions found at the Little Murun complex or with perovskites $(Na,REE)Ti_2O_6$. Natural analogues of perovskites containing actinides do not exist because the relevant actinides are artificial. However, these ternary actinide perovskite phases (Tabs. 1 and 2) were synthesized and investigated.

It is true for "synroc perovskite" that "These minerals are closely related to natural minerals known to have been stable in a wide range of geological and geochemical environments for periods of more than 10^9 years" [41], page 154. The diversity of stable crystal perovskite-type structures is large, but in case of hot-pressed synroc ceramics, polymorphs of ternary perovskites with different A- and B-site combinations are sufficient to incorporate many important waste ions. Since 1994, ANSTO began working with the US Lawrence Livermore National Laboratory to develop a synroc variant for plutonium disposition, which resulted in the development of a new pyrochlore-rich ceramic [150]. Recently, ANSTO plans to use synroc to immobilize the ILW liquid waste resulting from molybdenum-99 production, where perovskite phases play a minor role for the incorporation of sodium and uranium [151, 152].

References

[1] Nuraje N, Su K. Perovskite ferroelectric nanomaterials. Nanoscale 2013;5:8752–80.
[2] Cohen RE. Origin of ferroelectricity in perovskite oxides. Nature 1992;358:136–8.
[3] Hill NA. Why are there so few magnetic ferroelectrics ? J Phys Chem B 2000;104:6694–709.
[4] Lin Q, Armin A, Nagiri RCR, Burn PL, Meredith P. Electro-optics of perovskite solar cells. Nat Photon 2015;9:106–12.
[5] Almond DP, Bowen CR. An explanation of the photoinduced giant dielectric constant of lead halide perovskite solar cells. J Phys Chem Letters 2015;6:1736–40.
[6] Liao W-Q, Zhang Y, Hu C-L, Mao J-G, Ye H-Y, Li P-F, Huang SD, Xiong R-G. A lead-halide perovskite molecular ferroelectric semiconductor. Nat Commun 2015;6:7338.
[7] Shi C, Meidong L, Churong L, Yike Z, Da Costa J. Investigation of crystallographic and pyroelectric properties of lead-based perovskite-type structure ferroelectric thin films. Thin Solid Films 2000;375:288–91.
[8] Uchino K. Glory of piezoelectric perovskites. Sci Technol Adv Mater 2015;16:046001.
[9] López-Suárez FE, Bueno-López A, Illán-Gómez MJ, Trawczynski J. Potassium-copper perovskite catalysts for mild temperature diesel soot combustion. Appl Catal A 2014;485:214–21.
[10] Richter J, Holtappels P, Graule T, Nakamura T, Gauckler LJ. Materials design for perovskite SOFC cathodes. Monatsh Chem 2009;140:985–99.
[11] Lubini M, Chinarro E, Moreno B, de Sousa VC, Alves AK, Bergmann CP. Electrical properties of $La_{0.6}Sr_{0.4}Co_{1-y}Fe_yO_3$ ($y = 0.2$–1.0) fibers obtained by electrospinning. J Phys Chem C 2016;120:64–9.
[12] Wu Y, Cordier C, Berrier E, Nuns N, Dujardin C, Granger P. Surface reconstructions of $LaCo_{1-x}Fe_xO_3$ at high temperature during N_2O decomposition in realistic exhaust gas composition: impact on the catalytic properties. Appl Catal B 2013;140–1:151–63.
[13] Nie W, Tsai H, Asadpour R, Blancon J-C, Neukirch AJ, Gupta G, Crochet JJ, Chhowalla M, Tretiak S, Alam MA, Wang H-L, Mohite AD. High-efficiency solution-processed perovskite solar cells with millimeter-scale grains. Science 2015;347:522.

[14] Ferraris G, Makovicky E, Merlino S. Application of modularity to structure description and modelling. In: Ferraris G, Makovicky E, Merlino S, editors. Crystallography of modular materials. Oxford Scholarship Online; 2008:227–4.
[15] Mitchell RH. Perovskites: modern and ancient. Thunder Bay (ONT): Almaz Press; 2002.
[16] Glazer A. The classification of tilted octahedra in perovskites. Acta Crystallogr B 1972;28:3384–92.
[17] Glazer A. Simple ways of determining perovskite structures. Acta Crystallogr A 1975;31:756–62.
[18] Howard CJ, Stokes HT. Group-theoretical analysis of octahedral tilting in perovskites. Acta Crystallogr B 1998;54:782–9.
[19] Woodward P. Octahedral tilting in perovskites. I. Geometrical considerations. Acta Crystallogr B 1997;53:32–43.
[20] Woodward P. Octahedral tilting in perovskites. II. Structure stabilizing forces. Acta Crystallogr B 1997;53:44–66.
[21] Woodward P. POTATO- a program for generating perovskite structures distorted by tilting of rigid octahedra. J Appl Crystallogr 1997;30,:
[22] Shpanchenko RV, Nistor L, Van Tendeloo G, Van Landuyt J, Amelinckx S, Abakumov AM, Antipov EV, Kovba LM. Structural Studies on New Ternary Oxides $Ba_8Ta_4Ti_3O_{24}$ and $Ba_{10}Ta_{7.04}Ti_{1.2}O_{30}$. J Solid State Chem 1995;114:560–74.
[23] Battle D, Green MA, Laskey NS, Millburn JE, Murphy L, Rosseinsky MJ, Sullivan SP, Vente JF. Layered Ruddlesden-Popper manganese oxides: synthesis and cation ordering. Chem Mater 1997;9:552–9.
[24] Stoumpos CC, Cao DH, Clark DJ, Young J, Rondinelli JM, Jang JI, Hupp JT, Kanatzidis MG. Ruddlesden-Popper hybrid lead iodide perovskite 2D homologous semiconductors. Chem Mater 2016;28:2852–67.
[25] Kendall KR, Navas C, Thomas JK, zur Loye H-C. Recent developments in oxide ion conductors: Aurivillius phases. Chem Mater 1996;8:642–9.
[26] Hojamberdiev M, Bekheet MF, Zahedi E, Wagata H, Kamei Y, Yubuta K, Gurlo A, Matsushita N, Domen K, Teshima K. New Dion-Jacobson phase three-layer perovskite $CsBa_2Ta_3O_{10}$ and its conversion to nitrided $Ba_2Ta_3O_{10}$ nanosheets via a nitridation-protonation-intercalation-exfoliation route for water splitting. Crystal Growth Des 2016;16:2302–8.
[27] Levin I, Bendersky LA. Symmetry classification of the layered perovskite-derived $A_nB_nX_{3n+2}$ structures. Acta Crystallogr B 1999;55:853–66.
[28] Murakami M, Sinogeikin SV, Hellwig H, Bass JD, Li J. Sound velocity of $MgSiO_3$ perovskite to Mbar pressure. Earth Planet Sci Lett 2007;256:47–54.
[29] Murakami M, Hirose K, Kawamura K, Sata N, Ohishi Y. Post-perovskite phase transition in $MgSiO_3$. Science 2004;304:855.
[30] Buykx WJ, Hawkins K, Levins DM, Mitamura H, Smart RSC, Stevens GT, Watson KG, Weedon D, White TJ. Titanate ceramics for the immobilisation of sodium-bearing high-level nuclear waste. J Am Ceram Soc 1988;71:678–88.
[31] Hatch LP. Ultimate disposal of radioactive wastes. Am Sci 1953;41:410–21.
[32] Roy R. Nuclear waste: the Penn State connection. Earth Miner Sci 1980;50:13.
[33] McCarthy J G. High-level waste ceramics: materials considerations, process simulation, and product characterization. Nucl Technol 1977;32:92–105.
[34] Roy R. Rational molecular engineering of ceramic materials. J Am Ceram Soc 1977;60:350–63.
[35] McCarthy GJ. Crystalline and coated high-level forms. In: Conference on high-level radioactive solid waste forms, December 19–21,1978; held at Denver, Colorado. Washington (DC): Nuclear Regulatory Commission; 1979:623–50.
[36] McCarthy GJ, Davidson MT. Ceramic nuclear waste forms. I. Crystal chemistry and phase formation. Am Ceram Soc Bull 1975;54:782–6.
[37] McCarthy GJ, Davidson MT. Ceramic nuclear waste forms. II. A ceramic-waste composite prepared by hot pressing. Am Ceram Soc Bull 1976;55(2).

[38] Kennedy BJ, Hunter BA. High-temperature phases of $SrRuO_3$. Phys Rev B 1998;58:653–8.
[39] Vashook V, Nitsche D, Vasylechko L, Rebello J, Zosel J, Guth U. Solid state synthesis, structure and transport properties of compositions in the $CaRu_{1-x}Ti_xO_{3-\delta}$ system. J Alloys Compounds 2009;485:73–81.
[40] Ringwood AE, Kesson SE, Ware N, Hibberson W, Major A. The SYNROC process: a geochemical approach to nuclear waste immobilisation. Geochem J 1979;13:141–65.
[41] Ringwood AE. Safe disposal of high level nuclear wastes: a new strategy. Australia: Australian National University Press; 1978.
[42] Ringwood A E. Safe disposal of high-level radioactive wastes. Fortschr Miner 1980;58:149–68.
[43] Ringwood T. A rocky graveyard for nuclear waste. New Sci 1983;99:756–8.
[44] Ringwood AE, Oversby VM, Kesson SE, Sinclair W, Ware N, Hibberson W, Major A. Immobilisation of high-level nuclear reactor wastes in synroc: a current appraisal. Nucl Chem Waste Manage 1981;2:287–305.
[45] Pentinghaus H. To SYNROC through melting: thermal analysis, thermogravimetry and crystal chemical characterization of phases. In: Proceedings of the International Seminar on Chemical Process Engineering for High-Level Liquid Waste Solidification, June 1–5, 1981. Julichw: Kernforschungsanlage Jülich; 1981:713–32.
[46] Ringwood AE, Kesson SE, Ware N, Hibberson W, Major A. The SYNROC process: a geochemical approach to nuclear waste immobilisation. Geochem J 1979;13:141–65.
[47] Oversby VM, Ringwood AE. Leach testing of synroc and glass samples at 85 and 200°C. Nucl Chem Waste Manage 1981;2:201–6.
[48] Kesson SE, Ringwood AE. Immobilisation of sodium in synroc. Nucl Chem Waste Manage 1981;2:53–5.
[49] Mendel JE, Ross WA, Roberts FP, Yatayama YB, Westsik JH, Turcotte RP, Wall JW, Bradley DJ. Annual report on the characteristics of high-level waste glasses. BNWL-2252 1977, UC-70.
[50] Lumpkin GR, Smith KL, Blackford MG. Partitioning of uranium and rare earth elements in synroc: effect of impurities, metal additive, and waste loading. J Nuclear Mater 1995;224:31–42.
[51] Campbell J, Hoenig C, Bazan F, Ryerson F, Guinan M, Van Konynenburg R, Rozsa R. Properties of SYNROC-D nuclear waste form: a state-of-the-art review. United States 1982, 1982-01-01.
[52] Ringwood AE. Immobilisation of sodium and phosphorus-bearing PW-7a waste in SYNROC. Progress report. [SYNROC C]; 1982.
[53] Bernadzikowski TA, Allender JS, Gordon DE, Gould TH, Stone JA, Westberry CF. High-level nuclear waste form performance evaluation. Am Ceram Soc Bull 1983;62(12).
[54] Hench LL, Clark DE, Harker AB. Nuclear waste solids. J Mater Science 1986;21:1457–78.
[55] Ringwood AE, Kelly PM, Bowie SHU, Kletz TA. Immobilisation of high-level waste in ceramic waste forms [and discussion]. Philos Trans R Soc London A 1986;319:63.
[56] Ringwood AE. Mineral waste form development for US defense wastes. Progr Rep 1981;1–158.
[57] Hench LL, Clark DE, Campbell J. High level waste immobilisation forms. Nucl Chem Waste Manage 1984;5:149–73.
[58] Ringwood AE, Kesson SE, Ware N, Hibberson W, Major A. The SYNROC process: a geochemical approach to nuclear waste immobilisation. Geochem J 1979;13:141–65.
[59] Campbell J H, Hoenig C L, Bazan F, Ryerson F J, Rosza R, eds. Immobilisation of Savannah River high-level wastes in synroc: results from product performance tests. New York: Elsevier; 1982.
[60] Kesson SE, Ringwood AE. Immobilisation of HLW in Synroc-E. MRS Proc 1983;26.
[61] White TJ, Segall RL, Barry JC, Hutchison JL. Twin boundaries in perovskite. Acta Crystallogr B 1985;41:93–8.
[62] Kesson S E, Ringwood A E. Safe disposal of spent nuclear fuel. Radioactive Waste Manag Nucl Fuel Cycle 1983;4:159–74.

[63] Solomah AG, Richardson PG, McIlwain AK. Phase identification, microstructural characterization, phase microanalyses and leaching performance evaluation of SYNROC-FA crystalline ceramic waste form. J Nucl Mater 1987;148:157–65.
[64] Ringwood A E. Disposal of high-level nuclear wastes: a geological perspective. Mineral Mag 1983;49:159–76.
[65] Strunz H, Nickel E. Strunz mineralogical tables. Chemical-structural mineral classification system. Stuttgart: Schweizerbart 9th ed.; 2001.
[66] Rose G. Ueber einige neue Mineralien des Urals. J Prakt Chem 1839;19:459–60.
[67] Barth T. Die Kristallstruktur von Perowskit und verwandten Verbindungen. Norrsk Geol Tidsk 1925;8:201–6.
[68] Naray-Szabo S. Die Struktur des Perowskits $CaTiO_3$. Math Term Ert 1942;61:913–25.
[69] Bonshedt-Kupleckaya EM. Some data on the minerals of the perovskite group. Probl Min Geochim Petr Acad Sci 1946;43.
[70] Kay HF, Bailey PC. Structure and properties of $CaTiO_3$. Acta Crystallogr 1957;10:219–26.
[71] Koopmans HJA, van de Velde GMH, Gellings PJ. Powder neutron diffraction study of the perovskites $CaTiO_3$ and $CaZrO_3$. Acta Crystallogr C 1983;39:1323–5.
[72] Sasaki S, Prewitt CT, Liebermann RC. The crystal structure of $CaGeO_3$ perovskite and the crystal chemistry of the $GdFeO_3$-type perovskites. Am Mineral 1983;68:1189–98.
[73] Buttner RH, Maslen EN. Electron difference density and structural parameters in $CaTiO_3$. Acta Crystallogr B 1992;48:644–9.
[74] Liu X, Liebermann CR. X-ray powder diffraction study of $CaTiO_3$ perovskite at high temperatures. Phys Chem Minerals 1993;20:171–5.
[75] Beran A, Libowitzky E, Armbruster T. A single-crystal infrared spectroscopic and X-ray-diffraction study of untwinned San Benito perovskite containing OH groups. Can Mineral 1996;34:803.
[76] Yashima M, Ali R. Structural phase transition and octahedral tilting in the calcium titanate perovskite $CaTiO_3$. Solid State Ionics 2009;180:120–6.
[77] Liang L, Wencong L, Nianyi C. On the criteria of formation and lattice distortion of perovskite-type complex halides. J Solid State Chem 2004;65:855–60.
[78] Wu H, Zhou W, Udovic TJ, Rush JJ, Yildirim T. Crystal chemistry of perovskite-type hydride NaMgH3: implications for hydrogen storage. Chem Mater 2008;20(6):2335–42.
[79] Tilley RJD. Perovskites: structure-property relationships. Hoboke, New Jersey, USA: John Wiley & Sons; 2016.
[80] Smyth D M. Defects and structural changes in perovskite systems; from insulators to superconductors. Cryst Lattice Defects Amorphous Mater 1989;18:355–75.
[81] Scheunemann K, Müller-Buschbaum H. Zur Kristallstruktur von $Ca_2Nb_2O_7$. J Inorg Nucl Chem 1974;36:1965–70.
[82] Redhammer GJ, Tippelt G, Roth G, Amthauer G. Structural variations in the brownmillerite series $Ca_2(Fe_{2-x}Al_x)O_5$: single-crystal X-ray diffraction at 25°C and high-temperature X-ray powder diffraction (25°C ≤ T ≤ 1000°C). Am Miner 2004;89:405–20.
[83] Mitchell R H, Vladykin N V. Rare earth element-bearing tausonite and potassium barium titanates from the Little Murun potassic alkaline complex, Yakutia, Russia. Mineral Mag 1993;57:651–64.
[84] Vorob'yev YI, Konev AA, Malyshonok YV, Afonina GF, Sapozhnikov AN. Tausonite, $SrTiO_3$, A new mineral of the perovskite group. Int Geol Rev 1984;26:462–5.
[85] Haggerty SE, Mariano AN. Strontian-loparite and strontio-chevkinite: two new minerals in rheomorphic fenites from the Paraná Basin carbonatites, South America. Contrib Miner Petrol 1983;84:365–81.
[86] Goldschmidt V M. Die Gesetze der Krystallochemie. Naturwissenschaften 1926;14:477–85.
[87] Inorganic Crystal Structure Database (ICSD). Karlsruhe (Germany): Fachinformationszentrum (FIZ).
[88] Safiannikoff A. Un nouveau mineral de niobium. Bull Seances Acad R Sci Outre-Mer 1959;5:1251–5.

[89] Danø M, Sørensen H. An examination of some rare minerals from the nepheline syenites of South West Greenland. Grønlands Geol Undersøgelse Bull 1959.

[90] Nickel EH, McAdam RC. Niobian perovskite from Oka, Quebec; a new classification for minerals of the perovskite group. Can Mineral 1963;7:683.

[91] Chakhmouradian AR, Mitchell RH. Compositional variation of perovskite-group minerals from the carbonatite complexes of the Kola alkaline province, Russia. Can Mineral 1997;35:1293.

[92] Chakhmouradian AR, Mitchell RH. Lueshite, pyrochlore and monazite-(Ce) from apatite-dolomite carbonatite, Lesnaya Varaka complex, Kola Peninsula, Russia. Mineral Mag 1998;62:769.

[93] Chakhmouradian AR, Halden NM, Mitchell RH, Horváth L. Rb-Cs-rich rasvumite and sector-zoned "loparite-(Ce)" from Mont Saint-Hilaire (Québec, Canada) and their petrologic significance. Eur J Mineral 2007;19(4):533–546.

[94] Chakhmouradian AR, Mitchell RH. New data on pyrochlore- and perovskite-group minerals from the Lovozero alkaline complex, Russia. Eur J Mineral 2002;14:821.

[95] Vousden P. The structure of ferroelectric sodium niobate at room temperature. Acta Crystallogr 1951;4:545–51.

[96] Solov'ev S P, Venevtsev Y N, Zhdanov G S. An X-ray study of phase transitions in $NaNbO_3$. Kristallogr Sov Phys-Crystallogr 1961;6:218–24.

[97] Sakowski-Cowley AC, Lukaszewicz K, Megaw HD. The structure of sodium niobate at room temperature, and the problem of reliability in pseudosymmetric structures. Acta Crystallogr B 1969;25:851–65.

[98] Ahtee M, Glazer AM, Megaw HD. The structures of sodium niobate between 480° and 575°C, and their relevance to soft-phonon modes. Philos Mag 1972;26:995–1014.

[99] Glazer AM, Megaw HD. The structure of sodium niobate (T2) at 600°C, and the cubic-tetragonal transition in relation to soft-phonon modes. Philos Mag 1972;25:1119–35.

[100] Darlington CNW, Knight KS. High-temperature phases of $NaNbO_3$ and $NaTaO_3$. Acta Crystallogr B 1999;55:24–30.

[101] Mishra SK, Choudhury N, Chaplot SL, Krishna PSR, Mittal R. Competing antiferroelectric and ferroelectric interactions in $NaNbO_3$: neutron diffraction and theoretical studies. Phys Rev B 2007;76:024110.

[102] Jiang L, Mitchell DC, Dmowski W, Egami T. Local structure of $NaNbO_3$: a neutron scattering study. Phys Rev B 2013;88:014105.

[103] Mishra SK, Mittal R, Pomjakushin VY, Chaplot SL. Phase stability and structural temperature dependence in sodium niobate: a high-resolution powder neutron diffraction study. Phys Rev B 2011;83:134105.

[104] Peel MD, Thompson SP, Daoud-Aladine A, Ashbrook SE, Lightfoot P. New twists on the perovskite theme: crystal structures of the elusive phases R and S of $NaNbO_3$. Inorg Chem 2012;51:6876–89.

[105] Mitchell RH, Burns PC, Knight KS, Howard CJ, Chakhmouradian AR. Observations on the crystal structures of lueshite. Phys Chem Miner 2014;41:393–401.

[106] Brendan JK, Prodjosantoso AK, Christopher JH. Powder neutron diffraction study of the high temperature phase transitions in $NaTaO_3$. J Phys Condensed Matter 1999;11:6319.

[107] Mitchell R H, ed. Perovskites: a revised classification scheme for an important rare earth element host in alkaline rocks. In: Jones AP, Wall F, Williams CT, editors. Rare Earth Minerals: Chemistry, origin and ore deposits. London UK: Chapman & Hall; 1996;7:41–76.

[108] Ramsay W, Hackman V. Das Nephelinsyenitgebiet auf der Halbinsel Kola Fennia 1894;14.

[109] Kuznetsov IG. Loparite – a new rare earth mineral from the Khibina Tundra. Izv Geol Komit 1925;44:663–82.

[110] Hu M, Wenk H-R, Sinitsyna D. Microstructures in natural perovskites. Am Miner 1992;77:359.

[111] Wang Y, Liebermann RC. Electron microscopy study of domain structure due to phase transitions in natural perovskite. Phys Chem Miner 1993;20:147–58.

[112] von Gaertner H R. Die Kristallstrukturen von Loparit und Pyrochlor. N Jb Mineral Geol Palaeontol Bei Abt A 1930;61:30.
[113] Megaw HD. Crystal structures: a working approach. Philadelphia: WB Saunders Company; 1973.
[114] Mitchell RH, Burns PC, Chakhmouradian AR. The crystal structures of loparite-(Ce). Can Mineral 2000;38:145–52.
[115] Zubkova NV, Arakcheeva AV, Pushcharovskii DY, Semenov EI, Atencio D. Crystal structure of loparite. Crystallogr Reports 2000;45:210–4.
[116] Geller S. Crystal structure of gadolinium orthoferrite, $GdFeO_3$. J Chemical Phys 1956;24:1236–9.
[117] Mitchell RH, Chakhmouradian AR, Woodward PM. Crystal chemistry of perovskite-type compounds in the tausonite-loparite series, $(Sr_{1-2x}Na_xLa_x)TiO_3$. Phys Chem Miner 2000;27:583–9.
[118] Thomas N. A new global parameterization of perovskite structures. Acta Crystallogr B 1998;54:585–99.
[119] Aleksandrov KS. The sequences of structural phase transitions in perovskites. Ferroelectrics 1976;14:801–5.
[120] Mitchell RH, Chakhmouradian AR. Th-rich loparite from the Khibina alkaline complex, Kola Peninsula: isomorphism and paragenesis. Mineral Mag 1998;62:341–53.
[121] Mitchell RH, Chakhmouradian AR. Solid solubility in the system $NaLREETi_2O_6$-$ThTi_2O_6$ (LREE, light rare-earth elements): experimental and analytical data. Phys Chem Miner 1999;26:396–405.
[122] Mitchell RH, Liferovich RP. A structural study of the perovskite series $Na_{0.75}Ln_{0.25}Ti_{0.5}Nb_{0.5}O_3$. J Solid State Chem 2005;178:2586–93.
[123] Chakhmouradian A, Yakovenchuk V, Mitchell R H, Bogdanova A. Isolueshite: a new mineral of the perovskite group from the Khibina alkaline complex. Eur J Mineral 1997;9:483–90.
[124] Krivovichev SV, Chakhmouradian AR, Mitchell RH, Filatov SK, Chukanov NV. Crystal structure of isolueshite and its synthetic compositional analogue. Eur J Mineral 2000;12:597.
[125] Nickel EH. Latrappite; a proposed new name for the perovskite-type calcium niobate mineral from the Oka area of Quebec. Can Mineral 1964;8:121.
[126] Van Wambeke L. Lattrapite and ceriopyrochlore, new minerals for the Federal Republic of Germany. N Jb Mineral Mh 1980;171–4.
[127] Mitchell R H, Choi I K, Hawthorne F C, McGammon C A, Burns P C. Latrappite: a reinvestigation. Can Mineral 1998;36:107–16.
[128] McCammon C. Crystal chemistry of iron-containing perovskites. Phase Transitions 1996;58:1–26.
[129] Christoph GG, Larson AC, Eller PG, Purson JD, Zahrt JD, Penneman RA, Rinehart GH. Structure of barium plutonate by neutron powder diffraction. Acta Crystallogr B 1988;44:575–80.
[130] Geller S. Crystal structure of gadolinium orthoferrite, $GdFeO_3$. J Chemical Phys 1956;24:1236–9.
[131] Jacobson AJ, Tofield BC, Fender BEF. The structures of $BaCeO_3$, $BaPrO_3$ and $BaTbO_3$ by neutron diffraction: lattice parameter relations and ionic radii in O-perovskites. Acta Crystallogr B 1972;28:956–61.
[132] Chackraburtty DM, Jayadevan NC, Sivaramakrishan CK. Complex oxide systems of barium and plutonium. Acta Crystallogr 1963;16:1060–1.
[133] Russell LE, Harrison JDL, Brett NH. Perovskite-type compounds based on plutonium. J Nucl Mater 1960;2:310–20.
[134] Nakajima K, Arai Y, Suzuki Y, Yamawaki M. Vaporization behavior of $SrPuO_3$. J Nucl Mater 1997;248:233–7.
[135] Reynolds E, Kennedy BJ, Thorogood GJ, Gregg DJ, Kimpton JA. Crystal structure and phase transitions in the uranium perovskite, Ba2SrUO6. J Nucl Mater 2013;433:37–40.
[136] Keller C. Ueber die Festkoerperchemie der Actiniden-Oxide Nukleonik 1962;4:271–277.
[137] Lang S M, Knudson E P, Filmore C, Roth RS. High temperature reactions of uranium dioxide with various metal oxides. Nat Bur Stand Circ 1956;568.
[138] Degueldre C, Paratte JM. Concepts for an inert matrix fuel, an overview. J Nucl Mater 1999;274:1–6.

[139] Richter K, Fernandez A, Somers J. Infiltration of highly radioactive materials: a novel approach to the fabrication of targets for the transmutation and incineration of actinides. J Nucl Mater 1997;249:121–7.
[140] Konings RJM, Conrad R, Dassel G, Pijlgroms BJ, Somers J, Toscano E. The EFTTRA-T4 experiment on americium transmutation. J Nucl Mater 2000;282:159–70.
[141] Keller C, Walter KH. Ternäre oxide des americiums und einiger seltener erden vom typ $AX_{III}BO_3$. J Inorg Nucl Chem 1965;27:1247–51.
[142] Fullarton ML, Qin MJ, Robinson M, Marks NA, King DJM, Kuo EY, Lumpkin GR, Middleburgh SC. Structure, properties and formation of PuCrO3 and PuAlO3 of relevance to doped nuclear fuels. J Mater Chem A 2013;1:14633–40.
[143] Walter K H. Ternäre Oxide des drei-bis sechswertigen Americums. Karlsruhe: Gesellschaft für Kernforschung MBH;1965;581.
[144] Mosley WC. Self-radiation damage in curium-244 oxide and aluminate. J Am Ceram Soc 1971;54:475–9.
[145] Christopher JH, Brendan JK, Bryan CC. Neutron powder diffraction study of rhombohedral rare-earth aluminates and the rhombohedral to cubic phase transition. J Phys Condens Matter 2000;12:349.
[146] Megaw H. A note on the structure of lithium niobate, $LiNbO_3$. Acta Crystallogr A 1968;24:583–8.
[147] Moreau JM, Michel C, Gerson R, James WJ. Atomic displacement relationship to rhombohedral deformation in some perovskite-type compounds. Acta Crystallogr B 1970;26:1425–8.
[148] O'Keeffe M, Hyde BG. Some structures topologically related to cubic perovskite (E21), ReO_3 (D09) and Cu_3Au (L12). Acta Crystallogr B 1977;33:3802–13.
[149] Thomas NW, Beitollahi A. Interrelationship of octahedral geometry, polyhedral volume ratio and ferroelectric properties in rhombohedral perovskites. Acta Crystallogr B 1994;50:549–60.
[150] Zhang Y, Zhang Z, Thorogood G, Vance ER. Pyrochlore based glass-ceramics for the immobilisation of actinide-rich nuclear wastes: from concept to reality. J Nucl Mater 2013;432:545–7.
[151] Stewart MWA, Vance ER, Moricca SA, Brew DR, Cheung C, Eddowes T, Bermudez W. Immobilisation of higher activity wastes from nuclear reactor production of 99Mo. Sci Technol Nucl Install 2013;2013:16
[152] Cheung CKW, Vance ER, Stewart MWA, Brew DRM, Bermudez W, Eddowes T, Moricca S. The intermediate level liquid molybdenum-99 waste treatment process at the Australian Nuclear Science and Technology Organisation. Proc Chem 2012;7:548–53.

Reto Gieré, Gregory R. Lumpkin and Katherine L. Smith
11 Titanate ceramics for high-level nuclear waste immobilization

11.1 Introduction

Peaceful civilian uses of radioactive materials, such as nuclear power and the application of medical radioisotopes, generate radioactive by-products (wastes), some of which have to be isolated from the biosphere for long periods of time (10,000–1,000,000 years). Many countries have agreed that deep geologic disposal (Fig. 11.1) is the most viable long-term solution for high-level waste (HLW) [1, 2]. However, at present, there is only one geologic HLW repository under construction (in Finland [3]) and substantial amounts of HLW are stored in temporary facilities. Consequently, the management of HLW engenders a great deal of scientific, engineering, political and social interest.

Definitions of HLW vary [2, 4], but include spent nuclear fuel (SNF) and the residues from reprocessing of SNF. Several countries (e.g. Canada, Finland, Spain, Sweden, United States) have adopted a once-through nuclear fuel cycle with eventual direct disposal of SNF, whereas others (e.g. Belgium, China, France, India, Japan, Russia, United Kingdom) employ multiple cycles of fuel usage with intermediate reprocessing steps. Reprocessing generates small volumes (relative to SNF) of highly radioactive, heat-producing waste.

It is generally envisaged that SNF and/or HLW-bearing "waste forms" will be placed in geologic repositories [2], with a design similar to the one shown in Fig. 11.1. Borosilicate glass was the first waste form proposed for the immobilization of HLW [5]. It is currently the waste form of choice in many countries that reprocess their commercial SNF, primarily due to the simplicity of production and because of a well-established glass industry. The use of glass has been questioned, however, especially regarding its long-term thermodynamic stability and its aqueous durability, particularly at elevated temperature, under the conditions of a geologic repository, where the potential for higher dissolution rates can result in the release of radionuclides into the environment [6, 7]. Consequently, ceramic materials were proposed as alternative waste forms for HLW, and these are regarded as attractive materials for the long-term isolation of HLW in the geologic environment [5, 8, 9].

In this book chapter, we will review different types of ceramic waste forms, focusing on titanate ceramics, their design and fabrication, and their aqueous durability, and discuss radiation-damage effects in individual phases and how these can be modeled. In addition, we will review how the design and development of these ceramics have been influenced by important results obtained from studies of so-called "natural-analogue materials".

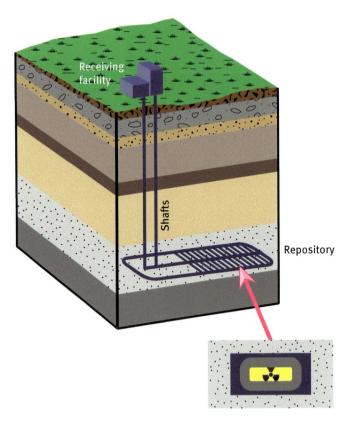

Fig. 11.1: Schematic diagram of a deep geologic repository. The waste canisters are received in an above-ground facility. From here, the canisters, other materials and workmen are transported down to the repository level via vertical shafts. The actual repository, located several hundred meters below the Earth's surface, consists of many disposal tunnels, connected by access tunnels. The inset displays the typical multibarrier concept: the high-level waste is encapsulated in a suitable waste form (yellow), which is enclosed in a metal container (dark gray). This container is surrounded by backfill material (dark blue). The host rock (stippled light gray) is the outermost barrier, which further delays the migration of radionuclides, possibly leached from the waste form, to the biosphere.

11.2 Multibarrier repository

During radioactive decay, radioactive isotopes release radiation, ejected particles (e.g. alphas, betas, neutrons and protons, which carry kinetic energy that degrades to heat), and a recoil nucleus. Both the amount of radiation and the amount of released heat die off with the half-life of the radioactive isotope.

HLW varieties, including SNF, have complex chemical and radionuclide compositions, are highly radioactive, and initially produce considerable heat [3]. Consequently, it is currently the practice, and is also planned for the future, that these wastes and/or their waste forms are kept in "interim" storage for periods of time (in some estimates, decades) prior to emplacement in deep underground repositories.

The repository concept is based around the idea of using multiple barriers, both engineered and geologic, each designed to delay, minimize, or prevent the release of radionuclides into the environment. A multibarrier HLW repository typically consists of the following barriers (from the center to the periphery; see Fig. 11.1):
- nuclear waste form, which initially hosts all the actual waste (e.g. SNF, borosilicate glass, ceramics);
- metal canister (e.g. steel, copper);
- backfill (e.g. bentonite, concrete);
- surrounding geologic formation (e.g. crystalline rocks, clay, rock salt).

For a review of geologic repository design and of different host-rock options, the readers are referred to individual contributions in a recent issue of *Elements* [10–14].

11.3 Design and fabrication of titanate ceramics

A suitable waste form for HLW should satisfy various requirements [8], including
- high durability in aqueous fluids;
- resistance to property changes resulting from radiation damage;
- high waste loading capability, and thus minimum volume
- long-term thermodynamic stability;
- suitable physical properties (e.g. mechanical strength, hardness, thermal conductivity, viscosity, electric resistivity);
- compatibility with the geologic environment;
- simple, safe, reliable, and cost-effective production procedure.

With these criteria in mind, waste forms can be optimized to account for the waste composition and/or the geologic environment. No HLW waste form can satisfy all of the mentioned requirements for emplacement in all geologic environments, so a whole range of crystalline waste forms, both single- and multiphase, have been developed, including titanate, zirconate, aluminate, and ferrite ceramics as well as glass ceramics [5, 8, 15–18]. In this short review, we will focus on the multiphase titanate ceramics.

The design of titanate ceramics, originally termed Synroc (short for "synthetic rock"), has been based primarily on the fact that a number of actinide (ACT)- and rare earth element (REE)–rich titanate minerals occur in nature, where they have survived over geologic time scales (up to 2 Ga). These minerals include zirconolite, pyrochlore, betafite, microlite, perovskite, brannerite, aeschynite, euxenite, polycrase, crichtonite, and fergusonite. Despite having been exposed to considerable irradiation doses due to the decay of their radioactive constituents, natural titanate phases have been relatively durable and have retained the actinides and their daughter products [19–27].

The design of the multiphase titanate ceramics also takes advantage of the range in crystallographic sites available in the different phases for the incorporation via solid-solution of a wide variety of waste species. Titanate waste forms typically contain some combination of the following phases as hosts for ACT and fission products: zirconolite, a pyrochlore-group phase, perovskite, Ti-based hollandite, and brannerite (Tab. 11.1). In some cases, murataite may also be a

Tab. 11.1: Structural and chemical characteristics of the most important constituent phases of HLW titanate ceramics.

Phase	Idealized formula	Crystal system	Structural site [coordination number]	Major elements[a]	Minor elements
Zirconolite	$CaZrTi_2O_7$	Monoclinic[b]	Ca [8]	Ca^{2+}, REE^{3+}, ACT^{4+}	Sr^{2+}
			Zr [7]	Zr^{4+}, Hf^{4+}	ACT^{4+}, REE^{3+}
			Ti [6], [5]	Ti^{4+}, Mg^{2+}, Fe^{2+}, Fe^{3+}, Al^{3+}	Mn^{2+}, Ta^{5+}, W^{6+}
Perovskite	$CaTiO_3$	Orthorhombic or cubic	Ca [12]	Ca^{2+}, Na^+, Sr^{2+}, REE^{3+}	K^+, Ba^{2+}, ACT^{4+}, ACT^{3+}, $\square$
			Ti [6]	Ti^{4+}, Nb^{5+}, Fe^{3+}, Fe^{2+}	Mg^{2+}, Al^{3+}, Ta^{5+}, Zr^{4+}
Hollandite	$A_{0-2}B_yC_{8-y}O_{16}$	Tetragonal or monoclinic	A [8]	Ba^{2+}, Cs^+, Rb^+, K^+, Na^+, Pb^{2+}	Sr^{2+}, H_2O, $\square$
			B, C [6]	Ti^{4+}, Fe^{2+}, Fe^{3+}, Mg^{2+}, Al^{3+}, Mn^{4+}, Mn^{3+}	Si^{4+}
Pyrochlore	$A_{2-m}B_2X_{6-w}Y_{1-n}$	Cubic	A [8]	Ca^{2+}, Na^+, Sr^{2+}, Pb^{2+}, Sn^{2+}, Sb^{3+}, REE^{3+}, ACT^{4+}, ACT^{5+}, ACT^{6+}, H_2O, $\square$	Ba^{2+}, Mn^{2+}, Fe^{2+}, K^+, Cs^+, Bi^{3+}, Sc^{3+}
			B [6]	Nb^{5+}, Ta^{5+}, Ti^{4+}, Sb^{5+}, W^{6+}	Zr^{4+}, V^{5+}, Sn^{4+}, Hf^{4+}, Fe^{3+}, Mg^{2+}, Al^{3+}, Si^{4+}
			X	O^{2-}	OH^-, F^-
			Y	O^{2-}, OH^-, F^-, H_2O, $\square$	K^+, Cs^+, Rb^+
Brannerite	UTi_2O_6	Monoclinic	U [6]	ACT^{4+}, ACT^{5+}, ACT^{6+}, REE^{3+}, Ca^{2+}, Pb^{2+}	Mn^{2+}
			Ti [6]	Nb^{5+}	Al^{3+}, Si^{4+}, Mn^{2+}, Fe^{2+}, Ni^{2+}
Murataite	$A_3B_2C_6O_{20-22}$[c]	Cubic	A [8]	Ca^{2+}, Mn^{2+}, REE^{3+}, ACT^{3+}, ACT^{4+}	Na^+
			B [5]	Mn, Ti^{4+}, Zr^{4+}, ACT^{4+}	Zn^{2+}, Fe^{2+}
			C [6]	Ti^{4+}, Al^{3+}, Fe^{3+}	Nb^{5+}
Rutile	TiO_2	Monoclinic	Ti [6]	Tc^{4+}	Ru^{4+}

[a]Abbreviations: ACT = actinides; REE = rare earth elements, including Y; $\square$ = vacancy.
[b]Most common polytype.
[c]Formula for natural murataite is approximately $A_4B_3C_6O_{18}(F,OH)_4$.

prominent host phase [8, 18]. Rutile is commonly an inert "buffer" phase in these ceramics, but can also be a host phase for Tc. Most titanate phases can incorporate a range of chemical elements in a variety of crystallographic sites [18, 26, 28–30], which makes titanate waste forms suitable for the incorporation of various types of nuclear waste (Tabs. 11.1 and 11.2; Fig. 11.2). Moreover, some of these phases can incorporate considerable amounts of neutron absorbers, such as Hf and Gd, which allows for enhanced ACT uptake by these phases, depending on partitioning behavior [31, 32].

Tab. 11.2: List of some common ceramic nuclear waste forms, their mineralogical composition, intended usage, and waste loading.

Waste form	Mineralogical composition	Intended usage (waste loading)
Synroc C	Zirconolite, perovskite, hollandite, rutile	Immobilization of HLW from reprocessing of SNF (up to 20 wt%)
Synroc D	Zirconolite, perovskite, spinel, nepheline	Immobilization of US defense wastes (60–70 wt%)
Synroc F	Pyrochlore, perovskite, uraninite	Conversion of spent fuel (~50 wt%)
Tailored ceramics	Magnetoplumbite, zirconolite, spinel, uraninite, nepheline	Immobilization of US defense wastes (≥60 wt%)
Zirconate (Pyrochlore)	Pyrochlore, zirconolite, brannerite, rutile	Immobilization of transuranic elements (up to 35 wt%)

An additional design aspect of titanate ceramics is that some of the phases will suffer more rapid and extensive radiation damage and associated swelling than others (see Chapter 11.4 below), and therefore grain size is kept small in order to minimize internal stress in the waste form (Fig. 11.3).

Waste-form development has also benefitted from, and was complemented by, numerous mineralogical and geochemical studies of analogous crystalline phases that occur in the natural environment [22, 26, 27, 33].

Various methods have been used to fabricate titanate ceramics, including hot pressing, sintering, and crystallization from a melt [8, 18, 34]. The method chosen or developed depends on a number of factors, including the desire to (a) limit the generation of by-products (e.g. off gases), and (b) optimize particular characteristics in the final product (e.g. high density, fine grain size) [34–36]. The original Synroc process involved using of reducing conditions during calcination and hot pressing in order to achieve the maximum retention of volatiles, and hot-pressing temperatures of at least ~1100°C to eliminate porosity and provide maximum aqueous durability.

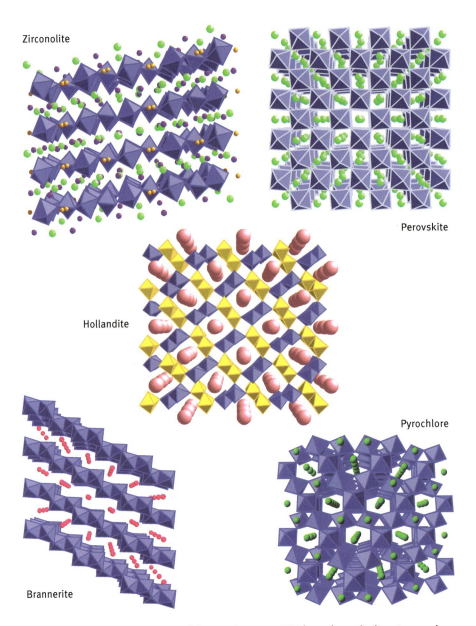

Fig. 11.2: Crystal structure drawings of the most important HLW host phases in titanate ceramic waste forms. The octahedra represent the sites occupied by Ti (blue) or similar cations (yellow). The spheres indicate the position, but not the exact size, of the other cations: green = Ca; purple in zirconolite = Zr; light brown in zirconolite = Ti in split site (coordination number 5); pink in hollandite = A-site cation; red in brannerite = U.

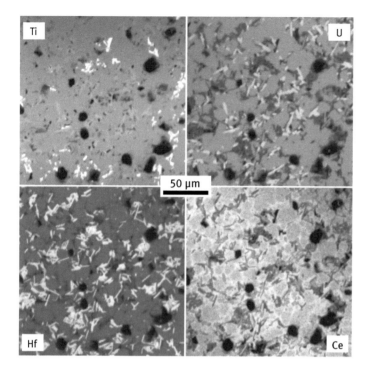

Fig. 11.3: Maps showing the distribution of Ti, U, Hf, and Ce in a pyrochlore ceramic, which consists of pyrochlore, zirconolite, brannerite, and minor rutile [31]. Note the needles of Hf-rich zirconolite (bright in Hf map) and the platy brannerite (bright in U map) embedded in the pyrochlore matrix (medium gray in U map, light gray in Ce map). Interstitial rutile appears bright in the Ti map. Areas that appear black in all maps represent pores.

11.4 Radiation resistance

A major concern regarding the performance of crystalline waste forms is that the constituent phases will undergo a crystalline-to-amorphous transformation as a result of alpha decay of the contained ACT [37, 38]. This transformation of titanate ceramic waste forms can affect their performance in two ways: (1) the aqueous durability can be considerably diminished and (2) the ACT host phases can swell as a result of radiation damage, which leads to uneven expansion and associated microcracking of the surrounding materials, thus increasing surface area and producing fluid pathways, which can enhance solubility [15, 26]. Alpha-decay damage accumulates over time and increases with increasing dose. Various techniques have been applied to study the dose-dependence of radiation damage in synthetic phases, i.e. doping with short-lived alpha-emitting isotopes, neutron irradiation, and heavy-ion implantation [15, 18]. In natural-analogue phases, on the other hand, accumulation of radiation damage is commonly studied by transmission electron microscopy, TEM [22, 39].

Investigations of radiation damage in natural titanate phases by TEM have shown that the onset of damage is marked by a mottled contrast in high-resolution TEM images and by the presence of local nanometer-sized areas of amorphous material [39]. At higher doses, larger amorphous areas form a matrix with embedded islands of crystalline material. At even higher doses, the material is fully amorphous, i.e. no lattice fringes and no electron-diffraction peaks can be observed (Fig. 11.4). This final stage is achieved once a phase has experienced alpha doses higher than what is known as the critical dose (D_c) for full amorphization [22, 40].

Fig. 11.4: TEM images of zirconolite from (a) Bergell (Italy) with an age of 32 Ma and a cumulative dose of 0.06×10^{16} α/mg; (b) Stavern (Norway) with an age of 300 Ma and a cumulative dose of 0.3×10^{16} α/mg; (c) Phalaborwa (South Africa) with an age of 2050 Ma and a cumulative dose of 2.8×10^{17} α/mg.

The susceptibility to radiation damage is different for different phases. For the phases of interest, zirconolite exhibits a smaller value of D_c than pyrochlore, i.e. pyrochlore is more radiation-resistant than zirconolite [22, 33]. Radiation resistance, however, is also different for different compositions of the same phase: Zr-rich members of the pyrochlore solid-solution series, for example, are much more radiation-resistant than the Ti-rich pyrochlores [15]. Similarly, volume expansion due to radiation dose varies according to host phase. ACT-doping experiments with zirconolite and pyrochlore have shown that swelling as a function of increased dose saturates at 5–6 vol%, which compares favorably with the much higher swelling of other phases, such as, zircon [26]. The swelling-induced microcracking of materials that surround or enclose highly radioactive titanate phases has been documented microscopically for natural samples (Fig. 11.5).

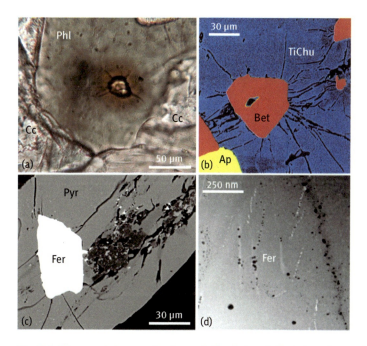

Fig. 11.5: Microscopic images showing radiation-induced effects in and around minerals that have experienced metamictization. (a) Optical microscope image (plane polarizers) of a metamict zirconolite grain with pleochroic halo (appearing brown in this view; alpha-particle damage) and radial cracks (due to volume expansion) in the enclosing green phlogopite. (b) False-color backscattered electron (BSE) image of metamict betafite (red) with radial cracks in the surrounding titanian clinohumite (blue). (c) BSE image of metamict fergusonite with radial cracks in enclosing pyrophanite. (d) High-angle annular dark-field image, obtained in scanning TEM mode, of metamict fergusonite with nanopores, which were interpreted as He bubbles [21]. Nanopores occur both isolated and in trails. The image further shows trails of U-rich nanoparticles (white). All samples from Adamello, Italy [21, 77]: those shown in (a) and (b) are from Ti-rich veins, those shown in (c) and (d) are from a granitic pegmatite. *Abbreviations:* Ap = apatite; Bet = betafite; Cc = calcite; Fer = fergusonite; Phl = phologopite; Pyr = pyrophanite; TiChu = titanian clinohumite.

The release of alpha particles during alpha decay can lead to accumulation of He in a mineral. Here, He is either accommodated interstitially, trapped at internal defects, or aggregated in bubbles [37, 41]. These He bubbles, which may occur isolated or agglomerated and/or aligned (Fig. 11.5d), can affect various physical properties of the host phase [37].

ACT-doping and irradiation experiments have shown that heating a phase can slow the progress of radiation damage and associated swelling and that radiation damage can be annealed if the phase is stored at elevated temperatures for a longer period of time [15], as also confirmed by natural-analogue studies. The latter document that D_c of geologically very old titanate minerals (>10^8 years) is considerably higher than that of young samples (e.g. [33]).

11.5 Durability of titanate ceramics

As water will be one of the principal agents that may adversely affect the survival of a waste form in a deep geologic repository for long time periods, the chemical durability of waste forms is of critical importance. Interactions between waste form and water may lead to dissolution or leaching, both processes that can be studied by laboratory investigations. Even though laboratory tests are restricted to relatively short time scales (up to several years at most), they can yield crucial thermodynamic data regarding the overall solubility of the waste form as well as data that can be used as input into theoretic models, which allow for stability assessments and/or extrapolation over time.

There are a number of experimental procedures for determining the chemical durability of waste forms. The methods may be static or dynamic, and they determine differential leach rate of individual constituents or cumulative leaching. Some of the most frequently used techniques are the Materials Characterization Center (MCC) and the single-pass flow-through (SPFT) tests as well as Soxhlet extraction. In these tests, either a monolithic sample or a powder is exposed to a leachant (deionized water or an aqueous solution that is representative of repository conditions, e.g. saline, carbonate or silicate solutions) for various durations at different temperatures and specimen surface-to-leachant volumes [8, 15, 18]. Most of the aqueous durability tests on polyphase ceramic nuclear waste forms have been conducted at temperatures between 25 and 200°C, whereby the fluid was replaced periodically. The dissolution data [34, 42–44] document that, under similar experimental conditions, the ceramic waste forms Synroc C and Synroc D exhibit lower release rates for their main chemical constituents than SNF and borosilicate glass (Tab. 11.3), implying that they are more durable.

Another strategy used to assess the durability of waste forms is to investigate the behavior of individual constituent phases in the geologic environment. These natural-analogue studies are focused on the characterization of the mineral analogues and their geochemical alteration [17, 22, 26, 27].

The data available in this context for both natural and synthetic phases [27, 44–47] show on the one hand that the main constituents in titanate ceramics, i.e. pyrochlore and zirconolite, are highly durable in aqueous systems at low temperature (see also [22, 26]). On the other hand, hollandite, brannerite, and especially perovskite are less durable (Tab. 11.3). Upon dissolution, brannerite and perovskite are converted to anatase and other alteration products [17, 26, 27]. It appears that amorphization decreases the aqueous durability of some of the titanate phases of interest, but the evidence for this relationship is currently relatively limited [26, 48].

Tab. 11.3: Examples of dissolution data for ceramic nuclear waste forms and some of their individual phases, as well as for spent nuclear fuel and borosilicate glass.

Waste form or phase	Condition	T (°C)	Release (g/cm²/d)	Element(s)[a]
Spent nuclear fuel[b]	Flow, pH 8–10[c]	19–78	5×10^{-4} to 5×10^{-3}	U
Borosilicate glass (SON68)[d]	SPFT	90	2×10^{-2} to 3×10^{-1}	Si
Synroc C[e]	MCC	95	$\sim 10^{-5}$ to 10^{-4}	Ti, Zr, Nd
Synroc D[f]	MCC, pH 6	90	$\sim 1 \times 10^{-4}$	U
Perovskite[g]	SPFT, pH 2–13	90	7×10^{-2} to 2×10^{-1}	Ca
Hollandite[f]	SPFT, pH 2–13	90	4×10^{-3} to 7×10^{-2}	Ba
Brannerite[f]	SPFT, pH 2–12	70	4×10^{-4} to 7×10^{-2}	U
Pyrochlore[f]	SPFT, pH 2–12	75	8×10^{-5} to 1×10^{-4}	U
Zircon[h]	Soxhlet	120–250	7×10^{-5} to 4×10^{-4}	Zr, Si
Zirconolite[f]	SPFT, pH 2–12	75	1×10^{-6} to 7×10^{-4}	U
Zirconia[i]	Static, pH 5.6	90	1×10^{-6} to 6×10^{-3}	Y, Zr, Ce, Nd, Sr
Monazite (natural)[j]	Flow, pH 1.5–10	70	6×10^{-7} to 6×10^{-4}	REEs, Th, U

[a]The listed elements are the *key elements* used for interpretation. The releases of U and Si are typically used for comparison with the behavior of spent fuel and borosilicate glass, respectively. Some studies, however, have quantified and listed more than one element measurable above detection limits (for details, the reader is referred to the original papers).
[b]Ref. [42].
[c]Suite of experiments with 0.2–20 mmol/L carbonate and either 0.3 or 2% oxygen.
[d]Ref. [43].
[e]Ref. [34].
[f]Ref. [44].
[g]Ref. [27].
[h]Ref. [46].
[i]Ref. [47].
[j]Ref. [45].

11.6 Atomistic modeling

Simulations of the structure, stability, defect behavior, and radiation damage of nuclear materials have become commonplace in materials science and engineering, including a range of existing and potential nuclear waste forms designed for the incorporation and safe storage of ACT and fission products. This is particularly true for the titanate and related oxide waste forms, which have evolved during a time when atomistic simulation methods have become increasingly advanced. Atomistic modeling primarily involves the use of empirical inter-atomic pair potentials that form the basis of molecular dynamics (MD) simulations on picosecond time scales. These simulations allow for the direct observation of the formation and evolution of alpha-recoil collision cascades and defects and can be performed as a function of temperature and pressure via the use of suitable boundary conditions. Empirical potentials are also used for static calculations of structural stability and the energetics of defect formation and migration.

Density functional theory (DFT) has been increasingly applied in studies of the stability and radiation effects of nuclear materials. DFT has the ability to more accurately describe the geometry, stability, and energetics of defect formation and migration in crystalline materials based on the appropriate use of a range of available density functionals (DFs) and approaches to take into account the effects of exchange and correlation. All MD and DFT simulations rely on careful validation of parameters via benchmarking against the known properties of crystalline compounds. Here, we take a brief look at some of the most important background studies and key papers on atomistic modeling of nuclear waste forms and related materials in studies that have used MD and/or DFT modeling.

A convenient starting point for atomistic modeling of titanate waste forms was to look into the atomic-scale behavior of the simplest titanate phases, i.e. the polymorphs of TiO_2 (e.g. [49, 50–53]). Although rutile is normally used as an inert component in nuclear waste forms, it has the capacity to serve as a host phase for certain radionuclides (Tab. 11.1) under appropriate conditions. Therefore, Kuo et al. [54] studied the possibility for incorporation of the long-lived fission product ^{99}Tc via direct substitution for Ti in the lattice (Tab. 11.1). They investigated the effects of defect clustering and the transmutation of Tc to Ru, demonstrating that Tc has a moderate (and temperature-dependent) solubility in rutile. However, the solubility increases when clustering is considered, with the preferred binary Tc-Tc pair having a similar distance as found in TcO_2. Importantly, the transmutation of Tc to Ru results in Ru-Ru pairs adopting a preferred second nearest neighbor configuration. As the calculated solubility of Ru in rutile is lower than that of Tc, this study suggested the possibility of second phase formation if Tc is present near the solubility limit.

In stepping up to more complex materials, Thomas et al. [55] investigated the defect behavior of the perovskite compound $SrTiO_3$, an important titanate host phase for ^{90}Sr, a key short-lived fission product in HLW. They quantified the energetics of defect formation and migration (Tab. 11.4) using MD and DFT methods, finding that O and Sr atoms prefer to form split interstitials and Ti atoms form interstitials in channel sites. Interstitials are more mobile than vacancies and Sr and O have low migration barriers, indicating the potential for mobility and recombination (see also [56]). MD simulations revealed that all three atom types are most easily displaced by direct replacement sequences on their own sub-lattices. Weighted average threshold displacement energies (TDE) of 70, 140, and 50 eV were obtained for Sr, Ti, and O, respectively. However, it was also shown that there exists strong dependence of the TDEs on crystallographic direction (see also [57]). Recently, Won et al. [58] used TEM and atomistic simulations to study radiation damage in $SrTiO_3$ containing Ruddlesden-Popper–type faults. Using empirical potentials and collision cascade simulations, they showed that the faults are more susceptible to amorphization and that this is related to a kinetically feasible, thermodynamic driving force for defects to migrate to the faults.

Tab. 11.4: Energetics of defect formation and migration in $SrTiO_3$.

Defect	Strontium (eV)	Titanium (eV)	Oxygen (eV)	Sr/Ti (eV)
Vacancy (isolated)	18.3[a]	41.7[a]	11.4[a]	—
Interstitial (isolated)	−1.8[a]	−23.5[a]	−1.3[a]	—
Frenkel (isolated)	16.5[a]	18.2[a]	10.1[a]	—
Anti-site (isolated)	—	—	—	9.3[a]
Frenkel	14.6[a]	—	9.8[a]	—
Frenkel (DFT)	9.0[a]	—	7.9[a]	—
Anti-site	—	—	—	9.8[a]
Anti-site (DFT)	—	—	—	13.2[a]
Vacancy (migration)	3.9[a]	11.0[a]	0.9[a]	—
	4.20[b]	None[b]	1.22[b]	—
Interstitial (migration)	0.3[a]	4.6[a]	0.3[a]	—
	0.52[b]	1.50[b]	0.43[b]	—

From [a]Thomas et al. [55] and [b]Bi et al. [53].

Even more complex perovskites containing La and A-site vacancies have been investigated by both experimental and atomistic modeling methods [59]. Thomas et al. [60] developed pair potentials for Sr, La, Ti, and O by fitting to experimental data and new *ab initio* data. This work provided the atomistic basis for understanding existing experimental data on the structure [61] and radiation effects [62] in the $Sr_{1-3x/2}La_xTiO_3$ system. In particular, this solid solution incorporates A-site vacancies according to the substitution $3Sr^{2+} \rightarrow 2La^{3+} + \square$ ($\square$ = vacancy), so the role of vacancies may have important consequences for defect behavior and recovery under irradiation. In fact, the solid-solution series exhibits a minimum in the critical temperature for amorphization near $x = 0.2$ [62]. Using both empirical pair potentials and DFT calculations, Thomas et al. [63] demonstrated that the A-site interactions follow electrostatic principles at low defect concentrations, leading to the association of La ions with vacancies and dissociation of vacancy-vacancy pairs. Once long-range A-site ordering is developed at higher values of x, the defect interactions are inverted due to strain forces arising from cooperative atomic relaxations (Fig. 11.6).

Zirconolite has been the subject of very few modeling and theoretical studies. Veiller et al. [64] reported both the TDEs for individual atoms and collision cascades introduced into the crystalline lattice, finding that the cascade region consists of a disordered core and a surrounding zone of isolated point defects. Radial distribution functions of the core zone are intermediate to those of crystalline and amorphous zirconolite. The simulations were stated to be consistent with the model of direct impact amorphization. A later DFT study [65] attempted to identify stable interstitial sites for intrinsic defects in the complicated zirconolite structure and reported significant dependence on charge states for vacancies and the possible formation of O_2 in Ti and Zr vacancies. Frenkel-defect energies of 1.1–3.2 eV for O (charge neutral), 1.97 eV for Ca, 8.22 eV for Zr, and 3.9–6.1 eV for

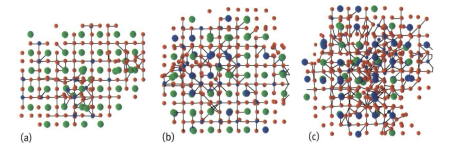

Fig. 11.6: Final damage states produced by MD collision cascades in La-Sr-Ti oxide perovskites with (a) no La, (b) 0.25 La atoms per formula unit, and (c) 0.5 La atoms per formula unit. Projection is down [100] of the cubic structure in (a) and [100] of the cubic subcell in (b,c). Color scheme: La is violet, Sr is green, Ti is blue, and O is red. Modified from Fig. 8.5 in [59], used with permission of the author.

Ti (all cation-vacancies are in the 2+/2− charge state) were calculated. It was concluded that the low Frenkel-defect energies, combined with their relative stability (high migration barriers?), provide an explanation for the ability to amorphize zirconolite at low irradiation or alpha-decay dose. Chappell et al. [66] recently developed new empirical potentials and studied zirconolite using MD to generate 70-keV single and double cascades within large simulation cells. Results of this study confirmed the damage "track" after about 15 ps to consist of a highly disordered region surrounded by a zone with scattered point defects. When normalized to stoichiometry, the relative numbers of defects produced by a single cascade are 542 O, 2106 Ca, 190 Ti, and 131 Zr.

The prototypical pyrochlore in nuclear waste forms is generally based on the composition $CaUTi_2O_7$; however, another generalized end-member component is $Ln_2Ti_2O_7$ which serves to accommodate REE fission products from the waste stream. Most simulation studies have actually been conducted on the latter type of pyrochlore composition containing a range of lanthanides or Y. Purton and Allan [67] conducted MD simulations of $Gd_2Ti_2O_7$ and $Gd_2Zr_2O_7$ with primary knock-on atom (PKA) energies up to 20 keV, showing that more defects were consistently produced in the titanate pyrochlore. Damage cascades in $Gd_2Zr_2O_7$ were smaller and more "ordered", whereas the larger cascades in $Gd_2Ti_2O_7$ appeared to be amorphous. Using MD simulations to determine the TDEs of $Gd_2Ti_2O_7$ and $Gd_2Zr_2O_7$ pyrochlores, Devanathan and Weber [68] found that the TDE for Ti is >170 eV for all directions that were examined. It was hypothesized that this result is due to the higher energy required for replacement of Gd by Ti in the structure relative to replacement of Gd by Zr. Xiao et al. [69] studied the TDEs of $Y_2Ti_2O_7$ using the DFT code SIESTA, modified to incorporate the generation of a PKA in the simulation cell. Average TDEs of 35.1 eV for Y, 35.4 eV for Ti, 17.0 eV for O on *48f*, and 16.2 eV for O on *8b* were determined for the three main cubic directions (100, 110, and 111). A cation anti-site energy of 2.32 eV was observed for the Y-Ti pair and a Frenkel-defect energy of 8.66 eV for Y (displaced to 8b). Other Frenkel defects were found to be unstable, suggesting the possibility of correlated recombination during relaxation. Low formation energies of cation interstitials (1.6–5.3 eV for Y; 1.2–2.2 eV for Ti) may contribute to the ability to render this pyrochlore amorphous under irradiation.

From these atomistic modeling studies, it is apparent that the ability of a pure crystalline solid to resist transformation to the amorphous state at the same temperature and pressure depends upon several basic properties, including the TDEs, defect-formation energies, and barriers to defect migration and recombination. Even in simple oxides, such as rutile, the ability to recover from cascade damage may be aided further by cooperative defect migration and recovery. A number of studies further indicate that volume-density and bonding considerations are also important in determining the ability to recover from cascade damage on short time scales. For pyrochlore systems in general, extensive literature covers in detail the above concepts [70–74] and the analysis of experimental data [75, 76].

11.7 Conclusions

Over the last ~50 years, various types of titanate ceramics have been developed to immobilize HLW before emplacement in geologic repositories. This development would not have been possible without contributions from scientists with expertise in a wide area of topics, including: mineralogy, crystallography, geochemistry, geology, materials science, physics, and radiochemistry. Using various methods, a large body of data has been collected from both natural environments and experimental investigations. These data provide the basis for the more recent simulations using atomistic modeling. Combining all the various techniques is essential, and it has led to considerable advances in our understanding how crystalline materials behave in intense radiation fields, under various environmental conditions, and over geologic time scales. This knowledge forms the basis for predicting the long-term behavior of titanate ceramics in deep geologic repositories for HLW.

References

[1] Sowder A, Kessler J, Apted M, Kozak M. What now for permanent disposal of used nuclear fuel and HLW in the United States? Radwaste Solutions 2013;20:26–39.
[2] IAEA. Classification of radioactive waste: general safety guide. In: IAEA Safety Standards Series. Vienna: International Atomic Energy Agency; 2009.
[3] Ewing RC, Whittleston RA, Yardley BWD. Geological disposal of nuclear waste: a primer. Elements 2016;12:233–7.
[4] NRC. High-Level Waste, United States Nuclear Regulatory Commission. (United States Nuclear Regulatory Commission); 2016.

[5] Lutze W, Ewing RC. Radioactive waste forms for the future. Amsterdam: North-Holland; 1988: 778 pp.
[6] McCarthy GJ. High-level waste ceramics. Trans Am Nucl Soc 1976;23:168–9.
[7] Ringwood AE, Kesson SE, Ware NG, Hibberson W, Major A. Immobilisation of high level nuclear reactor wastes in SYNROC. Nature 1979;278:219–23.
[8] Stefanovsky SV, Yudintsev SV, Gieré R, Lumpkin GR. Nuclear waste forms. In: Gieré R, Stille P, editors. Energy, waste, and the environment: a geochemical perspective, special publications. Geol Soc London Spec Publ 2004;236:37–63.
[9] Weber WJ, Ewing RC. Ceramic waste forms for uranium and transuranium elements. In: Burns PC, Sigmon GE, editors. Uranium – cradle to grave. Mineral Assoc Can Short Course Ser 2013;43:317–36.
[10] Grambow B. Geological disposal of radioactive waste in clay. Elements 2016;12:239–45.
[11] Hedin A, Olsson O. Crystalline rock as a repository for Swedish spent nuclear fuel. Elements 2016;12:247–52.
[12] Laverov NP, Yudintsev SV, Kochkin BT, Malkovsky CI. The Russian strategy of using crystalline rock as a repository for nuclear waste. Elements 2016;12:253–56.
[13] Swift PN, Bonano EJ. Geological disposal of nuclear waste in tuff: Yucca Mountain (USA). Elements 2016;12:263–8.
[14] von Berlepsch T, Haverkamp B. Salt as a host rock for the geological repository for nuclear waste. Elements 2016;12:257–62.
[15] Smith KL, Lumpkin GR, Blackford MG, Colella M. Titanate ceramics for the immobilisation of high-level nuclear waste, and their mineral analogues. In: White T, Stegemann JA, editors. International conference on materials for advanced technologies. Singapore: Materials Research Society; 2001:301–16.
[16] Harker AB. Tailored ceramics. In: Lutze W, Ewing RC, editors. Radioactive waste forms for the future. Amsterdam: North-Holland; 1988;335–92.
[17] Lumpkin GR. Ceramic waste forms for actinides. Elements 2006;2:365–72.
[18] Stefanovsky SV, Yudintsev SV. Titanates, zirconates, aluminates and ferrites as waste forms for actinide immobilization. Russ Chem Rev 2016;85:962–94.
[19] Gieré R, Buck EC, Guggenheim R, Mathys D, Reusser E, Marques J. Alteration of Uranium-rich microlite. In: Hart KP, Lumpkin GR, editors. Scientific basis for nuclear waste management XXIV. Mater Res Soc Symp Proc 2001:663:835–944.
[20] Gieré R, Swope RJ, Buck E, Guggenheim R, Mathys D, Reusser E. Growth and alteration of uranium-rich microlite. In: Smith R, Shoesmith D, editors. Scientific basis for nuclear waste management XXIII. Mater Res Soc Symp Proc 2000:608: 519–24.
[21] Gieré R, Williams CT, Wirth R, Ruschel K. Metamict fergusonite-(Y) in a spessartine-bearing granitic pegmatite from Adamello, Italy. Chem Geol 2009;261:333–45.
[22] Lumpkin GR. Alpha-decay damage and aqueous durability of actinide host phases in natural systems. J Nucl Mater 2001;289:136–66.
[23] Lumpkin GR, Ewing RC. Geochemical alteration of pyrochlore group minerals: Betafite subgroup. Am Mineral 1996;81:1237–48.
[24] Lumpkin GR, Ewing RC. Geochemical alteration of pyrochlore group minerals: Pyrochlore subgroup. Am Mineral 1995;80:732–43.
[25] Lumpkin GR, Ewing RC. Geochemical alteration of pyrochlore group minerals: microlite subgroup. Am Mineral 1992;77:179–88.
[26] Lumpkin GR, Geisler-Wierwille T. Minerals and natural analogues. In: Konings RJM, editor. Comprehensive nuclear materials. Vol 5. Amsterdam: Elsevier; 2012: 563–600.
[27] Lumpkin GR, Smith KL, Gieré R, Williams CT. Geochemical behaviour of host phases for actinides and fission products in crystalline ceramic nuclear waste forms. In: Gieré R, Stille P, editors. Energy, waste and the environment: a geochemical perspective, special publications. Vol. 236. London: The Geological Society; 2004:89–111.

[28] Gieré R, Williams CT, Lumpkin G, R. Chemical characteristics of natural zirconolite. Schweiz Mineral Petrogr Mitt 1998;78:433–59.
[29] Post JE, Von Dreele RB, Buseck PR. Symmetry and cation displacements in hollandites: structure refinements of hollandite, cryptomelane and priderite. Acta Crystallogr B38:1056–65. 1982.
[30] Puchkova EV, Bogdanov RV, Gieré R. Redox states of uranium in samples of microlite and monazite. Am Mineral 2016;101:1884–91.
[31] Gieré R, Hatcher C, Reusser E, Buck EC. Element partitioning in a pyrochlore-based ceramic nuclear waste form. In: McGrail BP, Cragnolino GA, editors. Scientific basis for nuclear waste management XXV. Mater Res Soc Symp Proc 2002;713:303–10.
[32] Lumpkin GR, Smith KL, Blackford MG. Partitioning of uranium and rare earth elements in Synroc: effect of impurities, metal additive, and waste loading. J Nucl Mater 1995;224:31–42.
[33] Lumpkin GR, Gao Y, Gieré R, Williams CT, Mariano AN, Geisler T. The role of Th-U minerals in assessing the performance of nuclear waste forms. Mineral Mag 2014;78:1071–95.
[34] Ringwood AE, Kesson SE, Reeve KD, Levins DM, Ramm EJ. SYNROC. In: Lutze W, Ewing RC, editors. Radioactive waste forms for the future. Amsterdam: North-Holland; 1988:233–334.
[35] Ramm EJ, Vance ER. In-bellows calcination of alkoxide precursor to synroc. Ceram Trans 1990;8:57–69.
[36] Buykx WJ, Levins DM, Smart RSC, Smith KL, Stevens GT, Watson KG, Weedon D, White TJ. Interdependence of phase chemistry, microstructure, and oxygen fugacity in titanate nuclear waste ceramics. J Am Ceram Soc 1990;73:1201–7.
[37] Weber WJ, Ewing RC, Catlow CRA, Diaz de la Rubia T, Hobbs LW, Kinoshita C, Matzke H, Motta AT, Nastasi M, Salje EKH, Vance ER, Zinkle SJ. Radiation effects in crystalline ceramics for the immobilization of high-level nuclear waste and plutonium. J Mater Res 1998;13(6):1434–84.
[38] Ewing RC, Meldrum A, Wang LM, Wang SX. Radiation-induced amorphisation. In: Ribbe PH, editor. Reviews in mineralogy and geochemistry. Vol 39. Washington (DC): Mineralogical Society of America; 2000: 319–61.
[39] Lumpkin GR, Smith KL, Gieré R. Application of analytical electron microscopy to the study of radiation damage in natural zirconolite. Micron 1997;28:57–68.
[40] Smith KL, Zaluzec NJ, Lumpkin GR. In situ studies of ion irradiated zirconolite, pyrochlore and perovskite. J Nucl Mater 1997;250:36–52.
[41] Ewing RC, Weber WJ, Clinard FW. Radiation effects in nuclear waste forms for high-level radioactive waste. Prog Nucl Energy 1995;29(2):63–127.
[42] Steward SA, Gray WJ. Comparison of uranium dissolution rates from spent fuel and uranium dioxide. Proc 5th Int High-Level Radioactive Waste Manage Conf 1994;4:2602–8.
[43] Icenhower JP, Steefel CJ. Dissolution rate of borosilicate glass SON68: A method of quantification based upon interferometry and implications for experimental and natural weathering rates of glass. Geochim Cosmochim Acta 2015;157:147–63.
[44] Hench LL, Clark DE, Campbell J. High level waste immobilization forms. Nucl Chem Waste Manage 1984;5:149–73.
[45] Oelkers EH, Poitrasson F. An experimental study of the dissolution stoichiometry and rates of a natural monazite as a function of temperature from 50 to 230°C and pH from 1.5 to 10. Chem Geol 2002;191:73–87.
[46] Ewing RC. Nuclear waste forms for actinides. Proc Natl Acad Sci USA 1999;86:3432–9.
[47] Kamizono H, Hayakawa I, Muraoka S. Durability of zirconium-containing ceramic waste forms in water. J Am Ceram Soc 1991;74:863–4.
[48] Williams CT, Bulakh AG, Gieré R, Lumpkin GR, Mariano AN. Alteration features in natural zirconolite from carbonatites. In: Hart KP, Lumpkin GR, editors. Scientific basis for nuclear waste management XXIV. Materials Research Society; 2001;845–52.
[49] Marks NA, Thomas BS, Smith KL, Lumpkin GR. Thermal spike recrystallisation: molecular dynamics simulation of radiation damage in polymorphs of titania. Nucl Instrum Methods Phys Res B 2008;266:2665–70.

[50] Lumpkin GR, Smith, KL, Blackford MG, Thomas BS, Whittle KR, Marks NA, Zaluzec NJ. Experimental and atomistic modeling study of ion irradiation damage in thin crystals of the TiO_2 polymorphs. Phys Rev B 2008;77:214201.

[51] Qin MJ, Kuo EY, Whittle KR, Middleburgh SC, Robinson M, Marks NA, Lumpkin GR. Density and structural effects in the radiation tolerance of TiO_2 polymorphs. J Phys Condens Matter 2013;25:355402.

[52] Robinson M, Marks NA, Lumpkin GR. Structural dependence of threshold displacement energies in rutile, anatase and brookite TiO_2. Mater Chem Phys 2014;147:311–8.

[53] Thomas BS, Marks NA, Corrales LR, Devanathan R. Threshold displacement energies in rutile TiO_2: a molecular dynamics simulation study. Nucl Instrum Methods Phys Res B 2005;239:191–201.

[54] Kuo EY, Qin MJ, Thorogood GJ, Whittle KR, Lumpkin GR, Middleburgh SC. Technetium and ruthenium incorporation into rutile TiO_2. J Nucl Mater 2013;441:380–9.

[55] Thomas BS, Marks NA, Begg BD. Defects and threshold displacement energies in $SrTiO_3$ perovskite using atomistic computer simulations. Nuclear Instruments and Methods in Physics Research B 2007;254:211–18.

[56] Bi Z, Uberuaga BP, Vernon LJ, Fu E, Wang Y, Li N, Wang H, Misra A, Jia QX. Radiation damage in heteroepitaxial $BaTiO_3$ thin films on $SrTiO_3$ under Ne ion irradiation. J Appl Phys 2013;113:023513.

[57] Liu B, Xiao HY, Zhang Y, Aidhy DS, Weber WJ. Ab initio molecular dynamics simulations of threshold displacement energies in SrTiO3. J Phys Condens Matter 2013;25:485003.

[58] Won B, Vernon LJ, Karakuscu A, Dickerson RM, Cologna M, Raj R, Wang Y, Yoo SJ, Lee SH, Misra A, Uberuaga BP. The role of non-stoichiometric defects in radiation damage evolution of SrTiO3. Journal of Materials Chemistry A 2013;1 (32):9235–45.

[59] Thomas BS. Atomistic simulation of Synroc-type titanates. PhD thesis. University of Sydney; 2008:231 pp.

[60] Thomas BS, Marks NA, Begg BD. Developing pair potentials for simulating radiation damage in complex oxides. Nucl Instrum Methods Phys Res B 2005;228:288–92.

[61] Howard CJ, Lumpkin GR, Smith RI, Zhang Z. Crystal structures and phase transition in the system $SrTiO_3-La_{2/3}TiO_3$. J Solid State Chem 2004;177:2726–32.

[62] Smith KL, Lumpkin GR, Blackford MG, Colella M, Zaluzec NJ. In situ radiation damage studies of $La_xSr_{1-3x/2}TiO_3$ perovskites. J Appl Phys 2008;103:083531.

[63] Thomas BS, Marks NA, Harrowell P. Inversion of defect interactions due to ordering in $Sr1-3x/2LaxTiO_3$ perovskites: an atomistic simulation study. Phys Rev B 2006;74:214109.

[64] Veiller L, Crocombette J-P, Ghaleb D. Molecular dynamics simulation of the alpha-recoil nucleus displacement cascade in zirconolite. J Nucl Mater 2002;306:61–72.

[65] Mulroue J, Morris AJ, Duffy DM. Ab initio study of intrinsic defects in zirconolite. Phys Rev B 2011;84:094118.

[66] Chappell HF, Dove MT, Trachenko K, McKnight REA, Carpenter MA, Redfern SAT. Structural changes in zirconolite under alpha-decay. J Phys Condens Matter 2013;25:055401.

[67] Purton JA, Allan NL. Displacement cascades in $Gd_2Ti_2O_7$ and $Gd_2Zr_2O_7$: a molecular dynamics study. J Mater Chem 2002;12:2923–6.

[68] Devanathan R, Weber WJ. Insights into the radiation response of pyrochlores from calculations of threshold displacement events. Journal of Applied Physics 2005;98:086110.

[69] Xiao HY, Gao F, Weber WJ. Threshold displacement energies and defect formation energies in $Y_2Ti_2O_7$. J Phys Condens Matter 2010;22:415800.

[70] Panero WR, Stixrude L, Ewing RC. First-principles calculation of defect-formation energies in the $Y_2(Ti, Sn, Zr)2O_7$ pyrochlores. Phys Rev B 2004;70:054110.

[71] Trachenko K, Pruneda JM, Artacho E, Dove MT. How the nature of the chemical bond governs resistance to amorphization by radiation damage. Phys Rev B 2005;71:184104.

[72] Pruneda JM, Artacho E. First-principles study of structural, elastic, and bonding properties of pyrochlores. Phys Rev B 2005;72:085107.
[73] Zhang ZL, Xiao HY, Zu XT, Gao F, Weber WJ. First-principles calculation of structural and energetic properties for $A_2Ti_2O_7$ (A = Lu, Er, Y, Gd, Sm, Nd, La). J Mater Res 2008;24:1335–41.
[74] Jiang C, Stanek CR, Sickafus KE, Uberuaga BP. First-principles prediction of disordering tendencies in pyrochlore oxides. Phys Rev B 2009;79:104203.
[75] Lumpkin GR, Smith KL, Blackford MG, Whittle KR, Harvey EJ, Redfern SAT, Zaluzec NJ. Ion irradiation of ternary pyrochlore oxides. Chem Mater 2009;21:2746–54.
[76] Lumpkin GR, Pruneda JM, Rios S, Smith KL, Trachenko K, Whittle KR, Zaluzec NJ. Nature of the chemical bond and prediction of radiation tolerance in pyrochlore and defect fluorite compounds. J Solid State Chem 2007;180:1512–8.
[77] Gieré R. Quantification of element mobility at a tonalite/dolomite contact (Adamello Massif, Provincia di Trento, Italy). No. 9141. PhD thesis. ETH Zürich; 1990.

Part III: **Biomineralization, biomimetics, and medical mineralogy**

E. Griesshaber, X. Yin, A. Ziegler, K. Kelm, A. Checa, A. Eisenhauer and W.W. Schmahl

12 Patterns of mineral organization in carbonate biological hard materials

12.1 Introduction

Organization of nanosized entities across many length scales poses a major challenge in the development and production of man-made materials with advanced functions. In contrast, in biologically formed hard tissues, this design feature and its formation principle are intrinsic. It began already with the emergence of first skeletal hard parts in late Precambrian and was diversified since then by evolutionary adaptation.

In this concept article, we describe nanoscale, mesoscale, and macroscale biocarbonate mineral organization in biological hard tissues such as shells, calcite teeth, and calcite intercalations into terrestrial isopod exocuticle. First, we highlight differences in extracellular matrix architectures and proceed to the interdigitation between biopolymer membranes/fibrils and biocarbonate mineral. Subsequently, we describe major types of fabrics and textures of mineral nanoparticles, crystallites and crystals in invertebrate hard tissues and discuss characteristic patterns of their organization on the different length scales.

In the last part of this article, we illustrate the utilization of carbonate mineral organization (texture) by the animal as a tool for habitat-adapted tailoring and functionalization of hard tissues.

12.2 Composite nature of biological hard tissues

Mineralized structures generated under biological control are widely recognized in materials science as prototypes for advanced materials. These skeletal elements and teeth are hierarchical composites with a large variety of structural design concepts (e.g. [1–7, 8]). They consist of two distinct types of materials: (1) a compliant biopolymer matrix that is reinforced by (2) stiff and hard components, such as Ca-carbonate or Ca-phosphate minerals.

In gastropod, bivalve, and brachiopod shells, the extracellular matrix is formed of two major components: a structural and a functional component (e.g. [9, 10, 11–17, 18, 19, 20, 21, 22]). Water-insoluble proteins form matrix membranes and constitute the scaffold of the hard tissue. These membranes are composites as well and consist of a core formed of chitin that is encased by proteins [9, 13, 15, 16, 23–25]. Matrix membranes are arranged in a characteristic grid or honeycomb structure (Figs. 12.1–3), subdividing space and influencing the shape and size of the basic

https://doi.org/10.1515/9783110497342-012

mineral units (Fig. 12.4) that comprise the biological hard tissue. The other biopolymer component in biological carbonate hard tissues consists of soluble polyanionic proteins, occluded within the basic mineral units as a foam-like network of fibrils (Fig. 12.2b, d–f). These polyanionic protein fibrils predominantly control crystallite attachment and orientation within the space that is defined by the matrix membranes [10, 13, 17, 18, 26, 27].

Scaffold (matrix) membranes are porous (Fig. 12.3 and [28, 29, 30, 31, 32, 33]) and enable the transfer of crystallographic orientation information from one compartment into the other. This facilitates the formation of mesoscale entities consisting of a few co-oriented mineral units, as, for example, is the case in both shell layers of the bivalve *Mytilus edulis* [34] or in the fibrous shell layer of the brachiopod *Megerlia truncata* [35].

There are two major routes of carbonate material infiltration into compartments defined by matrix membranes: (1) transport of ions (Ca^{2+}, HCO_3^-, H^+), that occurs with specific transport proteins, where carbonate ions cross from the cell interior into extracellular matrix compartments [36, 37, 38] or (2) vesicles loaded with mineral (amorphous calcium carbonate, ACC) which get exocytosed from cells. At the sites

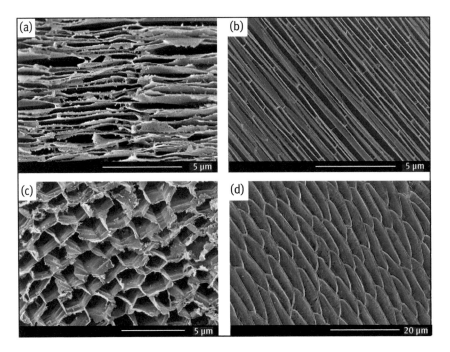

Fig. 12.1: FE-SEM micrographs of extracellular matrices present in modern gastropod (a, *Haliotis ovina*), bivalve (b, *Elliptio crassidens*; c, calcitic layer of *Mytilus edulis*), and brachiopod (d, *Liothyrella uva*) shells. Shell samples were microtome-cut, microtome-polished, decalcified (a–c) or etched (d), and critical point dried. Etching occurred for 90 and 180 seconds and was applied to remove the carbonate in order to visualize the spatial distribution of (glutaraldehyde-stabilized) biopolymers within the shell. Note the differences in space compartmentalization. Membranes in a and b surround nacre tablets, whereas the membranes presented in c and d envelope calcite fibers.

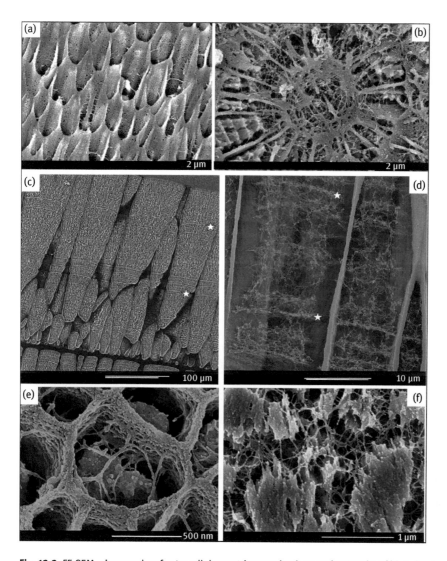

Fig. 12.2: FE-SEM micrographs of extracellular matrix organizations and networks of biopolymer fibrils in modern octopod (a, b, and e, *Argonauta argo*) and bivalve (c, d, *Elliptio crassidens*; f, *Arctica islandica*) shells. Shell samples were microtome-cut, microtome-polished, decalcified or etched, and critical point dried. Etching occurred for 90 and 180 seconds. With this procedure, the carbonate was removed and the distribution pattern of (glutaraldehyde-stabilized) biopolymers became apparent. Note the network of fibrils in d–f. The compartment, defined in size and morphology by biopolymer membranes, is filled with a foam-like network of fibrils. In the shell of the bivalve *Elliptio crassidens* (c, d), we observe localized fibril agglomerations, shown by white stars in c, d.

of crystallization amorphous carbonate becomes destabilized and crystallizes [39, 40–46]. However, as carbonate shells and calcite teeth are concerned, up to now, there is no definitive evidence for mineral transport from the cell into the extracellular matrix either via ACC-filled exocytosed vesicles or by protein-aided ion transport.

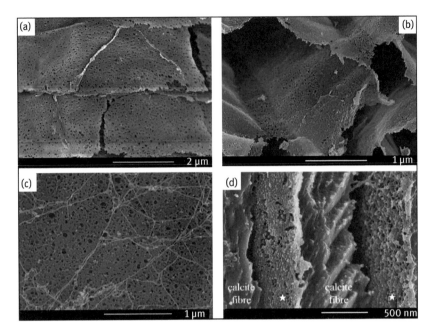

Fig. 12.3: FE-SEM micrographs visualizing the porosity of extracellular matrices in gastropod (Fig. 12.3a: *Haliotis ovina*), bivalve (Fig. 12.3b: calcitic shell layer of *Mytilus edulis*), Fig. 12.3c: *Elliptio crassidens*) and brachiopod (Fig. 12.3d: *Liothyrella uva*) shells. Samples were microtome cut, microtome polished, decalcified and critical point dried. Etching occurred for 90 and 180 seconds. White stars in Fig. 3D point to organic membranes intercalated between calcite fibres within the shell of the modern brachiopod *Liothyrella uva*. Extracellular membranes in carbonate hard tissues are highly porous, porosity not being in a regular arrangement.

12.3 Characteristic basic mineral units in gastropod, bivalve, and brachiopod shells

Biopolymer matrices are occluded within the hard tissues with organism-specific characteristic mesoscale grid structures and hence control the mesoscale architecture of the biologically formed mineral units (crystals). Basic biocarbonate mineral units are diverse: needles, laths, fibers, columns, tablets, irregularly shaped grains, and assemblages of granules (Figs. 12.4–7 and [18, 34, 47, 48, 49, 50, 51, 52, 53, 54, 55, 56]). Even though also consisting of calcium and/or magnesium carbonate, the size and outer morphology of all biologically formed mineral units (crystals) is in marked contrast to that of non-biological calcite and aragonite (Fig. 12.4a, b). This is due to the fact that biogenic carbonates are produced by organelles and under cellular control [45, 46, 57, 58], while non-biologic carbonates are precipitated from aqueous solution [59, 60, 61]. The latter exhibit typical rhombohedral, scalenohedral (calcite), and pseudo-hexagonal, prismatic (aragonite) morphologies, have smooth outer surfaces, and can grow to centimeter-size entities. Figures 12.4c–k and 7 display calcite and aragonite crystals (mineral units) of biological carbonate hard tissues: thin aragonite needles of the modern coral *Porites*

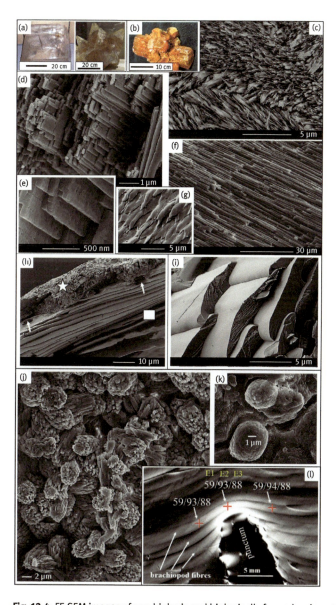

Fig. 12.4: FE-SEM images of non-biologic and biologically formed carbonate crystals, mineral units: calcite (a) and aragonite (b) single crystals precipitated from solution. (c–i) SEM images of fracture surfaces of modern coral (c, aragonite needles in *Porites* sp.), bivalve (d, e, aragonite laths in crossed-lamellar microstructure of *Propeamussium jeffreysii*; (f, g) calcite fibers in *Mytilus edulis* (in longitudinal and cross section), primary and fibrous layer calcite in the shell of the modern brachiopod *Liothyrella uva* (in longitudinal and cross section), aragonite granular prisms of the modern bivalve *Entodesma navicula*. White star in h, primary shell layer; white rectangle in h, fibrous shell layer; white arrows point to punctae. (l) Brachiopod fibers in *Megerlia truncata* that curve around a punctum. Even though fiber morphology is curved, the similarity of Euler angle (E1, E2, E3) values show that the calcite lattice in the fibers is coherent. Note the difference in crystal/mineral unit size, morphology, and curvature of outer surfaces between non-biologic and biologically formed carbonates.

sp. (Fig. 12.4c), thin aragonite laths in the shell of the modern bivalve *Propeamussium jeffereysii* (Fig. 12.4d, e), calcite fibers in the shell of the modern bivalve *M. edulis* (Fig. 12.4f in longitudinal section, Fig. 12.4g in cross section), calcite fibers in the shell of the modern brachiopod *Notosaria nigricans* (Fig. 12.4h, i) and aragonite granular prisms within the shell of the modern bivalve *Entodesma navicula* (Fig. 12.4j, k).

Biological calcite and aragonite fibers, laths and needles are frequently curved (Fig. 12.4l and [49, 53, 62]). Figure 12.4l shows a portion of the fibrous layer of the modern brachiopod *M. truncata*, where calcite fibers curve around a punctum and the morphological fiber axes are curved. However, as the similarity of Euler angles E1, E2, and E3 (Fig. 12.4l), determined at different spots with EBSD, shows, the pattern of calcite crystal orientation within this fiber does not change. Thus, throughout the curved fiber the calcite lattice is coherent – a feature intrinsic to biological hard materials. Curvature of geologic analogues would result in material failure.

Biological structural materials exhibit a hierarchical architecture. Thus, structural patterns of component assembly extend from the nanometer to the millimeter scale. The mesoscale mineral units (e.g. fibers, tablets, columns, laths) reveal internal structures (Fig. 12.5) as these are composed of primary particles with a size of a few tens of nanometers: the nanoparticles (Fig. 12.5b, d, f). Atomic force microscopy deflection images taken on conventionally polished (mildly etched) shell surfaces reveal the size, morphology and pattern of organization of nanoparticles in the shell of the modern brachiopod *Liothyrella uva* (Fig. 12.5a, b) and that of the modern bivalve *Mytilus galloprovincialis* (Fig. 12.5c–f). Calcite nanoscale particles sizes in the bivalve and the brachiopod shell scatter around 100 nanometers, while the size of aragonite nanoparticles in nacre tablets appears to be larger and might range up from 200 to even 300 nanometers (Fig. 12.5f). Figure 12.6 highlights bio-aragonite crystallite nanoparticles and crystal (basic mineral units) organization. The crystal lattices of the nanoparticles that compose a nacre tablet (Fig. 12.6a, b) are misoriented to each other (see the patchiness of the color-coded electron backscatter diffraction [EBSD] map in Fig. 12.6b and [34]). On the mesoscale, some pattern of crystallographic aragonite co-orientation is present. In the nacre of the bivalves *M. edulis* and *M. galloprovincialis*, the nacre crystals (tablets) aggregate to tower grains (white stars in Fig. 12.6c, marked with a c in Fig. 12.6b) while sharing one major crystallographic orientation [34].

The hard tissue of carbonate shells is structured also on the macroscale [63]. The shell might consist of only one carbonate polymorph that forms distinctly shaped assemblages of mineral units (Fig. 12.7a, b). These are stacked together in a well-organized, characteristic manner and constitute the different layers of the shell (Fig. 12.7a, b). In most cases, carbonate hard tissues consist of two carbonate polymorphs, with calcite and aragonite arranged in separate characteristic mineral units, e.g. the calcite often in fibers and columns, the aragonite in tablets, irregularly shaped grains, laths, and granules.

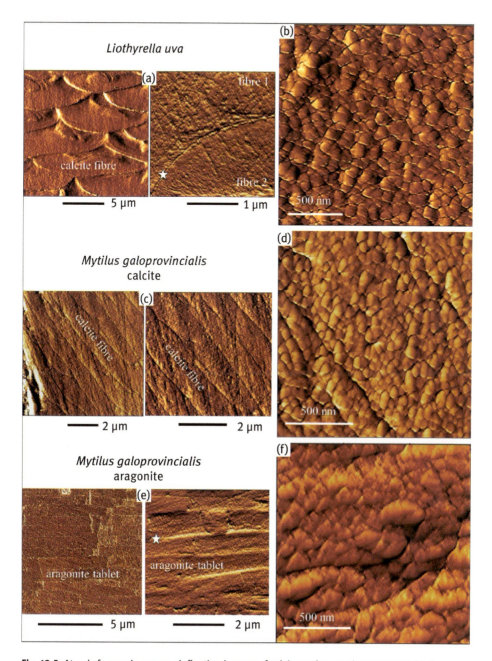

Fig. 12.5: Atomic force microscopy deflection images of calcite and aragonite nanoparticles in the shell of the modern brachiopod *Liothyrella uva* (a, b) and that of the modern bivalve *Mytilus galloprovincialis* (c–f). (a) Brachiopod calcite fibers in cross sections, with white star pointing to the organic membrane that separates two adjacent fibers. (c, 5) Nanoparticle assembly in the calcitic shell portion. (e, f) Nanoparticle organization in the aragonitic shell part of *M. galloprovincialis*. Calcite fibers in the bivalve are sectioned longitudinally. The white star in E points to matrix membranes that are present between nacre tablets.

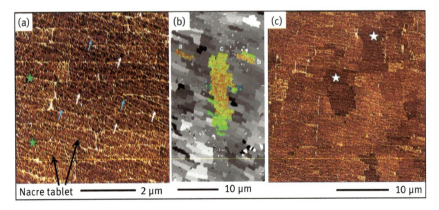

Fig. 12.6: Hierarchical arrangement of aragonite nanocrystals in the aragonitic shell portion of the bivalve *Mytilus edulis*. A nacre tablet is encased by matrix membranes (green stars in a) and consists of aragonite nanoparticles (white arrows in a) embedded in a network of biopolymer fibrils (blue arrows in a). Several tablets form a co-oriented mesoscale unit (white stars in c), the tower grain marked with c in the center of b. Within a tablet, aragonite nanocrystals are misoriented to each other, as highlighted by the patchiness of nacre tablets in the for orientation color-coded EBSD map in b. b is modified after [34].

The basic mineral unit of all nacreous hard tissues (Fig. 12.7c, d) is an aragonite tablet, enveloped by a biopolymer and a thin amorphous layer [64]. A nacre tablet is not a classical single crystal. It is a mesocrystal, as it is composed of slightly misoriented nanometer sized aragonite crystallites [34, 47]. Depending on the arrangement of tablets, two major nacreous microstructures prevail: (1) columnar nacre in gastropods (Fig. 12.7d), with tablets being arranged in columns and sheet nacre in bivalves (Fig. 12.7c), with tablets being stacked in a 'brick wall' organization (e.g. [10, 13, 14, 17, 26, 31, 33, 65]). However, not only the arrangement of tablets differs in these two mollusk classes but also tablet dimensions (see inserts in Fig. 12.7c, d). *Haliotis ovina* nacre tablets are thin (450 to 500 nm), while *M. edulis* nacre tablets are significantly thicker (1 to 2 µm).

12.4 Carbonate hard tissue microstructures and textures, difference in crystal co-orientation strength

Figures 12.8–11 present patterns of crystallite co- or misorientation in calcite and aragonite grown from solution and in biologically formed carbonate skeletons (shells and calcite teeth), respectively. Crystal orientation data are presented with color-coded EBSD maps and corresponding pole figures showing crystallographic *c*- and *a*-axes orientations, either as data points or as pole density distributions. Hard tissue microstructure is visualized with gray-scale EBSD band contrast measurement images.

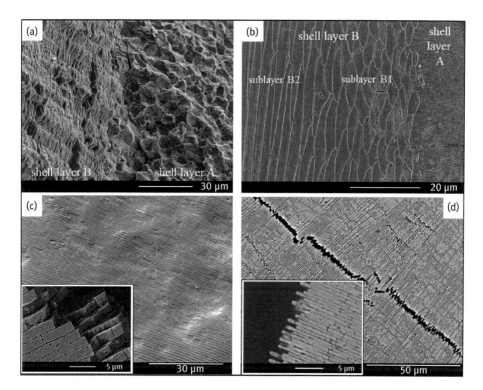

Fig. 12.7: Characteristic assemblies of mineral units for the formation of different shell layers. (a) FE-SEM image of a fracture surface in the shell of the gastropod *Haliotis ovina* showing aragonite mineral units in two different assemblages: irregular aragonite grains (shell layer A) at the seawater pointing side and the arrangement of nacre tablets to columns next to the soft tissue of the animal (shell layer B). (b) Polished surface showing a cross section through the shell of the modern brachiopod *Liothyrella uva*. Nanosized calcite crystallites comprise large interdigitating mineral units of the primary (shell layer A) and the fibers of the fibrous shell layer (shell layer B). Parallel arrays of calcite fibers form stacks, which are misoriented to each other by ~90°. A two-dimensional cut through the shell gives perpendicularly (sublayer B1) and transversely (sublayer B2) cut stacks of fibers. (c, d) FE-SEM images of polished shell surfaces depicting the two major modes of nacre tablet arrangements: in a bivalve (c, *Mytilus edulis*) and in a gastropod shell (d, *Haliotis ovina*).

EBSD band contrast is the signal strength in each measurement point. Thus, when a mineral is detected by the electron beam, a strong diffraction signal is obtained, while, when polymers or nanosized crystallites are encountered, that are too small to be differentiated by the electron beam, the diffraction signal is either absent or too weak to be detected automatically. Crystal co-orientation strength is assessed with MUD values (multiples of uniform distribution). A high MUD indicates high crystal co-orientation, whereas a low MUD value accounts for a high crystal misorientation (low crystal co-orientation) [66]. The uniform colors of the EBSD maps shown in Fig. 12.8 visualize the high co-orientation of calcite/aragonite crystallites in both the non-biological (Fig. 12.8a, b) and biological (Fig. 12.8d) carbonates. MUD values of geologic calcite and aragonite are above 700; the many measured *c*- and *a*-axes data points (1040 for calcite and 140 for aragonite) superimpose closely onto each other in the stereographic projections, resulting in no scatter of the *c*- and *a*-axes within the pole figures. Thus, these two carbonate crystals are single crystals. Figure 12.8c–e visualizes the mode and strength of calcite co-orientation in the tooth of the sea urchin *Paracentrotus lividus*. Even though the pattern of crystallite co-orientation resembles that present in non-biological calcite, the calcite in the sea urchin tooth is not a single crystal, as depicted by the slight variation of blue shades in the EBSD map of

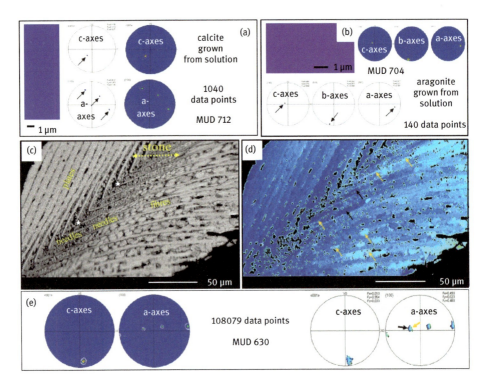

Fig. 12.8: Calcite crystal orientation pattern in geogene calcite and aragonite grown from solution (a, b) and in the calcitic tooth of the modern sea urchin *Paracentrotus lividus* (c–e). Crystal orientation data are given with EBSD measurements presented with gray-scale EBSD band contrast (c) and color-coded EBSD orientation maps (a, b, d) and corresponding pole figures (a, e) that show orientation data points and their density distributions. The strength of crystal co-orientation is given with MUD values. Geologic calcite (a) and aragonite (b) are single crystals. EBSD band contrast measurement image (c) visualizes well the microstructure of the calcite in the sea urchin tooth that consists of plates, needles, and fibers (the latter two shown in cross section in c). Even though crystal co-orientation is very high in this skeleton (see the uniform color of the crystal orientation map (d) and the close assemblage of *c*- and *a*-axes data points in the corresponding pole figures (e), the calcite in the sea urchin tooth, unlike the calcite grown from solution, is a mesocrystal and not a single crystal.

Fig. 12.8d and the slight scatter in *c*- and *a*-axes crystallographic orientation data in the corresponding pole figure. Crystallite co-orientation strength decreases, and the MUD value for the measurement shown in Fig. 12.8d declines to 630. The sea urchin tooth is a mesocrystal [52, 67] formed by an assembly of nanocrystallites with a remarkably high degree of crystallographic regularity [68–71].

Figure 12.9 shows the pattern and degree of aragonite co-orientation in the modern coral *Porites* sp. The crystal arrangement and strength of co-orientation here are markedly different from those in the sea urchin tooth. The microstructure of the modern coral *Porites* sp. is visualized with the band contrast measurement image shown in Fig. 12.9a. Aragonite crystallites nucleate on a thin organic template that lines the centers of calcification (Fig. 12.9b, white dots in Fig. 12.9a, c). When skeletal growth proceeds, aragonite crystallites increase in size and form thin fibers, which are more or less evenly developed. These grow outward from the centers of calcification (e.g. [72, 73]). The strength of aragonite co-orientation is weak in modern corals (Fig. 12.9c, d), and the MUD value of the measurement shown in Fig. 12.9 is 41. However, some aragonite crystal co-alignment is present in coral hard tissues as bundles of loosely co-oriented fibers form a mesoscale mineral unit (white star in Fig. 12.9c), with the mean orientation of this mineral unit being distinct from that of the neighboring unit.

Figures 12.10–12 highlight characteristic microstructures and textures of carbonate shells, e.g. that of the modern brachiopod *Magellania venosa* (Fig. 12.10), of the modern bivalve *M. edulis* (Figs. 12.11a, c and 12a, c) and that of the modern gastropod *H. ovina* (Figs. 12.11b, d and 12b, d). Depending on the species, the shells of modern rhynchonellide and terebratulide brachiopods consist of up to three calcite layers (e.g. [55, 74, 75, 76, 77]): the seaward primary, the inner fibrous, and the columnar layers. Figure 12.10 visualizes the microstructure and texture of the primary and the fibrous shell layer of the modern brachiopod *M. venosa*. Both shell layers consist of nanostructured calcite, that is assembled within the shell to distinctly shaped mineral units [53, 75, 76]. Nanosized calcite assemblages form the several micrometers large, irregularly shaped, interdigitating dendritic mineral units of the primary shell layer ([54], Fig. 12.10a, d). Nanoparticulate calcite constitutes the calcite fibers of the fibrous shell layer where parallel fibers form stacks. The morphological fiber axes of these arrays of fibers are rotated relative to each other by several tens of degrees, often by ~90°. In two dimensional sections longitudinal (Fig. 12.10e) and transverse (Fig. 12.10c) cut stacks emerge and constitute sublayers of the fibrous layer of the shell (Fig. 12.7b). In brachiopod shells calcite crystal co-orientation strength is moderate and scatters between MUD 80 and MUD 120. It is very similar for the primary and the fibrous shell layers (e.g. Fig. 12.10d, e [54]). This indicates that even though these two shell layers have highly different crystal morphologies and mineral unit boundaries, the preferred crystallographic orientation of the calcite is similar. However, the distinctness in mineral unit morphology indicates that at shell formation different control mechanisms are active [54]. For the measurements

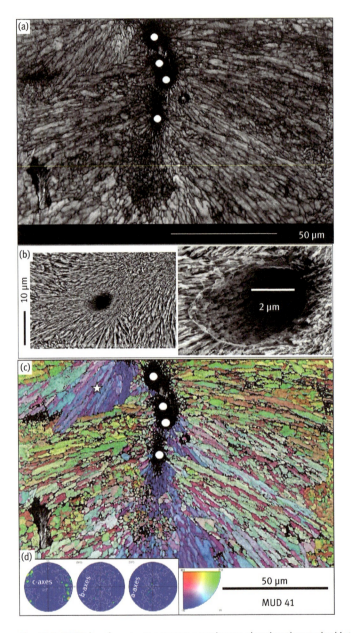

Fig. 12.9: EBSD band contrast measurement image showing the marked internal structuring of the skeleton of the modern coral *Porites* sp. White dots in a and c indicate centers of calcification where growth of aragonite crystallites starts (b). (b) Centers of calcification. When mineral growth proceeds further, aragonite crystals get co-oriented to some degree and form aragonite needles. Aragonite crystal organization in the skeleton of the modern coral *Porites* sp. is visualized with a color-coded EBSD map (c) and corresponding pole figure (d). The internal structuring of the aragonite skeleton is well visible within individual aragonite needles as well as in a mesoscale mineral unit (white star in c) consisting of loosely co-oriented aragonite needles. Note the distinctly low MUD value of 41 obtained for the EBSD measurement shown in Fig. 12.9.

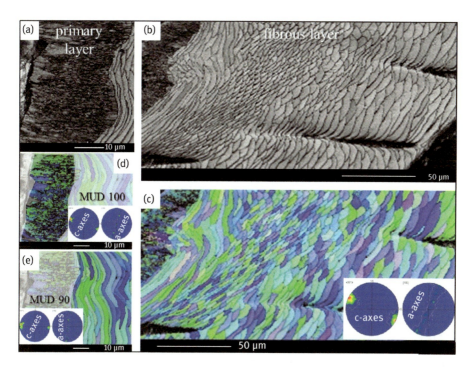

Fig. 12.10: Patterns of calcite orientation in the shell of the modern brachiopod *Magellania venosa* visualized with EBSD band contrast (a, b), color-coded orientation maps (c–e) and corresponding pole figures showing densities of crystal orientation distributions (c–e). Well visible is the microstructure and texture of the primary shell layer (a, d) that is highly distinct from that of the fibrous shell layer (c, e), containing stacks of transversely and some longitudinally cut fibers. The pole figures in d and e give orientation data of only one shell layer, the layer that is highlighted in the EBSD map: (d) primary and (e) fibrous shell layer. Even though the microstructure of the two shell layers differs, the strength of calcite crystallite co-orientation is very similar.

shown in Fig. 12.10d, e, we obtain an MUD value of 100 for the primary and an MUD value of 90 for the secondary shell layer, respectively. The difference in MUD is due to the fact that, in contrast to the fibrous layer, the primary shell layer is devoid of organic membranes and fibrils.

Even though nanoparticulate aragonite crystals comprise the entire shell of the gastropod *H. ovina* (Figs. 12.11b, d and 12b, 12d), their mode of assembly and strength of aragonite co-orientation are distinct in the two shell layers. The seaward layer consists of irregularly shaped aragonite mineral units that decrease in size toward the outside. The aragonite co-orientation strength is weak (MUD: 17, Fig. 12.12b) in this portion of the shell. The internal shell layer consists of nacreous aragonite; the nacre tablets are arranged in columns. In this shell portion aragonite nanocrystals are significantly more co-oriented (MUD: 102, Fig. 12.12d).

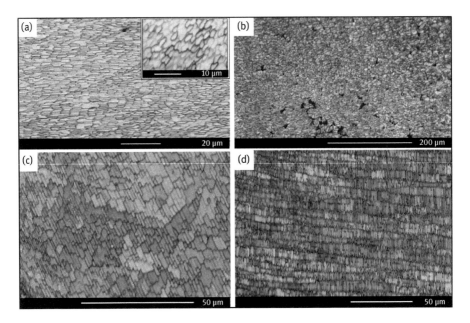

Fig. 12.11: EBSD band contrast measurement images visualizing the microstructures of the shell of the bivalve *Mytilus edulis* (a, c) and the gastropod *Haliotis ovina* (b, d). The outer shell layer of *Mytilus edulis* consists of co-oriented arrays of calcite fibers (insert in a gives a cross section through the fibers in higher magnification), the shell layer next to the soft tissue of the animal consists of nacreous aragonite in "brick-wall" arrangement (c). The outer shell layer of the gastropod *H. ovina* consists of irregularly shaped and sized aragonite mineral units (b), the shell layer next to the soft tissue of the animal consists of nacreous aragonite, with aragonite tablets forming interdigitating columns (d).

As the comparison of the two EBSD maps, and the corresponding pole figures as well as MUD values, shows (Figs. 12.11c, d and 12c, 12d), nacreous aragonite in the bivalve *M. edulis* is less nanostructured in comparison to that in the shell of *H. ovina*. Although the shells of both organisms contain a biomaterial type (nacre) formed with a very similar basic design principle (aragonite platelet embedded in a thin organic matrix), a large difference exists in the mode of assembly of both the mesoscale and macroscale structures and the strength of aragonite co-orientation (MUD of 102 in *H. ovina* nacre and MUD of 130 in *M. edulis* nacre). The outer shell layer of *M. edulis* is formed by parallel stacks of highly co-oriented calcite fibers (MUD of 331 for the measurement shown in Figs. 12.11a and 12a and [34, 78]). Here, we even see a strong tendency to a three-dimensional rather than to a simple cylindrical texture: compare the pole figures in Fig. 12.12a, where both the *c*- and *a*-axes have maxima, to the pole figures in Fig. 12.12b–d, where only the *c*-axes show a maximum, and the *a*-axes scatter on great circles, as it is the case for both shell layers of *H. ovina* and the nacreous shell portion of *M. edulis*.

Figure 12.13 shows the pattern of calcite crystal orientation in the egg shell of the goose *Tadorna ferruginea*. The shell is composed of large calcite units (crystals)

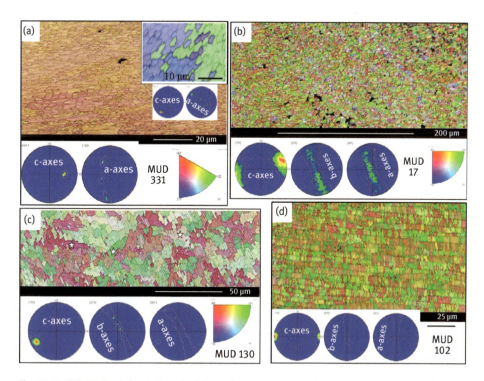

Fig. 12.12: Orientation information of calcite and aragonite mineral units in the shells of *Mytilus edulis* and *Haliotis ovina* visualized with color-coded orientation maps and corresponding pole figures. MUD values give the strength of carbonate crystal co-orientation of each EBSD map. Note the marked difference between the four shell layers in the pattern and strength of carbonate crystal co-orientation.

with irregular shapes (some marked with red letters in Fig. 12.13a). These are highly misoriented to each other (see the pole figure in Fig. 12.13a), interdigitate in three dimensions, and form a fractal (dendritic) microstructure (see the colored orientation map in Fig. 12.13a). For such a random pattern of orientation, an overall MUD value of 142 (Fig. 12.13a) is high. As Fig. 12.13b visualizes, calcite co-orientation strength in individual calcite units (crystals) is high and MUD values are well above 500. The high strength of crystal co-orientation within the individual units causes the high overall MUD of the measurement shown in Fig. 12.13. The misorientation versus distance diagrams (Fig. 12.13c) demonstrate that the large calcite units have an internal mosaic structure. These are composed of some (few) mosaic domains (indicated with red dashed lines and marked with red numbers in Fig. 12.13c). Misorientation within the large calcite units is between 3 and 5°, whereas misorientation between mosaic domains scatters between 1 and 2°.

Fractal morphologies of interdigitating crystals are present in other carbonate biological hard tissues as well, for example in the outer "primary" layer of modern brachiopod shells [54], pearl nacre and bivalve calcite.

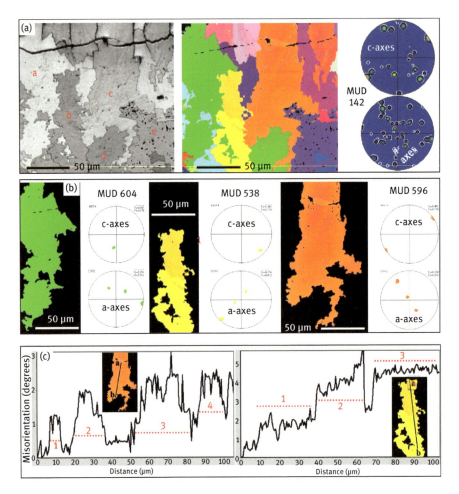

Fig. 12.13: Mode of calcite crystal assembly in the egg shell hard tissue of the goose *Tadorna ferruginea* visualized with EBSD band contrast measurement, color-coded orientation maps and pole figures. Note the irregular shape of large calcite units; their three-dimensional interdigitation causes the fractal microstructure of the hard tissue. However, within individual large calcite units, minerals are highly co-ordinated and are arranged in a mosaic structure.

12.5 Biomaterial functionalization through carbonate crystal orientation variation

The soft tissue of shelled organisms is surrounded by a hard and fracture-resistant skeleton, a material that is best produced when an organic matrix is reinforced by mineral [2, 35, 53]. In contrast to bioapatite-reinforced matrices (bones, phosphatic brachiopod shells), where adaptations to different functions are obtained by mineral/ biopolymer content gradations [1, 2, 63], this concept is not applicable to materials containing matrices reinforced by carbonate mineral, e.g. shells, calcitic teeth, and the exoskeleton of Crustacea. In the three latter biological hard materials, adaptation to various environments has to occur via other processes: for example, the formation of different carbonate mineral microstructures and their combination with a variety of biopolymer matrix fabrics. Hence, the diversification of carbonate hard tissue microstructure and the mode of arrangement of mineral units (e.g. grains, laths, fibers, tablets, columns, prisms) is of major importance for adaptation to various habitats, and is of evolutionary advantage [35, 49, 54, 79, 80, 81]. We discuss subsequently three different examples of environment-driven biomaterial adaptation: (1) calcitic teeth of sea urchins, (2) gastropod and bivalve shells, (3) calcite incorporation into isopod tergite and mandible cuticle.

12.5.1 Calcitic tooth of the sea urchin *P. lividus*

The masticator apparatus of a sea urchin has a set of five curved calcitic teeth (the Aristotle's lantern), that join at the end where the animal grazes on the sea floor substrate [82, 83]. Due to grazing, the tips of the teeth are continuously abraded, such that the teeth must be generated continuously at their posterior ends. The highly controlled structuring and crystal orientation of the tooth elements [52, 67] provides a predictable (reproducible) chipping-off at the very tip of the tooth which guarantees continuous self-sharpening and a constant shape of the five interlocking teeth.

Modern sea urchin teeth are composite materials consisting of calcite, high-Mg calcite and biopolymers [52, 84–87]. Sea urchin teeth combine two major structural elements: calcite plates that form the margins of the tooth and assemblages of high-Mg needles that accumulated mainly in the central tooth portion, the stone (Fig. 12.8c). Despite a marked ultrastructural and chemical differentiation and a curved outer morphology, calcite and Mg-calcite crystallites in the tooth are highly co-oriented; however, the tooth is not a single crystal. The tooth of the sea urchin *P. lividus* is a hierarchical multi-composite mesocrystal with calcite having a high degree of three-dimensional orientational coherence (MUD value of 630 for the EBSD measurement on the tooth portion shown in Fig. 12.8d, 8e; MUD value of 712 for the EBSD measurement of geogene calcite shown in Fig. 12.8a, [52]). At the cutting tip of the tooth we see two lattice orientations: one for the calcite plates and one for the high-Mg-calcite needles of the stone (see yellow and black arrows in Fig. 12.8d, e). Both orientations join at the tip of the tooth, the plates converging in a V-shaped manner, the very thin arrays of needles intercalating between the V-shaped plates. At self-sharpening, the plates are chipped off and the animal abrades material from the hard substrate with the highly co-oriented assemblage of high-Mg calcite needles [52, 67].

12.5.2 Shells of *H. ovina* and *M. edulis*

H. ovina lives in marine environments along rocky shores in water depths up to 60 m and is firmly attached to hard substrates by a powerful muscular foot. *M. edulis* lives in a surf environment, and it is attached firmly to rocks with byssal threads secreted by a mobile foot. Both mollusk species protect their soft tissue with a three-layered shell consisting of an outer organic layer, the periostracum, and a nacreous shell portion next to the soft tissue of the animal. Between these two shell layers is a region that is rigid in *H. ovina* and ductile in *M. edulis*. The central shell part of *H. ovina* consists of an ~60- to 70- and ~280- to 300-µm-thick layers of prismatic and nacreous aragonite, respectively. The central shell region of *M. edulis* comprises ~150 to 160 µm of fibrous calcite and a subjacent, ~80-µm-thick layer of nacreous aragonite. Thus, in *H. ovina* shell, nacreous aragonite prevails, while in the shell of *M. edulis*, fibrous calcite forms the major part of the hard tissue.

The outer shell portion of *H. ovina* with its irregularly shaped and sized aragonite mineral units being arranged with a low degree of co-orientation (MUD 17, Fig. 12.12b) provides the necessary high stiffness, while in *M. edulis*, the highly co-oriented (MUD 331, Fig. 12.12a) calcite fiber arrangement, with each fiber being sheathed by biopolymers, providing tensile strength and high ductility. In contrast to that of *H. ovina*, the shell of *M. edulis* is permanently exposed to friction by contact to sand particles and differences in pressure that are driven by wave motion. The living environment of *H. ovina* is not in the surf region; it is in deeper (subtidal) water levels. Thus, the different mechanical properties imparted by the mineral-biopolymer arrangement reflect the border conditions defined by the habitat.

12.5.3 Calcite in the tergite and mandible exocuticula of the isopod species *Porcellio scaber* and *Tylos europaeus*

The common rough woodlouse *P. scaber* and the sand burrowing species *T. europaeus* live in distinct habitats: *P. scaber* prefers mesic environments and the supralittoral *T. europaeus* lives on sandy substrate [88, 89]. Both isopod species incorporate mineral into their cuticle and reinforce the exocuticle and endocuticle layers with calcite, ACC, and amorphous calcium phosphate [90, 91]. Figure 12.14a–c visualizes the differences of calcite incorporation into the exocuticle of *P. scaber* (Fig. 12.13a) and that of *T. europaeus* (Fig. 12.14b, c). The amount of incorporated calcite differs significantly as well as the pattern of calcite organization. While the calcite layer within the exocuticle of *T. europaeus* is thick [90] and consists of a multitude of small calcite domains, which are highly misoriented to each other (see pole figure in Fig. 12.14b, MUD value of 8), the calcite in the exocuticle of *P. scaber* is significantly thinner [79] and is composed of a few large domains containing highly co-oriented calcite crystallites (see lower pole figure in Fig. 12.14a, MUD value of 148). The strength of calcite co-orientation increases

12.5 Biomaterial functionalization through carbonate crystal orientation variation — 263

when only one domain is regarded, in particular in *P. scaber* (upper pole figure in Fig. 12.14a), but also in *T. europaeus* (Fig. 12.14c, where as an example four domains, numbered 1 to 4, are highlighted). The above-described differences in mineral incorporation are relatable to function and evolutionary adaptations to the animals habitat and behavior. The tergite of *P. scaber* is thin and flexible. As upon predation the animal either runs away or clings still and firmly to the substrate [92], it needs a lightweight and highly flexible cuticle. In contrast, the beach-dwelling isopod *T. europaeus* rolls into a sphere upon threat, with the animal relying on its thick cuticle for protection of its soft body tissue [79].

Figure 12.14d and e shows the incorporation of a calcite band into the pars incisiva (PI) of the left mandible of the isopod *T. europaeus*. The beginning of the calcite layer starts a few hundred micrometers away of the incisive edge and extends into the corpus of the mandible [93]. It is fairly uniform in extent, is a few micrometers thick,

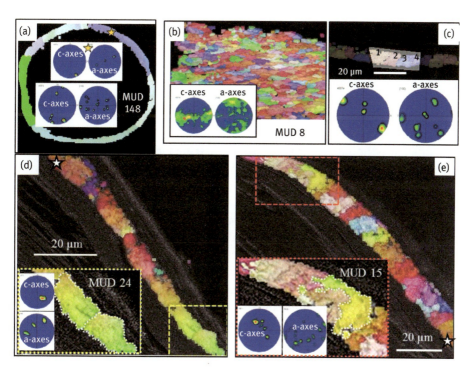

Fig. 12.14: Calcite crystallite orientation pattern and co-orientation strength variation in the tergite exocuticle of the terrestrial isopods: *Porcellio scaber* (a) and *Tylos europaeus* (b, c). Calcite in the cuticle of *P. scaber* is highly co-oriented (MUD: 148), while the calcite within the cuticle of *T. europaeus* is in almost random organization (MUD: 8). Incorporation of a calcite layer into partes incisivae of the mandibles of the isopod *Tylos europaeus* (d, e). Note the difference in microstructure, texture, and strength of calcite co-orientation within the intercalated calcite band that takes place in the hard tissue on short distances, within a few hundred micrometers. For further details, see [94]. Modified after [94].

and is subdivided into different domains (Fig. 12.14d, e [94]). The most remarkable feature of the microstructure and texture of the incorporated calcite is the distinctness in crystal co-orientation or misorientation between the tip of the calcite band (insert in Fig. 12.14d) and a portion located further to the base of the PI (insert in Fig. 12.14e). At the tip of the calcite band, calcite crystallites are aligned to each other in an almost perfect register and the first three mineral domains are graded mesocrystals (for further explanation see [93, 94]). Further toward the corpus, crystalline domains become irregular in shape, are smaller in size and contain calcite in a single crystalline organization. Misorientation between these calcite domains is highly increased [94]. It is remarkable that *T. europaeus* is able to vary calcite crystal organization not only within the same organism, but also on very small distances, as required for a particular skeletal functionality. This control of small-scale crystal organization construction (texture) appears to be intrinsic to isopods as, so far, it has not yet been observed in other carbonate hard tissues such as shells or calcite teeth. In addition, in *T. europaeus* and *P. scaber*, calcite crystal orientation appears to be independent of the orientation of organic fibrils of the twisted plywood structure, a characteristic not yet observed in shell materials where crystallite and mineral unit orientation and organization is strictly controlled by organic fibrils and membranes.

12.6 Concluding summary

Microstructure is the link between chemistry and technology of materials. In biological hard tissues, microstructure characteristics, biopolymer occlusion at all scale levels, and hierarchy are the keys for material optimization for environment adaptation and survival.

For biological carbonate hard tissues, we observe a vast diversity of microstructure and texture patterns that range from almost unaligned to highly co-aligned crystal assemblies (Fig. 12.15). As biological materials are highly evolved by evolution, this suggests that both a high order and a high disorder in crystallographic and mineral organization can be advantageous for the organism in different circumstances. In the case of biologically secreted hard materials, and in contrast to non-biological minerals, both the ordered and the disordered fabrics are tailored by the organism. In carbonate biological hard materials, we observe all intermediate stages of texture (Fig. 12.15) that range from three-dimensional to almost anisotropic mineral alignments (Tab. 12.1). The two classical end-members of crystal arrangement patterns, single crystals with perfect co-orientation and polycrystals with random orientational textures, are never developed. Pure Ca-carbonate single crystals are prone to brittle mechanical failure. The same is true for pure Ca-carbonate materials with random texture, where crack propagation is completely unpredictable. Current diffraction techniques in SEM and TEM prove clearly that sea urchin teeth, spines, coccolith scale units, and subunits are not single crystals but hierarchically structured composite entities (Yin et al., submitted for publication) [52]. On the other hand, the least textured biological carbonate materials found so far are the skeletons of stony corals (He et al., in preparation) [95] and calcite within the tergite cuticle of the conglobating isopod species *Armadillidium vulgare* and *T. europaeus* [79]. For conglobating isopods, isotropic properties are of

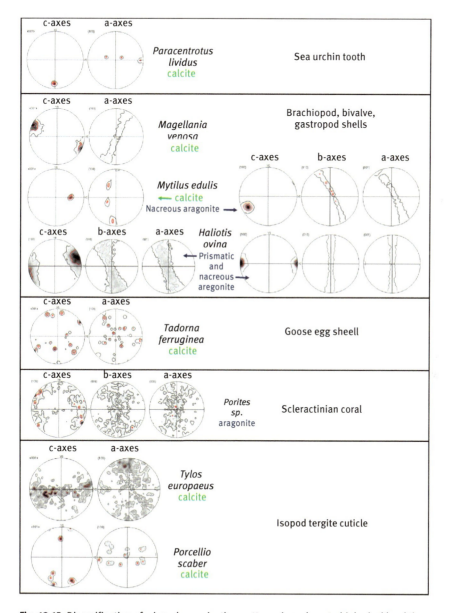

Fig. 12.15: Diversification of mineral organization patterns in carbonate biological hard tissues.

advantage since unexpected high mechanical loads, as exerted during predation, have no sharp directional preference [79].

EBSD measurements (in SEM and TEM) record the orientation of all crystallographic axes and prove that carbonate biological hard tissues, with the possible exception of coccoliths (Yin et al., submitted for publication) [87, 96], are hybrid mesocrystals on the submicron level, are crystallographically continuous, contain intercalated organic matrices, are morphologically well-defined in three dimensions, and where organic membranes define morphology, occur in a higher-order superstructure. This structural principle enables the biologic tailoring of shape and mechanical or other functional properties of the biological hard tissue, skeletal element, or tooth [69, 97, 98, 99]. Thus, mesocrystallinity is a characteristic that is worth to aim for when new synthetic materials or man-made materials for specific applications are developed [100].

Examples (Tab. 12.1) where biomaterial design principles are successfully transferred into biomimetic material systems are materials based on two- and three-dimensional colloidal assemblies of nanoparticles (e.g. gem, artificial opals [101–104]), materials with large and defect-free superstructures with developed orientational and translational order (maghemite nanocube superlattices [105], PbS nanocrystal

Tab. 12.1: Compilation of characteristic mineral assembly patterns in man-made materials and biological carbonate hard tissues.

	Single crystals	Collodial crystal arrays	Mesocrystals		Textured polycrystals	Polycrystals
Non-biological regimes	calcite	artificial opals	maghemite nanocubes PbS organic mesocrystals		goethite	marble

	Biological mesocrystals						
	Almost single crystals	Gradient crystals	Composite crystals	Multiplex composite crystals	Dendritic crystals	Axial co-oriented polycrystal	Almost untextured polycrystal
Biological regimes	Cidaris cidaris spine calcite	Tylos europaeus mandible calcite	Mytilus, edulis, Crassostrea gigas shell calcite	Paracentrotus lividus tooth calcite	Gryphus vitreus primary shell calcite, Tadorna ferruginea shell calcite	Mytilus edulis, Crassostrea gigas, Haliotis glabra, Elliptio eliptio nacreous aragonite	Tylos europaeus tergite calcite

superlattices [106]), polycrystalline materials that get textured by topotactic growth onto textured templates (formation of textured hematite from textured goethite [107]), and formation of nacre-like materials, hybrid and layered organic-inorganic composites [108, 109]. Contrasting to biological regimes, in synthetic and geological materials, single crystallinity and polycrystallinity are of similar importance as mesocrystalline behavior.

Acknowledgments

We are grateful to numerous colleagues for their fruitful collaborations in this work, particularly the following bachelor, master, and PhD students: L. Casella, M. Greiner, M. Heinig, F. Weitzel, F. Reidl, L. Reim, A. Goetz, F. Nindiyasari, K. Stevens, S. He, C. Beeken, M. Goos, M. Simonet Roda, S. Hahn, J. Frost, J. Huber, C. Reisecker, B. Seidl, and P. Alexa.

References

[1] Currey JD. Bones: structure and mechanics. Princeton (NJ): Princeton University Press; 2006.
[2] Currey JD. The design of mineralised hard tissues for their mechanical functions. J Exp Biol 1999;202:3285–94.
[3] Fratzl P, Weinkamer R. Nature's hierarchical materials. Prog Mater Sci 2007;52:1263–334.
[4] Dunlop JWC, Fratzl P. Biological composites. In: Clarke DR, Ruhle M, Zok F, editors. Annu Rev Mater Res. Vol 40. Palo Alto: Annual Reviews; 2010:1–24.
[5] Meyers MA, Chen PY, Lin AYM, Seki Y. Biological materials: structure and mechanical properties. Prog Mater Sci 2008;53:1–206.
[6] Meyers MA, Lin AYM, Chen PY, Muyco J. Mechanical strength of abalone nacre: role of the soft organic layer. J Mech Behav Biomed Mater 2008;1:76–85.
[7] Mayer G. Rigid biological systems as models for synthetic composites. Science 2005;310:1144–7.
[8] Addadi L, Joester D, Nudelman F, Weiner S. Mollusk shell formation: a source of new concepts for understanding biomineralization processes. Chemistry 2006;12:981–7.
[9] Levi-Kalisman Y, Falini G, Addadi L, Weiner S. Structure of the nacreous organic matrix of a bivalve mollusk shell examined in the hydrated state using cryo-TEM. J Struct Biol 2001;135:8–17.
[10] Rousseau M, Lopez E, Couté A, et al. Sheet nacre growth mechanism: a Voronoi model. J Struct Biol 2005;149:149–57.
[11] Nudelman F, Gotliv BA, Addadi L, Weiner S. Mollusk shell formation: mapping the distribution of organic matrix components underlying a single aragonitic tablet in nacre. J Struct Biol 2006;153:176–87.
[12] Cölfen H. Bio-inspired mineralization using hydrophilic polymers. In: Naka K, editor. Biomineralization II: Mineralization Using Synthetic Polymers and Templates. Berlin, Heidelberg: Springer Berlin Heidelberg;2007:1–77.
[13] Cartwright JHE, Checa AG. The dynamics of nacre self-assembly. J R Soc Interface 2007;4:491–504.

[14] Heinemann F, Launspach M, Gries K, Fritz M. Gastropod nacre: structure, properties and growth – biological, chemical and physical basics. Biophys Chem 2011;153:126–53.
[15] Marin F, Luquet G. Molluscan shell proteins. CR Palevol 2004;3:469–92.
[16] Marin F, Luquet G, Marie B, Medakovic D. Molluscan shell proteins: primary structure, origin, and evolution. Curr Top Dev Biol 2008;80:209–76.
[17] Rousseau M, Meibom A, Geze M, Bourrat X, Angellier M, Lopez E. Dynamics of sheet nacre formation in bivalves. J Struct Biol 2009;165:190–5.
[18] Checa AG, Macías-Sánchez E, Harper EM, Cartwright JHE. Organic membranes determine the pattern of the columnar prismatic layer of mollusc shells. Proc R Soc B 2016;283.
[19] Nakahara H. Nacre formation in bivalve and gastropod molluscs. In: Suga S, Nakahara H, editors. Mechanisms and phylogeny of mineralization in biological systems: Springer: Japan; 1991:343–50.
[20] Bevelander G, Nakahara H. An electron microscope study of the formation of the nacreous layer in the shell of certain bivalve molluscs. Calcif Tissue Res 1969;3:84–92.
[21] Gaspard D, Marin F, Guichard N, Morel S, Alcaraz G, Luquet G. Shell matrices of recent rhynchonelliform brachiopods: microstructures and glycosylation studies. Trans R Soc Edinburgh 2008;98:415–24.
[22] Nindiyasari F, Ziegler A, Griesshaber E, et al. Effect of hydrogel matrices on calcite crystal growth morphology, aggregate formation, and co-orientation in biomimetic experiments and biomineralization environments. Crystal Growth Des 2015;15:2667–85.
[23] Weiner S, Talmon Y, Traub W. Electron-diffraction of mollusk shell organic matrices and their relationship to the mineral phase. Int J Biol Macromol 1983;5:325–8.
[24] Weiner S, Traub W. Macromolecules in mollusk shells and their functions in biomineralization. Philos Trans R Soc Lond, Ser B: Biol Sci 1984;304.
[25] Worms D, Weiner S. Mollusk shell organic matrix – Fourier-transform infrared study of the acidic macromolecules. J Exp Zool 1986;237:11–20.
[26] Rousseau M, Lopez E, Stempflé P, et al. Multiscale structure of sheet nacre. Biomaterials 2005;26:6254–62.
[27] Osuna-Mascaró AJ, Cruz-Bustos T, Marin F, Checa AG. Ultrastructure of the interlamellar membranes of the nacre of the bivalve *Pteria hirundo*, determined by immunolabelling. Plos One 2015;10.
[28] Checa AG, Mutvei H, Osuna-Mascaró AJ, et al. Crystallographic control on the substructure of nacre tablets. J Struct Biol 2013;183:368–76.
[29] Checa AG, Cartwright JHE, Willinger M-G. Mineral bridges in nacre revisited. ArXiv e-prints 2012.
[30] Checa AG, Cartwright JHE, Willinger MG. Mineral bridges in nacre. J Struct Biol 2011;176:330–9.
[31] Checa AG, Cartwright JHE, Willinger MG. The key role of the surface membrane in why gastropod nacre grows in towers. Proc Natl Acad Sci USA 2009;106:38–43.
[32] Macías-Sánchez E, Checa AG, Willinger MG. The transport system of nacre components through the surface membrane of gastropods. Key Eng Mater 2015;672:103–12.
[33] Schäffer TE, Ionescu Zanetti C, Proksch R, et al. Does abalone nacre form by heteroepitaxial nucleation or by growth through mineral bridges? Chem Mater 1997;9:1731–40.
[34] Griesshaber E, Schmahl WW, Ubhi HS, et al. Homoepitaxial meso- and microscale crystal co-orientation and organic matrix network structure in *Mytilus edulis* nacre and calcite. Acta Biomater 2013;9:9492–502.
[35] Schmahl WW, Griesshaber E, Merkel C, et al. Hierarchical fibre composite structure and micromechanical properties of phosphatic and calcitic brachiopod shell biomaterials – an overview. Mineral Mag 2008;72:541–62.
[36] Ziegler A, Weihrauch D, Towle DW, Hagedorn M. Expression of Ca^{2+}-ATPase and $Na+/Ca^{2+}$-exchanger is upregulated during epithelial Ca^{2+} transport in hypodermal cells of the isopod *Porcellio scaber*. Cell Calcium 2002;32:131–41.

[37] Ziegler A, Weihrauch D, Hagedorn M, Towle DW, Bleher R. Expression and polarity reversal of V-type H+-ATPase during the mineralization-demineralization cycle in *Porcellio scaber* sternal epithelial cells. J Exp Biol 2004;207:1749–56.
[38] Ahearn GA, Mandal PK, Mandal A. Calcium regulation in crustaceans during the molt cycle: a review and update. Comp Biochem Physiol 2004;137:247–57.
[39] Weiss IM, Tuross N, Addadi L, Weiner S. Mollusc larval shell formation: amorphous calcium carbonate is a precursor phase for aragonite. J Exp Zool 2002;293:478–91.
[40] Beniash E, Aizenberg J, Addadi L, Weiner S. Amorphous calcium carbonate transforms into calcite during sea urchin larval spicule growth. Proc R Soc B 1997;264:461–5.
[41] Addadi L, Raz S, Weiner S. Taking advantage of disorder: amorphous calcium carbonate and its roles in biomineralization. Adv Mater 2003;15:959–70.
[42] Politi Y, Arad T, Klein E, Weiner S, Addadi L. Sea urchin spine calcite forms via a transient amorphous calcium carbonate phase. Science 2004;306:1161–4.
[43] Politi Y, Levi-Kalisman Y, Raz S, et al. Structural characterization of the transient amorphous calcium carbonate precursor phase in sea urchin embryos. Adv Funct Mater 2006;16:1289–98.
[44] Weiner S, Addadi L. Crystallization pathways in biomineralization. In: Clarke DR, Fratzl P, editors. Annu Rev Mater Res. Vol. 41. Palo Alto: Annual Reviews; 2011:21–40.
[45] Vidavsky N, Addadi S, Schertel A, et al. Calcium transport into the cells of the sea urchin larva in relation to spicule formation. Proc Natl Acad Sci USA 2016;113:12637–42.
[46] Vidavsky N, Addadi S, Mahamid J, et al. Initial stages of calcium uptake and mineral deposition in sea urchin embryos. Proc Natl Acad Sci USA 2014;111:39–44.
[47] Almagro I, Drzymała P, Berent K, et al. New crystallographic relationships in biogenic aragonite: the crossed-lamellar microstructures of mollusks. Crystal Growth Des 2016;16:2083–93.
[48] Willinger MG, Checa AG, Bonarski JT, Faryna M, Berent K. Biogenic crystallographically continuous aragonite helices: the microstructure of the planktonic gastropod *Cuvierina*. Adv Funct Mater 2016;26:553–61.
[49] Checa A, Harper EM, Willinger M. Aragonitic dendritic prismatic shell microstructure in *Thracia* (Bivalvia, Anomalodesmata). Invertebrate Biol 2012;131:19–29.
[50] Checa AG, Esteban-Delgado FJ, Ramírez-Rico J, Rodríguez-Navarro AB. Crystallographic reorganization of the calcitic prismatic layer of oysters. J Struct Biol 2009;167:261–70.
[51] Checa AG, Esteban-Delgado FJ, Rodríguez-Navarro AB. Crystallographic structure of the foliated calcite of bivalves. J Struct Biol 2007;157:393–402.
[52] Goetz AJ, Griesshaber E, Abel R, Fehr T, Ruthensteiner B, Schmahl WW. Tailored order: the mesocrystalline nature of sea urchin teeth. Acta Biomater 2014;10:3885–98.
[53] Schmahl WW, Griesshaber E, Kelm K, et al. Hierarchical structure of marine shell biomaterials: biomechanical functionalization of calcite by brachiopods. Z Kristallogr 2012;227:793–804.
[54] Goetz AJ, Steinmetz DR, Griesshaber E, et al. Interdigitating biocalcite dendrites form a 3-D jigsaw structure in brachiopod shells. Acta Biomater 2011;7:2237–43.
[55] Griesshaber E, Schmahl WW, Neuser R, et al. Crystallographic texture and microstructure of terebratulide brachiopod shell calcite: an optimized materials design with hierarchical architecture. Am Mineral 2007;92:722–34.
[56] Schmahl WW, Griesshaber E, Neuser R, Lenze A, Job R, Brand U. The microstructure of the fibrous layer of terebratulide brachiopod shell calcite. Eur J Mineral 2004;16:693–7.
[57] Ziegler A. Ultrastructural changes of the anterior and posterior sternal integument of the terrestrial isopod *Porcellio scaber* Latr. (Crustacea) during the moult cycle. Tissue Cell 1997;29:63–76.
[58] Ziegler A. Ultrastructural evidence for transepithelial calcium transport in the anterior sternal epithelium of the terrestrial isopod *Porcellio scaber* (Crustacea) during the formation and resorption of CaCO3 deposits. Cell Tissue Res 1996;284:459–66.

[59] Meldrum FC, Cölfen H. Controlling mineral morphologies and structures in biological and synthetic systems. Chem Rev 2008;108:4332–432.
[60] Park RJ, Meldrum FC. Synthesis of single crystals of calcite with complex morphologies. Adv Mater 2002;14:1167–9.
[61] Kim YY, Schenk AS, Ihli J, et al. A critical analysis of calcium carbonate mesocrystals. Nat Commun 2014;5:14.
[62] Schmahl WW, Kelm K, Griesshaber E, et al. The hierachical organization in biomaterials: from nanoparticles vio mescrystals to functionality. Semin Soc Esp Mineral 2010;7:05–21.
[63] Merkel C, Deuschle J, Griesshaber E, et al. Mechanical properties of modern calcite- (*Mergerlia truncata*) and phosphate-shelled brachiopods (*Discradisca stella* and *Lingula anatina*) determined by nanoindentation. J Struct Biol 2009;168:396–408.
[64] Nassif N, Pinna N, Gehrke N, Antonietti M, Jager C, Cölfen H. Amorphous layer around aragonite platelets in nacre. Proc Natl Acad Sci USA 2005;102:12653–5.
[65] Checa AG, Rodríguez-Navarro AB. Self-organisation of nacre in the shells of Pterioida (Bivalvia : Mollusca). Biomaterials 2005;26:1071–9.
[66] Engler O, Randle V. Introduction to texture analysis: macrotexture, microtexture, and orientation mapping. CRC Press London, UK; 2009.
[67] Griesshaber E, Goetz AJ, Howard L, Ball A, Ruff S, Schmahl WW. Crystal architecture of the tooth and jaw bone (pyramid) of the sea urchin *Paracentrotus lividus*. Bioinspired Biomim Nanobiomat 2012;1:133–9.
[68] Seto J, Ma YR, Davis SA, et al. Structure-property relationships of a biological mesocrystal in the adult sea urchin spine. Proc Natl Acad Sci USA 2012;109:3699–704.
[69] Cölfen H, Antonietti M. Mesocrystals and nonclassical crystallization. John Wiley & Sons Hoboken, New Jersey, USA; 2008.
[70] Song RQ, Colfen H. Mesocrystals-ordered nanoparticle superstructures. Adv Mater 2010;22:1301–30.
[71] Cölfen H. Biomineralization: a crystal-clear view. Nat Mater 2010;9:960–1.
[72] Meibom A, Cuif JP, Hillion FO, et al. Distribution of magnesium in coral skeleton. Geophys Res Lett 2004;31.
[73] Meibom A, Mostefaoui S, Cuif JP, et al. Biological forcing controls the chemistry of reef-building coral skeleton. Geophys Res Lett 2007;34.
[74] Griesshaber E, Neuser R, Schmahl W. The application of EBSD analysis to biomaterials: microstructural and crystallographic texture variations in marine carbonate shells. Semin Soc Esp Mineral 2010;7:22–34.
[75] Goetz AJ, Griesshaber E, Neuser RD, et al. Calcite morphology, texture and hardness in the distinct layers of rhynchonelliform brachiopod shells. Eur J Mineral 2009;21:303–15.
[76] Griesshaber E, Kelm K, Sehrbrock A, et al. Amorphous calcium carbonate in the shell material of the brachiopod *Megerlia truncata*. Eur J Mineral 2009;21:715–23.
[77] Pérez-Huerta A, Cusack M, Zhu WZ, England J, Hughes J. Material properties of brachiopod shell ultrastructure by nanoindentation. J R Soc Interface 2007;4:33–9.
[78] Maier BJ, Griesshaber E, Alexa P, Ziegler A, Ubhi HS, Schmahl WW. Biological control of crystallographic architecture: hierarchy and co-alignment parameters. Acta Biomater 2014;10:3866–74.
[79] Seidl BHM, Reisecker C, Hild S, Griesshaber E, Ziegler A. Calcite distribution and orientation in the tergite exocuticle of the isopods *Porcellio scaber* and *Armadillidium vulgare* (Oniscidea, Crustacea) – a combined FE-SEM, polarized SC mu-RSI and EBSD study. Z Kristallogr 2012;227:777–92.
[80] Morton JB, Harper SN. Bilinguals show an advantage in cognitive control – the question is why. Dev Sci 2009;12:502–3.
[81] Harper EM, Peck LS, Hendry KR. Patterns of shell repair in articulate brachiopods indicate size constitutes a refuge from predation. Mar Biol 2009;156:1993–2000.

[82] Märkel K, Röser U, Mackenstedt U, Klostermann M. Ultrastructural investigation of matrix-mediated biomineralization in echinoids (Echinodermata, Echinoida). Zoomorphology 1986;106:232–43.

[83] Wang RZ, Addadi L, Weiner S. Design strategies of sea urchin teeth: structure, composition and micromechanical relations to function. Philos Trans R Soc Lond B 1997;352:469–80.

[84] Ma Y, Cohen SR, Addadi L, Weiner S. Sea urchin tooth design: an "all-calcite" polycrystalline reinforced fibre composite for grinding rocks. Adv Mater 2008;20(8):1555–9.

[85] Ma YR, Aichmayer B, Paris O, et al. The grinding tip of the sea urchin tooth exhibits exquisite control over calcite crystal orientation and mg distribution. Proc Natl Acad Sci USA 2009;106:6048–53.

[86] Killian CE, Metzler RA, Gong YUT, et al. Mechanism of calcite co-orientation in the sea urchin tooth. J Am Chem Soc 2009;131:18404–9.

[87] Hoffmann R, Kirchlechner C, Langer G, Wochnik AS, Griesshaber E, Schmahl, WW, Scheu C. Insight into *Emiliania huxleyi* coccospheres by focused ion beam sectioning. Biogeosciences 2015;12:825–34.

[88] Warburg MR. Water relation and internal body temperature of isopods from mesic and xeric habitats. Physiol Zool 1965;38:99–109.

[89] Schmalfuss H, Vergara K. The isopod genus Tylos (Oniscidea: Tylidae) in Chile, with bibliographies of all described species of the genus. Stutt Beitr Naturke A 2000;162:1–42.

[90] Seidl B, Huemer K, Neues F, Hild S, Epple M, Ziegler A. Ultrastructure and mineral distribution in the tergite cuticle of the beach isopod *Tylos europaeus* Arcangeli, 1938. J Struct Biol 2011;174:512–26.

[91] Neues F, Ziegler A, Epple M. The composition of the mineralized cuticle in marine and terrestrial isopods: a comparative study. Crystengcomm 2007;9:1245–51.

[92] Schmalfuss H. Eco-morphological strategies in terrestrial isopods. In: Sutton SL, Holdich DM, editors. The Biology of Terrestrial Isopods. London: Oxford University Press;1984:49–63.

[93] Huber J, Fabritius HO, Griesshaber E, Ziegler A. Function-related adaptations of ultrastructure, mineral phase distribution and mechanical properties in the incisive cuticle of mandibles of *Porcellio scaber* Latreille, 1804. J Struct Biol 2014;188:1–15.

[94] Huber J, Griesshaber E, Nindiyasari F, Schmahl WW, Ziegler A. Functionalization of biomineral reinforcement in crustacean cuticle: calcite orientation in the partes incisivae of the mandibles of *Porcellio scaber* and the supralittoral species *Tylos europaeus* (Oniscidea, Isopoda). J Struct Biol 2015;190:173–91.

[95] Casella LA, Griesshaber E, Yin X, et al. Experimental diagenesis: insights into aragonite to calcite transformation of *Arctica islandica* shells by hydrothermal treatment. Biogeosciences 2017;14:1461–92.

[96] Hoffmann R, Wochnik AS, Heinzl C, Betzler SB, Matich S, Griesshaber E, Schulz H, Kucera M, Young JR, Scheu C, Schmahl WW. Nanoprobe crystallographic orientation studies of isolated shield elements of the coccolithophore species *Emiliania huxleyi*. Eur J Mineral 2014;26:473–83.

[97] Penn RL, Banfield JF. Morphology development and crystal growth in nanocrystalline aggregates under hydrothermal conditions: insights from titania. Geochim Cosmochim Acta 1999;63:1549–57.

[98] Mann S. Self-assembly and transformation of hybrid nano-objects and nanostructures under equilibrium and non-equilibrium conditions. Nat Mater 2009;8:781–92.

[99] Cölfen H, Mann S. Higher-order organization by mesoscale self-assembly and transformation of hybrid nanostructures. Angew Chem 2003;42:2350–65.

[100] Sturm EV, Cölfen H. Mesocrystals: structural and morphogenetic aspects. Chem Soc Rev 2016;45:5821–33.

[101] Bergström L, Sturm EV, Salazar-Alvarez G, Colfen H. Mesocrystals in biominerals and colloidal arrays. Acc Chem Res 2015;48:1391–402.
[102] Fritsch E, Gaillou E, Rondeau B, Barreau A, Albertini D, Ostroumov M. The nanostructure of fire opal. J Non-Crystall Solids 2006;352:3957–60.
[103] Gaillou E, Fritsch E, Aguilar-Reyes B, et al. Common gem opal: an investigation of micro- to nano-structure. Am Mineral 2008;93:1865–73.
[104] Velev OD, Kaler EW. Structured porous materials via colloidal crystal templating: from inorganic oxides to metals. Adv Mater 2000;12:531–4.
[105] Ahniyaz A, Sakamoto Y, Bergstrom L. Magnetic field-induced assembly of oriented superlattices from maghemite nanocubes. Proc Natl Acad Sci USA 2007;104:17570–4.
[106] Simon P, Rosseeva E, Baburin IA, et al. PbS-organic mesocrystals: the relationship between nanocrystal orientation and superlattice array. Angew Chem 2012;51:10776–81.
[107] Uchikoshi T, Nakamura N, Sakka Y. Fabrication of textured hematite via topotactic transformation of textured goethite. Appl Phys Exp 2009;2(10):101601.
[108] Antonietti M, Ozin GA. Promises and problems of mesoscale materials chemistry or why meso? Chem Eur J 2004;10:28–41.
[109] Zhang TJ, Ma YR, Qi LM. Bioinspired colloidal materials with special optical, mechanical, and cell-mimetic functions. J Mater Chem B 2013;1:251–64.

Klaus G. Nickel, Katharina Klang, Christoph Lauer and Gerald Buck
13 Sea urchin spines as role models for biological design and integrative structures

13.1 Introduction

In October 2014, the Collaborative Research Center (Transregio-SFB) TRR 141 started its activities under the name "Biological Design and Integrative Structures – Analysis, Simulation and Implementation in Architecture". As the name implies, it deals with a biomimetic approach to study biological role models, which are investigated and functionally abstracted in order to be potentially implemented in the context of the building and construction technologies. This includes "reverse biomimetics", i.e. the transfer of knowledge from the engineering and architectural sciences to understand the biological background.

Since the numerous projects are outlined on the website of the center (www.trr141.de), we will concentrate here on a few aspects of those projects in which applied mineralogy plays a significant part. Our aim is to show how applied mineralogy is connected to the materials science of non-metallic inorganic substances and their composites. We therefore deal with natural mineral-based materials and structures as well as technical artificial materials and constructions, which are usually addressed as ceramics and/or building materials such as mortar and concrete. Even though there are naturally formed phases within these material classes (e.g. the autogenous binder phases in some Roman buildings [1] or the "natural concrete" found there [2]), the characteristic phases of concrete have definitely artificial character. Hence, applied mineralogists often cross the border from traditional mineralogy with its limitation to naturally occurring materials toward a more general mineralogical materials science.

Sea urchins may be seen as a bonanza for biomimetics. Looking at 500 million years of successful survival in the record of life on Earth, they have developed many particularities, which predestined them as role models. These include the sophisticated tooth system ("Aristotle's lantern"), a self-sharpening rock-grinding construction [3, 4], connective tissues with controllable stress-transfer capacity [5], underwater adhesives [6], and segmented shell skeletons akin to modern buildings [7].

Naturally, the most prominent feature of sea urchins, their spines, became subject of interest as well. In our studies, we concentrate on spines of *Heterocentrotus mammillatus* and *Phyllacanthus imperialis* (Fig. 13.1), because they represent extreme types in terms of size (adult spines can exceed 10 cm in length and 1 cm in diameter), composition ($MgCO_3$-contents can exceed 13 mol%), habitat (reefs with high hydrodynamic forces), and other properties discussed below.

Fig. 13.1: *Phyllacanthus imperialis* (background) and two *Heterocentrotus mammillatus*.

13.2 Properties related to composition and nanostructure

Sea urchin spines are made of biogenic Mg-calcite. The content of $MgCO_3$ in the spine varies between about 2 and 13 mol% among different species, which, on the one hand, correlates positively with the water temperature of the habitat, but, on the other hand, varies within the spines, causing variations in the hardness of the calcite [8]. It is known for quite some time that at least those from some species possess unusual mechanical strength [9], and it was puzzling that they were single crystals despite their high porosity [10]. Nowadays, it is known that the spines are indeed formed by mesocrystals of nanoscale size [11], which are largely coherent. The slight misfit between individual mesocrystals is evident not only in transmission electron microscopy (TEM) studies but can be inferred already from X-ray microdiffraction studies, revealing the degree of misorientation being less than 2° [12].

This ultrastructural design is typical for biogenically formed minerals of which many examples of well-oriented crystal growth and its relation to cell biology exist [13]. However, unlike nacre with several percent of organic matter serving as interface material [14], sea urchin spines contain much less organic material, usually well below 1 wt% [15]. However, organic phases may play a role in the formation of amorphous calcium carbonate (ACC), which is much more soluble as calcite and may contain up to 15 wt% water. ACC is formed during spine growth and repairing processes [16] and its amounts may suffice to act as an interface material. Furthermore, ACC can develop by dehydration into a transient phase, transforming to the crystalline calcite near or even below room temperature [17].

The appearance of ACC in sea urchin spines as a residual interface phase may be a key factor for another peculiarity of its behavior: fracturing of the spine in compression or tension is not akin to the fracture behavior of inorganically formed single crystals of calcite, which fail along their plane of perfect cleavage. The spines have a macroscopically conchoidal fracturing (Fig. 13.2), which has previously been attributed to the Mg content of the calcite alone [8]. Theoretical reasoning and the observation that a dispersion of small secondary phases or defects can cause conchoidal fracturing [18] lead to the assumption that organic molecules [19] may be responsible for the fracture mode of spines. However, evidence from HRTEM analysis is mounting that ACC may in fact play the dominant role [20] and, in case of an interfacial character, could have a toughening effect similar to nacre. It is interesting to note that another part of the same sea urchin, their teeth, do possess a toughening mechanism akin to nacre with fibrous calcite surrounded by organic matrix, which leads to a very tough material indeed [21].

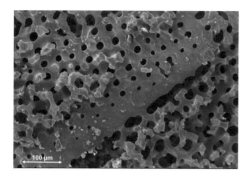

Fig. 13.2: Conchoidal fracturing of the stereom in *Heterocentrotus mammillatus*, both on a more compact growth line and the trabeculae of the stereom.

Thus, these nanostructural effects are important for the animal as they empower it to use the mechanically difficult brittle material calcite for spines and the whole skeleton. Even though carbonates have been used as implants or as templates for scaffolds converted to apatite [22, 23], calcite is not really viewed as an important source for biomimetics, because biomimetic materials and constructions will most probably not be made from a relatively soft carbonate and alternative polycrystalline ceramic materials do not have the problem of perfect cleavage at all. Thus, here we are rather addressing the "reverse biomimetics" question, i.e. how nature does turn a poor but ubiquitous material into a useful one, with which a large number of environments can be conquered.

The same is probably true if we turn to the chemical composition of the spines. The fact that they are made mainly of Mg-calcite is favorable for increased hardness but creates problems of thermal stability. Certainly, a sea urchin will never face the problem to survive higher temperatures, but a biomimetic building construction may well have to be qualified to withstand at least a moderate fire.

Therefore, the question of the thermal stability is a "reverse biomimetics" issue with significance not only for the possibilities of fossil preservation but for basic chemical and mineralogical research as well. This is because the unusually high level of Mg in the biogenic calcites points to a debate, which is almost as old as geoscience is: the problem to explain the low-temperature formation of dolomites and high-Mg calcites. Even in very modern theoretical and experimental investigations, the maximum equilibrium content of $MgCO_3$ in calcite at low temperatures is not a well fixed value, but probably very low below 300°C [24]. Important clues to the problem come from the investigation of the thermal treatment of sea urchin material [25], supporting other studies, in which the presence, fluctuation, and content of water and other impurities are identified as key to the stabilization and formation kinetics [26]. Regardless of whether the water content of several percent is located in calcite, ACC, or another water-bearing carbonate phase: in sea urchin spines, heating to temperatures of 300°C or higher results in pore formation within the calcitic trabeculae. The consequences of their mechanical properties are currently investigated.

13.3 Properties related to design

The internal design of the sea urchin spines is mainly related to the organization of porosity. Looking at the cross section of a spine (Fig. 13.3), the porosity can vary from a low percentage in the growth lines to more than 80% in the central stereom ("medulla"). Not only the degree of porosity is varying, but also, at close inspection, an ordering of the trabeculae roughly perpendicular to the growth lines becomes visible. This goes along with a pore shape change toward elongated pores or channels. Some consequences are immediately clear: this design makes the construction light. An overall density of 1 to 1.3 g/cm³ helps the animal to have little effort to carry and move spines in the water despite their length, which keeps predators at distance. It also makes the construction permeable, because its open porosity, filled by living tissue ("stroma"), in turn may allow skeleton assembly, re-shaping, and repair. In some sea urchins, poison is delivered to aid protection. Another strategy for survival is to hide in reef holes. The two species we consider wedge themselves in such holes, which protect against hydrodynamic wave activity or water-jet emitting predators like triggerfishes. The attack thus executes mechanical stresses to the spines, pressing them at high stress rates against the rocks in compressive or bending modes.

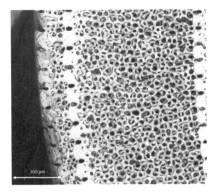

Fig. 13.3: Partial cross section of a spine of *Heterocentrotus mammillatus* with relatively dense growth lines separating a meshwork ("stereom") with varying degrees of ordering.

The unusually high strength of the spines despite the high degree of porosity was noted quite some time ago [9], but detailed investigation followed only recently [27]. Several aspects were striking. First, defining the compressive strength as the stress at which a spine segment loaded in direction of the c-axis (i.e. along the length of the spine) develops the first large cracks accompanied by a force/stress drop, we find a very significant increase in a stress-porosity plot relative to other porous calcites (Fig. 13.4).

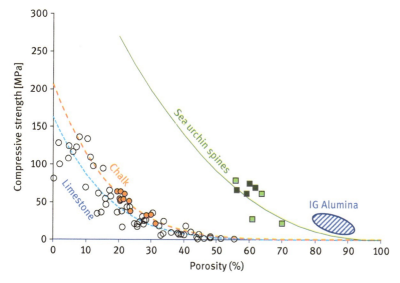

Fig. 13.4: Comparison of compressive strength of natural limestones (open circles), chalk (solid circles), and sea urchin spines (squares) relative to their porosity level. Data sources see text. The hatched area is the approximate range of values now obtained by the biomimetic material (IG alumina = ionotropic gelled Al_2O_3) described in Section 4 of this chapter.

The regression lines in Fig. 13.4, in which the data of Weber [9], Palchik [28], and Presser [27] were used, follow a general mixing rule model with compressive strength σ and porosity P of the type [29]

$$\sigma_{sample} = \sigma_0 \cdot (1-P)^X.$$

Second, the material is not behaving like an ordinary brittle material with catastrophic failure after reaching a critical stress. Instead, it behaves either like a brittle foam with a characteristic plateau at lower stresses after the first breakdown [30] or even in a cascading manner, in which the level of compressive stresses are repeatedly increasing and dropping during compression over a range of several tens of percent of compression [31] (Fig. 13.5a).

This in turn opens the bionic potential for a crushing zone material, because the processes depicted in Fig. 13.5a translate into a high energy, which is dissipated in the crushing process (Fig. 13.5b). A most interesting feature is the approximately constant high level of dissipation, which is a direct engineering figure of merit as it shows how much energy may be converted by a piece of a certain geometry.

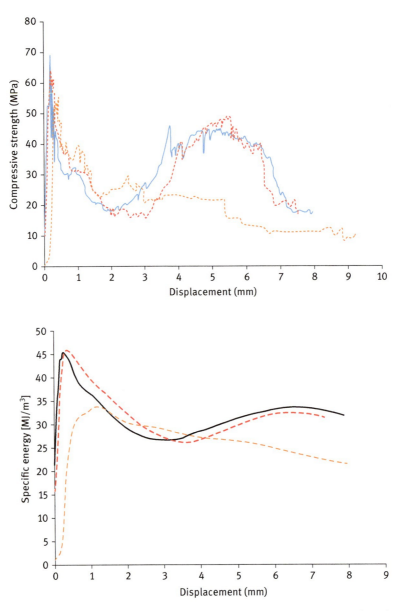

Fig. 13.5: (a) Typical plots from the compression of sea urchin spine segments from three specimen segments of *Phyllacanthus imperialis* (PI). The deviations are due to natural variations of the internal microstructure. The 10-mm displacement corresponds here to roughly 50% shrinkage. (b) Conversion of the data of (a) into specific energy used to conduct the crushing.

The physical process behind this behavior is difficult to evaluate in these crushing experiments because the samples are destroyed during the test. Therefore, we have developed another type of experiment, in which a blunt pin penetrates the stereom in a controlled manner. The advantage of the method is that we are able to evaluate a cross section through the penetrated path, which reveals the microstructure of the penetrated material and at the same time allows following the forces and displacements in a more restricted volume. This is desirable because of a plentitude of microstructural variations and inhomogeneities present in the naturally grown samples.

Figure 13.6 shows an example from such a penetration experiment. The penetration was from the top. The penetration path appears blurred, because it is outlined by powdery debris, which, under normal conditions, obviously cannot enter easily extended lateral areas of the stereom. It tends to cover the first lines of pores and to hold back the powder. However, just above the growth line, which is a denser section of the stereom, the debris was piled up and was injected into the stereom after buildup of sufficient pressure. With further increase in pressure, the breakthrough eventually occurred.

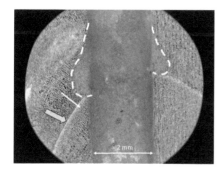

Fig. 13.6: Penetrated sea urchin spine segment after removal of the pin and cutting the sample alongside the indentation. The bright central line (arrow pointing toward it) is a growth line; the white dashed lines mark an area of debris ingression. The pin path has a diameter of 2 mm.

It follows that the mechanism of energy dissipation is related to the formation of new surfaces. If we know the amount of powder and its grain size, we may compare the energy from a calculation utilizing the surface energy of calcite (in the order of 0.3 to 0.8 J/m² [32]) to the mechanical energy input. In detail, the issue will be more complicated as the mechanical input energy consumption has also a frictional component and is furthermore complicated by the self-densification beneath the zone of sample destruction. This is currently under investigation and is discussed elsewhere [33].

An immediate conclusion is the awareness that a multitude of parameters govern the process. The diameter of the pores and their connectivity will determine whether debris is either transported or piled up. The thickness and shape of the trabeculae will control in which mode they fail (buckling, shearing, bending tension) and hence the number of cracks that will form and grow to yield debris of a specific size distribution. The surface energy of the material at hand and its strength will both contribute to the energy dissipated.

13.4 Biomimetic materials for energy dissipation

From the aim of the SFB to investigate biomimetic approaches toward use in the building and construction realm, it is clear that the studies here would not be complete by looking at the natural role models alone. The abstraction and application of the principles in appropriate materials needs to be part of it. One direction is to improve classic building materials such as concrete. This effort is headed by partner groups and discussed elsewhere [34]. Here, we describe only briefly the studies aimed at developing technical ceramic materials based on the principles discovered in the organisms dealt with.

In order to do so, there are two main tasks: (i) to create a material with porosity exceeding or at least approaching that of the natural role models and (ii) to be able to control the porosity distribution (gradation, layering, anisotropy, etc.). In Tübingen, this is done on the basis of alumina, because both surface energy (effective values between 2 and 4 J/m^2 [35]) and compressive strength of alumina are much higher than those of calcite, and well-characterized powders are commercially available.

In earlier studies, we have shown that a simple stacking of layers of different porosity by slip casting techniques can already be successful to reproduce a "graceful failure" mode [36]. In order to come much closer to our role models in terms of pore diameter, porosity level and channel-like pore geometries, we have combined the technique of freeze casting and ionotropic gelation similar to the approach by Xue et al. [37].

Ionotropic gelation uses alginates that form gels with channels in the 10- to 500-µm diameter range by a self-ordering process when crosslinked by ions of appropriate size and valence [38]. Self-organization within a slurry of ceramic powder forces the particles into the walls of the porous gel structure [39]. After a strengthening step with gluconolactone, the preforms can be freeze-dried, keeping the delicate channel structure intact and creating intra-channel porosity.

Figure 13.7 shows a product of the process after sintering at high temperatures. These ceramics have a compressive strength of about 5 to 20 MPa at a porosity exceeding 80%, which compares well with values obtained by Xue et al. [37]. This is indeed a strength level for this porosity grade, which is higher than that of natural spines as indicated by the hatched region in Fig. 13.4.

Fig. 13.7: Fracture surface of an alumina with approximate 80% porosity produced via a freeze drying/ionotropic gelation process.

13.5 Conclusions and outlook

The biomimetic approach to materials science and applied mineralogy includes the investigation of natural organisms to discover the principles behind their properties and attempt to convert them to materials and constructions of economic, ecologic, or technical interest. Sea urchin spines of *H. mammillatus* and *P. imperialis*, which live in environments where high forces are exerted on these parts, are biomimetic role models for energy-dissipating mechanisms of failure under compression in overload or impact events. The principle behind an astonishingly graceful failure behavior of such spines under compression is the layer-by-layer dissection of a porous structure into a powder, thus dissipating mechanical energy. It rests on the stress distribution obtained by their porous microstructure, enhanced by the orientation and gradation of their pores. In biogenic calcitic structures, this mechanism does only work because they are constructed of mesocrystals with amorphous and organic interfaces.

By combining ionotropic gelation and freeze-drying, we are able to manufacture highly porous anisotropic ceramics, which serve as models for materials implementing the effect. Efforts to implement this behavior in technical ceramics and concrete show promising results as indicated in Fig. 13.4 and described in Toader et al. [40], where we demonstrate that a functionally graded concrete containing several layers of high porosity already could carry loads after a shrinkage of up to 10% and dissipated energy at least 8 to 10 times the amount an ordinary concrete of normal density could.

For implementation in building constructions, a number of steps have to follow, in particular, upscaling of selected materials to appropriate sizes and developing processing methods to introduce this type of porosity in classic building materials. We see good chances to implement further advantages of this construction, e.g. for improving heat insulation, utilizing it for media transport or noise dampening.

Acknowledgments

We gratefully acknowledge the funding for these studies from DFG (SFB TRR 141 "Biologic Design and Integrative Structures") and Baden-Württemberg Foundation (within "Neue Materialien aus der Bionik" and "Biomimetische Materialsynthese").

References

[1] Jackson MD, Landis EN, Brune PF, Vitti M, Chen H, Li Q, Kunz M, Wenk H-R, Monteiro PJM, Ingraffea AR. Mechanical resilience and cementitious processes in Imperial Roman architectural mortar. Proc Natl Acad Sci USA 2014;111(52):18484–9.

[2] Vanorio T, Kanitpanyacharoen W. Rock physics of fibrous rocks akin to Roman concrete explains uplifts at Campi Flegrei Caldera. Science 2015;349(6248):617–21.

[3] Killian CE, Metzler RA, Gong Y, Churchill TH, Olson IC, Trubetskoy V, Christensen MB, Fournelle JH, De Carlo F, Cohen S, Mahamid J, Scholl A, Young A, Doran A, Wilt FH, Coppersmith SN, Gilbert PUPA. Self-Sharpening Mechanism of the Sea Urchin Tooth. Adv Funct Mater 2011;21:682–690.

[4] Ma Y, Cohen SR, Addadi L, Weiner S. Sea urchin tooth design: an "all-calcite" polycrystalline reinforced fiber composite for grinding rocks. Adv Mater 2008;20(5):1555–9.

[5] Trotter JA, Tipper J, Lyons-Levy G, Chino K, Heuer AH, Liu Z, Mrksich M, Hodneland C, Dillmore WS, Koob TJ, Koob-Emunds MM, Kadler K, Holmes D. Towards a fibrous composite with dynamically controlled stiffness: lessons from echinoderms. Biochem Soc Trans 2000;28(4):357–62.

[6] Santos R, Costa Gd, Franco C, Gomes-Alves P, Flammang P, Coelho AV. First insights into the biochemistry of tube foot adhesive from the sea urchin Paracentrotus lividus (Echinoidea, Echinodermata). Mar Biotechnol 2009;11:686–98.

[7] Grun TB, Dehkordi LK. F, Schwinn T, Sonntag D, Scheven Mv, Bischoff M, Knippers J, Menges A, Nebelsick JH. The skeleton of the sand dollar as a biological role model for segmented shells in building construction: a research review. In: Knippers J, Nickel KG, Speck T, editors. Biomimetic research for architecture and building construction: biological design and integrative structures. Springer: Berlin; 2016:217–242.

[8] Magdans U, Gies H. Single crystal structure analysis of sea urchin spine calcites: systematic investigations of the Ca/Mg distribution as a function of habitat of the sea urchin and the sample location in the spine. Eur J Mineral 2004;16(2):261–8.

[9] Weber J, Greer R, Voight B, White E, Roy R. Unusual strength properties of echinoderm calcite related to structure. J Ultrastruct Res 1969;26(5–6):355–66.

[10] Su X, Kamat S, Heuer AH. The structure of sea urchin spines, large biogenic single crystals of calcite. J Mater Sci 2000;35(22):5545–51.

[11] Blake DF, Peacor DR, Allard LF. Ultrastructural and microanalytical results from echinoderm calcite: implications for biomineralization and diagenesis of skeletal material. Micron Microsc Acta 1984;15(2):85–90.

[12] Eiberger J. Röntgenographische Spannungsmessung durch Mikrodiffraktion, Diploma thesis, Institut für Geowissenschaften. Tübingen: Eberhard-Karls-Univeristät Tübingen; 2007:169pp.
[13] Märkel K, Röser U, Mackenstedt U, Klostermann M. Ultrastructural investigation of matrix-mediated biomineralization in echinoids (Echinodermata, Echinoida). Zoomorphology 1986;106(4):232–43.
[14] Barthelat F. Nacre from mollusk shells: a model for high-performance structural materials. Bioinspir Biomimet 2010;5(3):035001.
[15] Benson SC, Benson NC. Wilt F. The organic matrix of the skeletal spicule of sea urchin embryos. J Cell Biol 1986;102(5):1878–86.
[16] Weiner S, Levi-Kalisman Y, Raz S, Addadi L. Biologically formed amorphous calcium carbonate. Connect Tissue Res 2003;44(Suppl 1):214–8.
[17] Politi Y, Metzler RA, Abrecht M, Gilbert B, Wilt FH, Sagi I, Addadi L, Weiner S, Gilbert P. Transformation mechanism of amorphous calcium carbonate into calcite in the sea urchin larval spicule. Proc Natl Acad Sci USA 2008;105(50):20045–5.
[18] Lange FF, The interaction of a crack front with a second-phase dispersion. Philos Mag 1970;22(179):0983–92.
[19] Berman A, Addadi L, Kvick A, Leiserowitz L, Nelson M, Weiner S. Intercalation of sea urchin proteins in calcite: study of a crystalline composite material. Science 1990;250(4981):664–7.
[20] Seto J, Ma Y, Davis SA, Meldrum F, Gourrier A, Kim YY, Schilde U, Sztucki M, Burghammer M, Maltsev S, Jager C, Colfen H. Structure-property relationships of a biological mesocrystal in the adult sea urchin spine. Proc Natl Acad Sci USA 2012;109(10):3699–704.
[21] Wang R. Fracture toughness and interfacial design of a biological fiber-matrix ceramic composite in sea urchin teeth. J Am Ceram Soc 1998;81(4):1037–40.
[22] Clarke SA, Walsh P, Maggs CA. Buchanan F. Designs from the deep: marine organisms for bone tissue engineering. Biotechnol Adv 2011;29(6):610–7.
[23] Vecchio KS, Zhang X, Massie JB, Wang M, Kim CW. Conversion of bulk seashells to biocompatible hydroxyapatite for bone implants. Acta Biomater 2007;3(6):910–8.
[24] Anovitz LM. Essene EJ. Phase equilibria in the system $CaCO_3$-$MgCO_3$-$FeCO_3$. J Petrol 1987;28(2):389–414.
[25] Dickson JAD. Transformation of echinoid Mg calcite skeletons by heating. Geochim Cosmochim Acta 2001;65(3):443–54.
[26] Lopez O, Zuddas P, Faivre D. The influence of temperature and seawater composition on calcite crystal growth mechanisms and kinetics: implications for Mg incorporation in calcite lattice. Geochim Cosmochim Acta 2009;73(2):337–47.
[27] Presser V, Schultheiß S, Berthold C, Nickel KG. Sea urchin spines as a model-system for permeable, light-weight ceramics with graceful failure behavior. Part I. Mechanical behavior of sea urchin spines under compression. J Bionic Eng 2009;6(3):203–13.
[28] Palchik V, Hatzor YH. The influence of porosity on tensile and compressive strength of porous chalks. Rock Mech Rock Eng 2004;37(4):331–41.
[29] Pabst W, Gregorová E. Young's modulus of isotropic porous materials with spheroidal pores. J Eur Ceram Soc 2014;34(13):3195–207.
[30] Toda H, Phgaki T, Uesugi K, Kobayashi M, Kuroda N, Kobayashi T, Kuroda N, Kobayashi T, Niinomi M, Akahori T, Makii K, Aruga Y. Quantitative assessment of microstructure and its effect on compression behavior of aluminum foams via high-resolution synchrotron X-ray tomography. Metall Mater Trans A 2006;37A:1211–9.
[31] Presser V, Schultheiß S, Kohler C, Berthold C, Nickel KG, Vohrer A, Finckh H, Stegmaier T. Lessons from nature for the construction of ceramic cellular materials for superior energy absorption. Adv Eng Mater 2011;13(11):1043–9.

[32] Røyne A, Bisschop J, Dysthe DK. Experimental investigation of surface energy and subcritical crack growth in calcite. J Geophys Res 2011;116:B4.
[33] Schmier S, Lauer C, Schäfer I, Klang K, Bauer G, Thielen M, Termin K, Berthold C, Schmauder S, Speck T, Nickel KG. Developing the experimental basis for an evaluation of scaling properties of brittle and quasi-brittle biological materials. In: Knippers J, Nickel KG, Speck T, editors. Biomimetic research for architecture and building construction. Berlin: Springer; 2016:277–94.
[34] Klang K, Bauer G, Toader N, Lauer C, Termin K, Schmier S, Haase W, Berthold C, Nickel KG, Speck T, Sobek W. Plants and animals as source of inspiration for energy dissipation in load bearing systems and facades. In: Knippers J, Nickel KG, Speck T, editors. Biomimetic research for architecture and building construction. Berlin: Springer; 2016:109–32.
[35] Davidge RW. Tappin G. The effective surface energy of brittle materials. J Mater Sci 1968;3:165–73.
[36] Presser V, Kohler C, Zivcova Z, Berthold C, Nickel KG, Schultheiß S, Gregorova E, Pabst W. Sea urchin spines as a model-system for permeable, light-weight ceramics with graceful failure behavior. Part II. Mechanical behavior of sea urchin spine inspired porous aluminum oxide ceramics under compression. J Bionic Eng 2009;6(4):357–64.
[37] Xue W, Huang Y, Xie Z, Liu W. Al2O3 ceramics with well-oriented and hexagonally ordered pores: the formation of microstructures and the control of properties. J Eur Ceram Soc 2012;32(12):3151–9.
[38] Thiele H, Hallich K. Kapillarstrukturen in ionotropen Gelen. Colloids Polym Sci 1957;151(1):1–12.
[39] Dittrich R, Despang F, Bernhardt A, Mannschatz A, Hanke T, Tomandl G, Pompe W, Gelinsky M. Mineralized scaffolds for hard tissue engineering by ionotropic gelation of alginate. Adv Sci Technol 2006;49:159–64.
[40] Toader N, Sobek W, Nickel KG. Energy absorption in functionally graded concrete bioinspired by sea urchin spines. J Bionic Eng 2017;14:369–378.

Marthe Rousseau
14 Nacre: a biomineral, a natural biomaterial, and a source of bio-inspiration

14.1 Introduction

Nacre, or mother of pearl, is a calcium carbonate structure produced by bivalves, gastropods, and cephalopods as an internal part of the shell. Because of its highly organized and hierarchical structure, chemical complexity, mechanical properties, and optical effects, which create a characteristic and beautiful luster, the formation of nacre is among the best-studied examples of calcium carbonate biomineralization. In this chapter, we will detail the composition and structure of nacre and its potential as biomaterial. We will summarize the recent studies of new materials inspired by nacre.

14.2 Nacre: a biomineral

14.2.1 Composition of nacre

Nacre is an organo-mineral composite. The mineral part is mainly calcium carbonate as aragonite form. For the inorganic fraction (about 95–97%), various elements are detected in *Pinctada* nacre [1, 2]. Despite the presence of Ca and Na, most elements exist usually as a trace element and show temporal and spatial variations [3]. Although it is established that the doubly charged ions Mn, Ti, Zn, and particularly Mg take part in the regulation of aragonite-calcite polymorphism and even favor the precipitation of $CaCO_3$ as aragonite rather than the more stable calcite [4]. Sr contents in nacreous layer and prismatic layer are similar, indicating the strong biological control of the biomineralization process [2]. The water in nacre is measured at 0.34 wt% [1] and contributes significantly to nacre's viscoelasticity [5].

For organic fraction (about 3–5%), it is a mixture of proteins, peptides, glycoproteins, chitin, polysaccharides, lipids, pigments, and many other small molecules lower than 1000 Da [6, 7], and most of the known organic molecules has been reviewed [8]. To date, more than 50 proteins and 50 peptides from nacre have been identified (UniProt protein database, see www.uniprot.org). The organic molecules can be extracted with aqueous and organic solvents [9]. The remarkable bioactive properties of nacre may relate to the abundant diffusible factors.

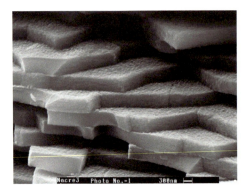
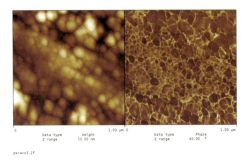

Fig. 14.1: The characteristic brick and mortar structure of the nacreous layer of Pinctada. AFM picture in Phase Contrast (1 × 1 µm²). At the nanometer (nm) length scale, the aragonite component inside individual tablets is embedded in a crystallographically oriented foam-like structure of intra-crystalline organic materials in which the mean size of individual aragonite domains is around 50 nm.

14.2.2 Multiscale structure of an iridescent biomineral

The interdigitating brickwork array of nacre tablets (Fig. 14.1), specifically in bivalves ("sheet nacre"), is not the only interesting aspect of nacre structure. Nacre is an organo-mineral composite at microscale and nanoscales. The biocrystal itself is a composite. The primary structural component is a pseudohexagonal tablet, about 0.5 µm thick and about 5 to 10 µm in width, consisting primarily (97%) of aragonite, a polymorph of $CaCO_3$, and of organics (3%).

Transmission electron microscopy performed in the dark-field mode evidences that a highly crystallized intracrystalline matrix. The organic matrix is continuous inside the tablet, and mineral phase is thus finely divided but behaves at the same time as a single crystal [10].

The tablet of nacre, the biocrystal, diffracts as a crystal but is made up of a continuous organic matrix (intracrystalline organic matrix) which breaks the mineral up into coherent nanograins (~45 nm mean size, flat-on) which share the same crystallographic orientation. The single crystal-like mineral orientation of the tablet is supposedly created by the heteroepitaxy of the intracrystalline organic matrix. This is a strong hypothesis because this work demonstrates at the same time that this intracrystalline matrix is well crystallized (i.e. periodic) and diffracts as a "crystal" too (dark-field TEM mode). However, these "organic crystals" do not show the same orientation in adjacent tablets.

Neighboring tablets above and below can maintain a common orientation, which again raises the issue of the transmission of mineral orientation from one row to the next. Bridges are well identified in *Pinctada* between successive rows in the pile. This implies that an organic template is controlling the orientation of the aragonite [11].

Intermittent-contact atomic force microscopy with phase detection imaging reveals a nanostructure within the tablet (Fig. 14.1B). A continuous organic framework divides each tablet into nanograins. Their mean extension is 45 nm. It is proposed that each tablet results from the coherent aggregation of nanograins keeping strictly the same crystallographic orientation, owing to a heteroepitaxy mechanism [11, 12].

It is well known for almost 10 years that biological carbonates are mesocrystals. This has been shown for *Mytilus edulis* nacre [13], as well as for *Paracentrotus lividus* teeth and spine [14] and red corals [15].

14.2.3 Nacre tablet formation

Bevelander and Nakahara [16] introduced decisive insights to establish the compartment theory. During mollusk shell formation, the mineral phase forms within an organic matrix composed of β-chitin, silk-like proteins, and acidic glycoproteins rich in aspartic acid. The matrix is widely assumed to play an important role in controlling mineralization [17]. Nacre tablets do not form in the presence of an aqueous solution, but in a hydrated gel-like medium [18]. The ESEM and cryo-SEM results are consistent with this space-filling material being the silk-like proteins that create the appropriate microenvironment where nacre tablets form.

The nacre of gastropod molluscs is intriguingly stacked in towers. It is covered by a surface membrane, which protects the growing nacre surface from damage when the animal withdraws into its shell. The surface membrane is supplied by vesicles that adhere to it on its mantle side and secretes interlamellar membranes from the nacre side [19].

14.3 Nacre: a natural biomaterial

A major breakthrough was done in 1992, when Lopez et al. [20] discovered that nacre from the pearl oyster *Pinctada maxima* is simultaneously biocompatible and osteoinductive. Nacre shows osteogenic activity after implantation in human bone environment [21]. Raw nacre pieces designed for large bone defects were used as replacement bone devices in the femur of sheep. Over a period of 12 months, the nacre blocks show persistence without alteration of the implant shape. A complete sequence of osteogenesis resulted from direct contact between newly formed bone and the nacre, anchoring the nacre implant [22]. Furthermore, when nacre is implanted in bone, new bone formation occurs, without any inflammatory reaction or fibrous formation. We observed an osteoprogenitor cellular layer lining the implant, resulting in a complete sequence of new bone formation (Fig. 14.2). Results showed calcium and phosphate ions lining the nacre within the osteoprogenitor tissue [23].

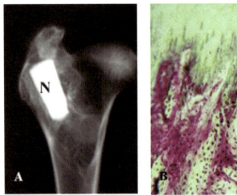
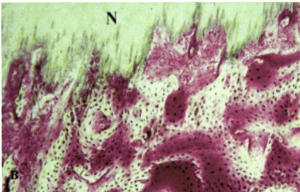

Fig. 14.2: Nacre pieces: assays on sheep model bone regeneration induced *in vivo* by nacre: A. Nacre piece (N) grafted in the sheep thighbone (Radiography) original magnification x2 B. Interface bone/nacre. (N)(optical microscopy × 350). [23, 24]

Osteogenesis is thought to begin with the recruitment of mesenchymal stem cells, which differentiate and form osteoblasts in response to one or more osteogenic factors. Previous studies [21, 24], as reviewed by Westbroek and Marin [25], has been shown through in vivo and in vitro experiments that the nacre can attract and activate bone marrow stem cells and osteoblasts.

Other authors have demonstrated the same activity of nacre. Liao et al., in 2000 [26], have published results on the implantation of nacre pieces from the shell of the freshwater *Margaritifera* to the back muscles and femurs of rats. They concluded that their nacre was biocompatible, biodegradable, and osteoconductive. They confirmed the results obtained with *Pinctada* nacre.

14.3.1 Nacre implant

Inspired by the nacre teeth found in Mayan skulls, nacre, as a bone graft substitute, was first designed many centuries later to restore dental defects in dogs and humans [27]. For the last 20 years, nacre has been designed more ingeniously and tested, in vivo, at various implantation sites for different uses (Fig. 14.2). New bone formation stimulated by nacre has been observed in humans, rats, sheep, rabbits, and pigs, as reviewed by Zhang et al. [28].

14.3.2 Nacre powder

A process of nacre powder preparation has been developed. The obtained nacre powder has been experimented in injectable form in vertebral and maxillary sites of sheep [29, 30]. This work has demonstrated that nacre powder is resorbable and that this resorption induced the formation of normal bone. The nacre powder filled the whole experimental cavity 1 week post-surgery. There was no inflammatory or foreign body reaction in the cavity area. Samples taken at 8 weeks after injection showed dissolution of the nacre within the cavity. Angiogenesis had begun by that time and the cavity was invaded by a network of capillaries. The cavity contained newly formed woven bone (Fig. 14.3). The vertebral bone adjacent to the cavities contained interconnected bone lamellae. They were also bone remodeling units with central lacunae rich in bone marrow. This new formed bone was functional as normal bone.

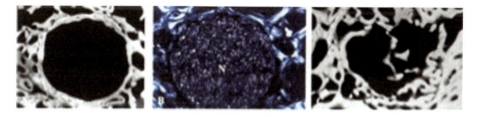

Fig. 14.3: Injection of nacre powder in endo-osseous vertebral site in sheep original magnification x25 for all three images:
A: vertebral cavity and trabecular bone (Microradiography).
B: vertebral cavity filled with nacre powder (N) (polarised light).
C: vertebral cavity and new formed bone after three months (Microradiography) [21–22].

Based on the studies done on nacre bioactivity on bone, development of medical devices made of nacre for bone repair in orthopedic surgery, for example, appears possible.

14.4 Nacre: a source of bio-inspiration for new materials

Materials scientists have done many studies on the mechanical behavior of nacre such as its strength and ductility [31]. Nacre is characterized by a layered structure consisting of strong platelets embedded in a soft, ductile organic matrix. Despite the inherently weak inorganic constituents (here, calcium carbonate in aragonite form), the high strength of the inorganic building blocks is ensured by limiting at least one of their dimensions to the nanoscale [10]. These tiny building blocks are organized into a hierarchical structure spanning over various length scales. Using principles found in natural composites such as nacre, materials scientists have, in the last 10 years, tried to improve the mechanical properties of polymers and composites.

14.4.1 Layered structure

Some material scientists have chosen to mimic the layered structure. Guided by bio-design principles, Bonderer et al. [32] have selected inorganic platelets and an organic polymer with features that should lead to artificial hybrid materials exhibiting high strength and ductility. Alumina platelets with estimated tensile strength of 2 GPa and a chitosan polymer with a yield shear strength around 40 MPa were chosen. The combination of these materials should lead to strength values higher than that of nacre while ensuring that fracture occurs by platelet pull-out. In order to maximize strength without impairing the polymer's ductility, 200-nm-thick platelets were used (Fig. 14.4). The use of high-strength artificial platelets as mechanical reinforcement imposes less stringent microstructural requirements to the composites in comparison to the natural hybrid structures. Artificial composites reinforced with strong platelets show remarkable mechanical properties despite their less elaborate microstructure.

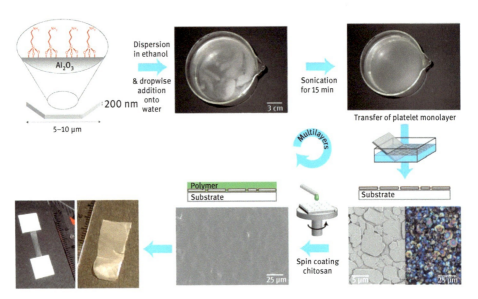

Fig. 14.4: Bottom-up colloidal assembly of multilayered hybrid films. Surface modified platelets are assembled at the air-water interface to produce a highly oriented layer of platelets after ultrasonication. The 2D assembled platelets are transferred to a flat substrate and afterwards covered with a polymer layer by conventional spin-coating [25].

Bai et al. [33] used ice-templated assembly and UV-initiated cryopolymerization to fabricate a novel kind of composite hydrogel which has both aligned macroporous structure at micrometer scale and a nacre-like layered structure at nanoscale. Such hydrogels are macroporous, are thermoresponsive, and exhibit excellent mechanical performance (i.e. they are tough and highly stretchable). They have successfully fabricated a novel type of macroporous composite hydrogel (composed of poly (N-isopropylacrylamide) and clay platelets) (Fig. 14.5).

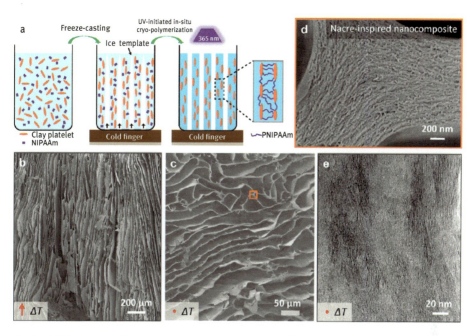

Fig. 14.5: (a) Schematic illustration of the fabrication method of macroporous composite hydrogels by freeze-casting. A solution, composed of monomer (NIPAAm), initiator (DEAP), and cross-linker (clay platelet) at a given concentration was placed on a coldfinger connected to a liquid nitrogen reservoir. During the cooling process, ice crystals grew from the coldfinger and templated the assembly of monomer and clay platelets into a nacre-like layered nanocomposite. After freezing, the sample was placed under UV light to initiate cryopolymerization. The as-prepared nanocomposite hydrogels have an anisotropically aligned structure at micrometer scale, as shown by the SEM images in both the (b) parallel and (c) perpendicular directions to the temperature gradient (ΔT). In the wall of the aligned structure, clay platelets and PNIPAAm are assembled into nacre-like layered nanocomposites, as shown by (d) SEM and (e) TEM images [26].

The process used by Bouville et al. [34] is based on ice-templating. However, they did not rely on ice crystals to form the elementary bricks, but instead took advantage of their growth as a driving force for the local self-assembly of anisotropic particles (platelets, Figs. 14.6–14.8). Those particles have the same dimensions as nacre

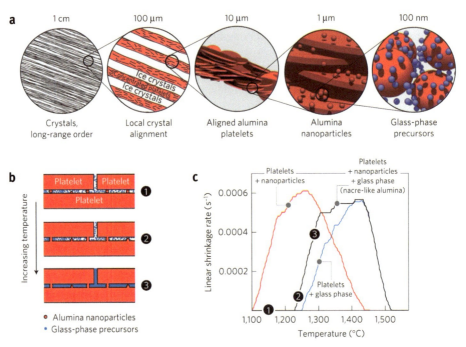

Fig. 14.6: Design strategy describing the control at multiple scales of structural self-organization, and densification strategy. (a) Self-organization of all the structural features occurs during the freezing stage. The growth of ordered-ice crystals triggers the local alignment of platelets. Alumina nanoparticles and liquid-phase precursors are entrapped between the platelets. (b) Schematic representation of the densification scenario. (c) The linear shrinkage rate illustrates that the densification of the composition comprising either one of the building blocks (3 vol.% nanoparticles or 5 vol.% liquid phase) occurs at different temperatures. In the composition comprising all the building blocks, the densification starts at a temperature between the two other compositions (liquid phase only and nanoparticles only), showing an interaction between nanoparticles and liquid-phase precursors [27].

platelets (500 nm thickness, 7 µm diameter) and constitute the elementary building blocks of the structure. Alumina nanoparticles (100 nm) incorporated in the initial suspension serve as a source of both inorganic bridges between the platelets and the nanoasperities at the surface of platelets, similar to that observed in nacre. Finally, smaller nanoparticles (20 nm) of liquid-phase precursors (silica and calcia) are added to aid filling the remaining gaps during the sintering stage. The entire process thus consists of three steps: preparation of an aqueous colloidal suspension containing all the required building blocks and processing additives, ice templation of this suspension, and a pressure-assisted sintering step at 1,500°C. The bioinspired, material-independent design presented by the study of Bouville et al. [34] is a specific but relevant example of a strong, tough, and stiff material, in great need for structural, transportation, and energy-related applications where catastrophic failure is not an option.

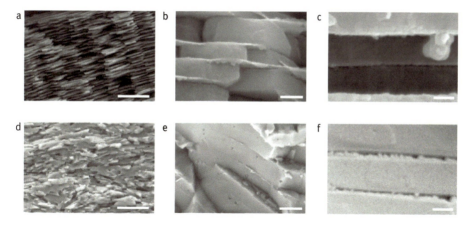

Fig. 14.7: Comparison of microstructures. Panels (a–c) correspond to nacre; panels (d–f) to nacre-like alumina. (a,d) SEM micrographs showing the short- and long-range order of platelets. The nacre-like alumina shows relatively high organization of the platelets. (b,e) Local stacking of platelets. A liquid-phase film is present even when the platelets are close, mimicking the protein layer in the nacre structure. (c,f) Closer views of the platelet interface, revealing the presence of the inorganic bridges and nano-asperities along with the glassy (f), or organic (c), phase filling the space between adjacent platelets. Some residual pores are also visible. Scale bars, 10 µm (a,d); 500 nm (b,e); 250 nm (c,f) [27].

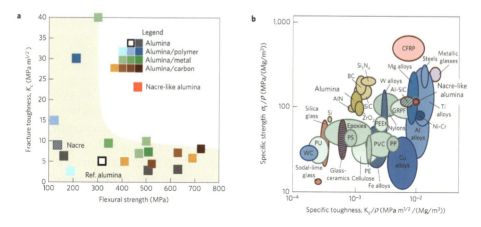

Fig. 14.8: Comparison of the relative materials performance. (a) The conflict between strength and toughness: fracture toughness versus flexural strength for alumina-based materials. Improvements in toughness are obtained by the introduction of a ductile phase, such as a metallic phase, a polymer, or carbon reinforcements such as graphene, carbon nanotubes, or whiskers. Improvements in strength are achieved through a texturation of the grains, hot isostatic pressing or the addition of strong reinforcements. (b) Ashby diagram of specific strength versus specific toughness for a range of engineering and natural materials. The nacre-like aluminas have specific strength/toughness properties similar to those of titanium or magnesium metallic alloys. CFRP, carbon-fibre-reinforced polymers; GRFP, glass-fibre-reinforced polymers; PEEK, polyether ether ketone [27].

14.4.2 Interfacial/interlocking structure

Most current synthetic composites have not exhibited their full potential of property enhancement compared to the natural prototypes they are mimicking. One of the key issues is the weak junctions between stiff and compliant phases, which need to be optimized according to the intended functions of the composite material. Motivated by the geometrically interlocking designs of natural biomaterials, Zhang et al. [35] propose an interfacial strengthening strategy by introducing geometrical interlockers on the interfaces between compliant and stiff phases (Fig. 14.9). The tensile strength of the composites with proper interlocker design can reach up to 70% of the ideal value.

Using a new compositing route, we replicated nacre's three-dimensional (3D) interlocking selection in engineering composites. The interlocked and mutually supported skeleton architecture consists of well-arranged ceramic bridges between parallel Al_2O_3 lamellae. Such Al_2O_3-CE composite (25 vol% Al_2O_3) with 3D interlocking skeleton (3D IL) possesses the specific strength of 162 MPa/(g/cm^3), the highest among Al_2O_3 composites, while reducing the weight by more than 25% and 50% compared to nacre and other nacre-like composites, respectively; this ceramic composite achieved a high flexural strength of 300 MPa and failure strain of 5%. With the unique 3D interlocking skeleton, the composite has exceptionally high shock resistance (Fig. 14.10) [36].

14.4.3 Innovative methods

Of the processing techniques employed to date, freezecasting (ice-templating) of aqueous suspensions of ceramic powders to produce unidirectional lamellar scaffolds has shown most success; subsequent cold pressing of the scaffolds to create brick-like structures, followed by infiltration of a polymer or metallic compliant ("mortar") phase, has generated several nacre-like ceramic hybrid materials with exceptional damage-tolerant properties. The advantage of freeze-casting is that it can produce bulk material, whereas most "bottom-up" approaches, e.g., layer-by-layer deposition and self-assembly, are incapable of processing macroscopic samples. However, because of the void space necessary for complete infiltration, ice-templating cannot achieve high ceramic volume fractions (typically >80%), resulting in too much mortar phase for optimum properties. Additionally, most results have involved polymeric mortars, whereas theoretical modeling suggests that metallic mortars can realize better combinations of strength and toughness. However, infiltrating metal into ceramic scaffolds is inherently difficult due to poor wetting between most ceramic-metal combinations. Correspondingly, fine-scale, "nacre-like" brick-and-mortar structures with high ceramic content and a metallic compliant phase, which are predicted to display optimal damage tolerance, have yet to be made

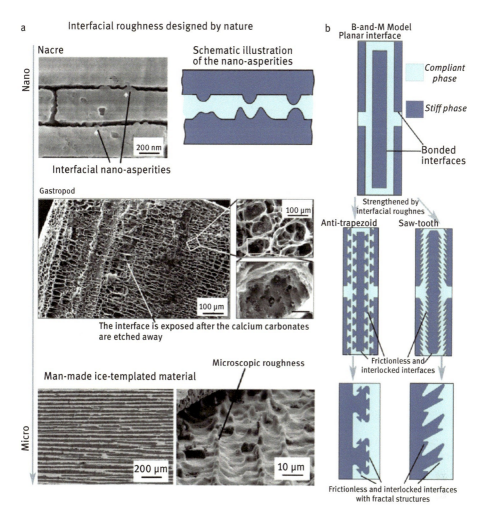

Fig. 14.9: Interfacial interlocking designs adopted in natural and synthetic B-and-M composites. (a) The interfacial interlocking designs at different length scales in natural materials. Top: the nanoasperities on aragonite surface in nacre; middle: the organic layer surface exposed after removal of the adjacent mineral layer by etching; bottom: the bumpy interface in a layered synthetic composite; (b) Schematics of the original (plain) B-and-M model (top), modified B-and-M structure with interfacial interlocker (middle), and modified B-and-M structure with fractal interfacial interlocker designs (bottom) [28].

using freeze-casting techniques. Freeze-casting, as a method for the development of bioinspired materials, is well developed, described, and used as a technique by Wegst and her group [37, 38].

Wilkerson et al. [39] present an alternative bulk processing technique, that of coextrusion, to make bioinspired, brick-and-mortar structures comprising high (≈90 vol%) ceramic volume fractions with a metallic compliant phase. A feed rod of material is

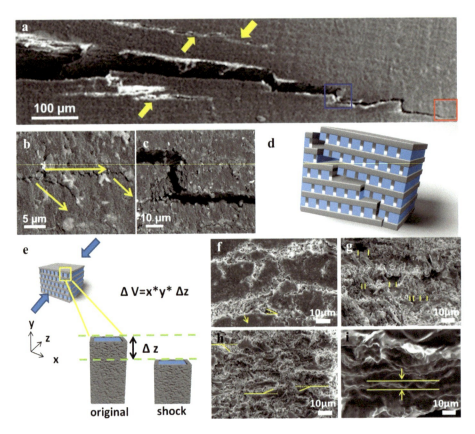

Fig. 14.10: Reinforcing mechanisms of 3D interlocking in composites. (a) SEM images taken from in situ three-point bending tests. (b,c) Enlarged images taken, respectively, from the red boxed area and the blue boxed area in (a), showing long-range crack deflection, multiple cracks, crack bridging, and the sliding between lamellae. (d) Schematic illustration of crack propagation. (e) Strain rate sensitivity in 3D composites. SEM images taken from samples dynamically loaded by SHPB at the strain rates of (f) 4000 s−1, (g) 7300 s−1, (h) 9700 s−1, and (i) 12000 s−1, respectively [29].

produced with specific core and shell diameters; this aspect ratio is preserved during extrusion as the rod's cross section is reduced down to a filament (Fig. 14.11). The filament is then sectioned into individual pieces, representing individual "bricks" in a brick-and-mortar structure (Fig. 14.11). Coextrusion allows the bricks to be individually coated with mortar with precise volume control, such that high-ceramic/low-mortar volume fractions can be attained. The mechanical properties of their compliant-phase ceramics were either comparable or superior to traditional ceramic-metal composites.

Mao et al. [40] describe a mesoscale "assembly and mineralization" approach inspired by the natural process in mollusks to fabricate bulk synthetic nacre that

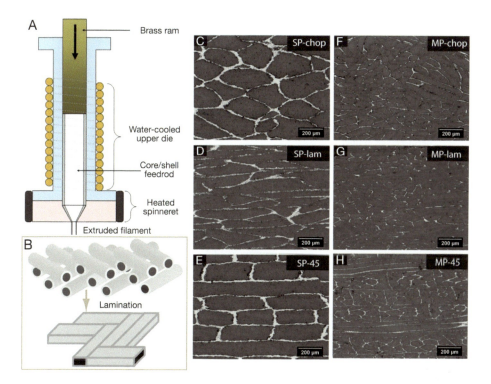

Fig. 14.11: (A) Schematic of the coextrusion assembly used to produce the NiO-coated Al_2O_3 filament. A feed rod is extruded through a heated spinneret to reduce the viscosity of the polymeric binder. The final filament has a diameter of ≈300 μm and is collected on a spool as it is extruded. (B) Coextruded filaments are chopped into short sections and laminated with heat and pressure in a die; this allows the filaments to deform laterally and become more brick-like before they are bonded together to form a "brick-and-mortar" architecture. Optical microscopy images of the resulting Al_2O_3/10Ni metal compliant-phase ceramics: (C–E) Single-pass coextrusions to 300 μm, specifically the Sp-chop structure (chopped and laminated in constrained die), SP-lam (chopped and pre-laminated into sheets), and SP-45 (oriented filament laminated into sheets and consecutively offset 45° from the previous layer); (F–H) multipass coextrusions counterparts MP-chop, MP-lam, and MP-45, that were extruded to 6 mm, then bound together and extruded again to 300 μm. The brighter phases are nickel metal while darker phases are the alumina ceramic [30].

highly resembles both the chemical composition and the hierarchical structure of natural nacre (Fig. 14.12). The millimeter-thick synthetic nacre consists of alternating organic layers and aragonite platelet layers (91 wt%) and exhibits good ultimate strength and fracture toughness. This predesigned matrix-directed mineralization method represents a rational strategy for the preparation of robust composite materials with hierarchically ordered structures, where various constituents are adaptable, including brittle and heat-labile materials.

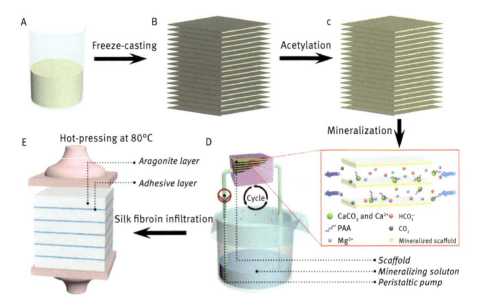

Fig. 14.12: Fabrication scheme of the synthetic nacre. (A) Starting solution, chitosan/acetic acid solution. (B) Freeze-casted laminated chitosan matrix. (C) Matrix after acetylation where chitosan is converted to β-chitin. (D) Mineralization of the matrix. Fresh mineralizing solution is pumped to flow through the space between the layers in the matrix, bringing in Ca^{2+}, Mg^{2+}, HCO_3^- and PAA for mineralization, and taking out excess CO_2. $CaCO_3$ precipitates onto the layers and CO2 diffuses into the air through the pin holes in the paraffin film. (E) Laminated synthetic nacre is obtained after silk fibroin-infiltration and hot-pressing [31].

14.5 Conclusion

Nature has created beautiful skeletal structures with the help of biominerals, such as nacre. Such biominerals are an unlimited source of inspiration for biomaterials and materials with improved mechanical properties.

References

[1] Bellaaj-Zouari A, Chérif K, Elloumi-Hannachi I, Slimane N, Habib Jaafoura M. Characterization of mineral and organic phases in nacre of the invasive pearl oyster Pinctada radiata (Leach, 1814). Cah Biol Mar 2011;52(3):337–48.

[2] Farre B, Brunelle A, Laprévote O, Cuif J-P, Williams CT, Dauphin Y. Shell layers of the black-lip pearl oyster Pinctada margaritifera: matching microstructure and composition. Comp Biochem Physiol B 2011;159:131–9.

[3] Pourang N, Richardson CA, Chenery SRN, Nasrollahzedeh H. Assessment of trace elements in the shell layers and soft tissues of the pearl oyster Pinctada radiata using multivariate analyses: a potential proxy for temporal and spatial variations of trace elements. Environ Monit Assess 2014;186:2465–85.

[4] Blackwelder PL, Weiss RE, Wilbur KM. Effects of calcium, strontium, and magnesium on the coccolithophorid Cricosphaera (Hymenomonas) carterae. I. Calcification. Mar Biol 1976;34:11–6.

[5] Verma D, Katti K, Katti D. Nature of water in nacre: a 2D Fourier transform infrared spectroscopic study. Spectrochim Acta A 2007;67:784–8.

[6] Weiss IM, Kaufmann S, Mann K, Fritz M. Purification and characterization of perlucin and perlustrin, two new proteins from the shell of the mollusc Haliotis laevigata. Biochem Biophys Res Commun 2000;267:17–21.

[7] Bédouet L, Rusconi F, Rousseau M, Duplat D, Marie A, Dubost L, Le Ny K, Berland S, Péduzzi J, Lopez E. Identification of low molecular weight molecules as new components of the nacre organic matrix. Comp Biochem Physiol B 2006;144:532–43.

[8] Marin F, Le Roy N, Marie B. The formation and mineralization of mollusk shell. Front Biosci 2012;4:1099–125.

[9] Rousseau M, Pereira-Mouries L, Almeida MJ, Milet C, Lopez E. The water-soluble matrix fraction from the nacre of Pinctada margaritifera produces earlier mineralization of MC3T3-E1 mouse pre-osteoblasts. Comp Biochem Physiol B 2003;135:1–7.

[10] Rousseau M, Lopez E, Stempflé P, Brendlé M, Franke L, Guette A, Naslain R, Bourrat X. Multiscale structure of sheet nacre. Biomaterials 2005;26:6254–62.

[11] Rousseau M, Lopez E, Couté A, Mascarel G, Smith DC, Naslain R, Bourrat X. Sheet nacre growth mechanism: a Voronoi model. J Struct Biol 2005;149:149–57.

[12] Rousseau M, Meibom A, Gèze M, Bourrat X, Angellier M, Lopez E. Dynamics of sheet nacre formation in bivalves. J Struct Biol 2009; 165: 190–5.

[13] Griesshaber E, Schmahl WW, Ubhi HS, Huber J, Nindiyasari F, Maier B, Ziegler A. Homoepitaxial meso- and microscale crystal co-orientation and organic matrix network structure in Mytilus edulis nacre and calcite. Acta Biomater 2013;9(12):9492–502.

[14] Goetz AJ, Griesshaber E, Abel R, Fehr T, Ruthensteiner B, Schmahl WW. Tailored order: the mesocrystalline nature of sea urchin teeth. Acta Biomater 2014;10(9):3885–98.

[15] Vielzeuf D, Garrabou J, Baronnet A, Garuaby O, Marschal C. Nano to macroscale biomineral architecture of red coral. Am Mineral 2008;93:1799–815.

[16] Bevelander G, Nakahara H. An electron microscope study of the formation of nacreous layer in the shell of certain bivalve molluscs. Calc Tiss Res 1969;3:84–92.

[17] Levi-Kalisman Y, Falini G, Addadi L, Weiner S. Structure of the nacreous organic matrix of a bivalve mollusk shell examined in the hydrated state using cryo-TEM. J Struct Biol 2001;135:8–17.

[18] Nudelmann F, Shimoni E, Klein E, Rousseau M, Bourrat X, Lopez E, Addadi L, Weiner S. Forming nacreous layer of the shells of the bivalves Atrina rigida and Pinctada margaritifera: an environmental- and cryo-scanning electron microscopy study. J Struct Biol 2008;162:290–300.

[19] Checa AG, Cartwright JHE, Willinger MG. The key role of the surface membrane in why gastropod nacre grows in towers. Proc Natl Acad Sci USA 2009;106: 38–43.

[20] Lopez E, Vidal B, Berland S, Camprasse S, Camprasse G, Silve C. Demonstration of the capacity of nacre to induce bone formation by human osteoblasts maintained in vitro. Tissue Cell 1992;24:667–79.

[21] Silve C, Lopez E, Vidal B, Smith DC, Camprasse S, Camprasse G, Couly G. Nacre initiates biomineralization by human osteoblasts maintained in vitro. Calcif Tissue Int 1992;51:363–9.

[22] Atlan G, Balmain N, Berland S, Vidal B, Lopez E. Reconstruction of human maxillary defects with nacre powder: histological evidence for bone regeneration. CR Acad Sci Sér III Sci Vie 1997;320:253–8.

[23] Atlan G, Delattre O, Berland S, LeFaou A, Nabias G, Cot D, Lopez E. Interface between bone and nacre implants in sheep. Biomaterials 1999;20:1017–22.

[24] Berland S, Delattre O, Borzeix S, Catonné Y, Lopez E. Nacre/bone interface changes in durable nacre endosseous implants in sheep. Biomaterials 2005;26:2767–73.
[25] Westbroek P, Marin F. A marriage of bone and nacre. Nature 1998;392:861–2.
[26] Liao H, Mutvei H, Sjöström M, Hammarström L, Li J. Tissue responses to natural aragonite (Margaritifera shell) implants in vivo. Biomaterials 2000;21:457–68.
[27] Bobbio A. The first endosseous alloplastic implant in the history of man. Bull Hist Dent 1972;20:1–6.
[28] Zhang G, Brion A, Willemin AS, Piet MH, Moby V, Bianchi A, Mainard D, Galois L, Gillet P, Rousseau M. Nacre, a natural, multi-use, and timely biomaterial for bone graft substitution. J Biomed Mater Res A. 2017 Feb;105(2):662–71. doi: 10.1002/jbm.a.35939. Epub 2016 Nov 7.
[29] Lamghari M, Almeida MJ, Berland S, Lamghari M, Almeida MJ, Berland S, Huet H, Laurent A, Milet C, Lopez E. Stimulation of bone marrow cells and bone formation by nacre: in vivo and in vitro studies. Bone 1999;25:91S-4S.
[30] Lamghari M, Huet H, Laurent A, Berland S, Lopez E. A model for evaluating injectable bone replacements in the vertebrae of sheep: radiological and histological study. Biomaterials 1999;20:2107–14.
[31] Jackson AP, Vincent JFV, Turner RM. The mechanical design of nacre. Proc R Soc Lond B 1988;234:415–40.
[32] Bonderer L, Studart A, Gauckler L. Bioinspired design and assembly of platelet reinforced polymer films. Science 2008;319:1069–73.
[33] Bai H, Polini A, Delattre B, Tomsia A. Thermoresponsive composite hydrogels with aligned macroporous structure by ice-templated assembly. Chem Mater 2013;25:4551–6.
[34] Bouville F, Maire E, Meille S, Van de Moortèle B, Stevenson A, Deville S, Strong, tough and stiff bioinspired ceramics from brittle constituents. Nat Mater 2014;13(5):508–14.
[35] Zhang Y, Gao Y, Ortiz C, Xu J, Dao M. Bio-inspired interfacial strengthening strategy through geometrically interlocking designs. J Mech Behav Biomed Mater 2012;15:70–7.
[36] Zhao H, Yue Y, Guo L, Wu J, Zhang Y, Li X, Mao S, Han X. Cloning nacre's 3d interlocking skeleton in engineering composites to achieve exceptional mechanical properties. Adv Mater 2016;28:5099–105.
[37] Wegst UG, Schecter M, Donius AE, Hunger PM. Biomaterials by freeze casting. Philos Trans A Math Phys Eng Sci 2010;368(1917):2099–121.
[38] Hunger PM, Donius AE, Wegst UG. Platelets self-assemble into porous nacre during freeze casting. J Mech Behav Biomed Mater 2013;19:87–93.
[39] Wilkerson RP, Gludovatz B, Watts J, Tomsia AP, Hilmas GE, Ritchie RO. A novel approach to developing biomimetic ("nacre-like") metal-compliant-phase (nickel-alumina) ceramics through coextrusion. Adv Mater 2016;28(45):10061–7.
[40] Mao LB, Gao HL, Yao HB, Liu L, Cölfen H, Liu G, Chen SM, Li SK, Yan YX, Liu YY, Yu SH. Synthetic nacre by predesigned matrix-directed mineralization. Science 2016;354(6308):107–10.

Robert B. Heimann

15 Hydroxylapatite coatings: applied mineralogy research in the bioceramics field

15.1 Introduction

Bioceramics are inorganic non-metallic compounds designed to replace a part or a function of the human body in a safe, reliable, economic, and physiologically and aesthetically acceptable manner. During the last decades, bioceramic materials have developed into a powerful driver of advanced ceramics, and thus, are an important arena of activity in the field of applied mineralogy today. As result of intense scientific endeavor, bioceramics, both bioinert materials, such as alumina and zirconia, and bioactive materials such as hydroxylapatite (HAp), tricalcium phosphate (TCP), and tetracalcium phosphate (TTCP), are being applied successfully in the clinical practice. Nevertheless, considering the complexity of their interaction with living tissue and their inherent socio-economic importance, there is still much need of fundamental research.

It is not surprising then that development of novel and optimization of existing bioceramics are the vanguard of health-related research effort in many countries worldwide. Large proportions of an aging population rely increasingly on repair or replacement of body parts, or restoration of lost body functions. These applications range from artificial dental roots, to alveolar ridge, iliac crest and cheek augmentation, to spinal implants, hip and knee endoprostheses, to ocular implants.

Consequently, the number of patients in need of receiving large-joint reconstructive hip and knee implants to repair the ambulatory knee-hip kinematic are constantly increasing. In addition, dental, small-joint, and spine implants are target areas of biomedical implantology today. Presently, the worldwide sales of hip and knee orthopedic surgical joint replacement products are US$ 16.7 billion, anticipated to double by reaching US$ 33 billion in 2022 [1].

15.2 Geological HAp vs. biological HAp

Calcium phosphates, in particular HAp, are of overwhelming importance to sustain life. HAp constitutes the inorganic component of the biocomposite materials "bone" and "tooth", designed by nature to provide the mechanical strength and resilience of the gravity-defying bony skeletons as well as the resistance of teeth to masticatory stresses of all vertebrates. In 1926, De Jong [2] was first to identify the structure of the calcium phosphate phase in bone as being akin to geological apatite. The biological HAp-collagen composites provide not only strength but also flexibility, their porous structure allowing exchange of essential nutrients and a biologically compatible resorption and precipitation behavior under appropriate physical and chemical conditions that control the build-up of bony matter by osteoblasts and its resorption by osteoclasts in response to varying stress levels. In addition, bone-like, i.e. calcium-deficient defect HAp is a reservoir of phosphorus that can be delivered to the body on demand.

15.2.1 Structure of geological HAp

HAp, $Ca_{10}(PO_4)_6(OH)_2$ is a member of a large group of chemically different but structurally identical compounds, many of them found as minerals. Their general formula is $M_{10}(ZO_4)_6X_2$, whereby M = Ca, Pb, Cd, Sr, La, Ce, K, Na; Z = P, V, As, Cr, Si, C, Al, S; X = OH, Cl, F, CO_3, H_2O, □, obeying the hexagonal space group $P6_3/m$. Ca polyhedra share faces to form chains parallel to the crystallographic c-axis [00.1]. These chains are linked into a hexagonal array by sharing edges and corners with [PO_4] tetrahedra. The OH⁻ ions are located in wide hexagonal channels parallel to the 6_3 screw axis [3]. Since the 16 OH⁻ positions in the unit cell are statistically occupied to only 50%, there exist on average 8 vacancies/unit cell along the c-axis. Hence, there are direction-dependent differences in the mobility of OH⁻ ions that are extremely relevant when considering structural transformation from amorphous calcium phosphate (ACP) to crystalline HAp as well as stepwise dehydroxylation of HAp to form oxyapatite (OAp) (see Section 15.2.4.2). OAp, $Ca_{10}O(PO_4)_6$ is the product of complete dehydroxylation of HAp. In its structure, there exists a linear chain of O^{2-} ions parallel to the c-axis, each one followed by a vacancy. This means that the character of the crystallographic c-axis changes from a screw axis 6_3 to a polar axis $\bar{6}$, thus lowering the overall symmetry.

15.2.2 Structure of biological apatite

Although the structure of geological HAp is well known, it is much less so for biological apatite. Natural bone is a composite material, the water-free substance of which consists of about 70 mass% apatite and 30 mass% collagen I. The apatite platelets of about 40 × 25 × 3 nm³ size are orderly arranged along the triple-helical strands of collagen I. Hence, this abundant protein acts as a structural template for crystallization of nano-HAp, presumably mediated by carboxylate terminal groups and osteocalcin. Because of its open channel structure, HAp is able to incorporate other ions by substituting Ca^{2+} cations as well as OH⁻ and PO_4^{3-} anions without large distortion of the lattice. In biological apatite, some Ca^{2+} is substituted by Na^+, Mg^{2+}, Sr^{2+}, K^+ and some trace elements such as Pb^{2+}, Ba^{2+}, Zn^{2+} and Fe^{2+}. The PO_4^{3-} groups are being partially replaced by CO_3^{2-}, whereas CO_3^{2-}, Cl^-, O^{2-} and, in particular F^- can substitute for OH^-. This compositional variability of HAp is the root cause of its high biocompatibility and osteoconductivity.

The substitution by other ions reduces the theoretical stoichiometric Ca/P ratio of biological HAp from 1.67 to values as low as 1.4 [4]. The non-stoichiometry of biological apatite can be described by the approximate formula:

$$Ca_{10-x}(HPO_4)_x(PO_4)_{6-x}(OH,O,Cl,F,CO_3,\square)_{2-x} \cdot nH_2O; \; 0 < x < 1; \; n = 0 - 2.5.$$

The fact that the OH^- positions can be occupied by mobile O^{2-} ions and vacancies $\square$ is of vital importance for understanding the kinetics of the dehydroxylation reaction of HAp to oxyhydroxyapatite (OHAp) and OAp, respectively that occurs during plasma-spraying (see Section 15.2.4.2).

While the behavior of OH^- ions in the HAp lattice has been recognized and widely studied in natural and synthetic inorganic materials, it came as a complete surprise that biological apatite was found to be essentially free of hydroxyl ions [5]. Hence, contrary to the general medical nomenclature, bone-like apatite appears to be non-hydroxylated. There are suggestions that the specific state of atomic order, imposed biochemically by the body, is essential for cell metabolism and the ability of the body to carry out tissue-specific functions. Since the lack of hydroxyl ions causes a high density of vacancies lined up along the c-axis, high mobility of Schottky-type defects may influence HAp solubility and thus may provide a mechanism for fast and efficient bone reorganization by dissolution (osteoclastesis) and precipitation (osteoblastesis) during bone reconstruction in response to changing stress and load levels controlled by Wolff's law [6].

15.2.3 Biocompatibility of bioceramics

Any artificial material incorporated into the human organism has to abide by certain properties that will assure that there are no negative interactions with living tissue. Consequently, one of the key properties of bioceramics is biocompatibility. Biocompatibility is not an individual property *per se* but relates to the various interactions the inorganic material is part of at the cell and tissue levels. Hence, a systemic approach is required. According to modern view, biocompatibility refers to the ability of a material to perform with an appropriate host response, in a specific application [7]. The interaction of biomaterials with living tissue can be defined by the increasing degree of biocompatibility, ranging from incompatible, biotolerant (e.g. austenitic stainless steel, polymethylmethacrylate), bioinert (e.g. titanium alloys, alumina, zirconia, titania, diamond-like carbon) to bioactive (e.g. calcium phosphates, bioglasses) materials [8].

In contrast to bioinert materials, bioactive materials show positive interaction with living tissue. This interaction can be characterized by two system-relevant terms: osseoconductivity and osseoinductivity. On the one hand, osseoconductivity is the ability of a material to foster the in-growth of bone cells, blood capillaries, and perivascular tissue into the gap between implant and existing bone. On the other hand, osseoinductivity refers to the transformation of undifferentiated mesenchymal precursor stem cells into osseoprogenitor cells preceding enchondral ossification. There is chemical bonding to the bone along the interface triggered by the adsorption of bone growth-mediating proteins at the biomaterial surface. Hence, there will be a biochemically mediated strong bonding osteogenesis. The bioactivity of calcium phosphates is associated with the formation of hydroxyl carbonate apatite, similar to biological, i.e. bone-like apatite [9].

The clinical applications of bioceramics range from bioinert structural ceramics such as alumina and zirconia [10–13] for femoral balls and insets of acetabular cups of total hip endoprosthesis systems [14] to bioactive (osseoconducting) bone growth-stimulating ceramics such as HAp and TCP applied as coatings to the metallic stem of hip endoprostheses and dental root implants [15, 16]. The state of the art of applying HAp coatings is still atmospheric plasma spraying (APS) [17, 18], despite the fact that HAp melts incongruently and thus decomposes in the extremely hot plasma jet into TCP, TTCP, and cytotoxic CaO. To alleviate this problem, low-temperature techniques are being explored to deposit HAp or other calcium phosphates on suitable implant substrates without decomposition. Such techniques include electrochemical deposition (ECD), plasma electrolytic oxidation, electrophoretic deposition, pulsed laser deposition, and ion-beam assisted deposition [19], as well as biomimetic treatment of titanium surfaces with alkalis and acids at elevated temperature [20].

15.2.4 Work in the area of bioceramics

The work of my research group focused on four issues related to the improvement of the performance of plasma-sprayed bioceramic coatings for medical implants. These were
- strengthening the adhesion of coatings to the metallic stem of endoprosthetic implants by designing appropriate adhesion-mediating bond coats;
- exploring the reactions HAp undergoes during exposure to the extremely hot plasma jet;
- studying the alteration of the chemistry and morphology of HAp coatings during exposure to simulated body fluid *in vitro*;
- understanding the development and distribution of residual coating stresses.

Most of the following will relate to the development of bond coats as well as the thermal decomposition of HAp during plasma spraying.

My own involvement with bioceramics started in the early 1990s at a time when I acted as co-supervisor of a master's thesis [21] on plasma-sprayed HAp coatings, conducted at the University of Alberta in Edmonton, Alberta, Canada. The goal of this work was to strengthen the notoriously weak adhesion of HAp to Ti alloy substrates by designing appropriate bond coats [22].

Later, at the Department of Mineralogy, Technische Universität Bergakademie Freiberg, this work continued with my students and co-workers. We recognized the need to improve the adhesion of plasma-sprayed osseoconductive HAp coatings to a Ti alloy implant surface. This research program was based on the awareness that in an alarmingly high number of clinical cases, coating failure occurred by chipping, spalling, delamination, and dissolution, observed on explanted hip endoprostheses consistently close to the implant/coating interface. This effect could be attributed to the existence of a thin layer of amorphous calcium phosphate (ACP) adjacent to the Ti/coating interface, formed by rapid quenching of molten droplets of calcium phosphate with extremely high cooling rates typical for plasma-sprayed coatings [17, 23]. We thought that such a continuous ACP layer (Fig. 15.1A, B) might act as a low-energy fracture path, guiding the propagation of an incipient crack along the interface. ACP will preferentially dissolve *in vivo* on contact with body fluid thus further weakening the mechanical integrity of the interface.

Our medical research partners working at the university hospitals in Essen and Aachen confirmed by clinical studies of retrieved hip endoprostheses coated with HAp that micromorphological features of plasma-sprayed coatings might play a decisive role in implant longevity. Micromorphology critically influences adhesion, crack initiation and propagation and, eventually fragmentation of the coatings, dissolution of the amorphous phase of the coatings in areas of high loading as well as osteoclastic resorption. In particular, failure of coating adhesion to the implant metal causes formation of a gap into which acellular connective tissue will inevitably invade. This pliable tissue layer formed to encapsulate the foreign implant will prevent solid attachment of the latter to the cortical bone tissue; eventually causing aseptic loosening that in many cases will require a remediation operation. Hence, this deleterious gap formation must be counteracted to guarantee long-term functionality and performance of the biomedical implant.

15.2.4.1 Role of adhesion-mediating bond coats

One way to achieve better coating adhesion is the application of mechanically and thermodynamically stable bioinert bond coats. Bond coats are workhorses of thermal spray technology and thus, well documented in the relevant literature [17, 24, 25]. An ideal bioinert bond coat should have several key characteristics that include uncompromised biocompatibility, good adhesion to both the metal substrate underneath and the osseoconductive HAp top coat, matching coefficient of thermal expansion, and a well-defined melting point to allow application by thermal spray technology.

In addition, such bond coats prevent direct contact between metal and HAp since there is experimental evidence that the metal catalyzes thermal decomposition of the latter [4]. Furthermore, it should prevent the release of metal ions from the Ti alloy substrate to the surrounding living tissue. Such ion release has been found to cause massive hepatic degeneration in animals as well as impaired development of human osteoblasts. Heavy metal ions, in particular vanadium, are thought to affect negatively the transcription of RNA in cell nuclei and, in addition, influence the activity of enzymes by replacing Ca or Mg ions at binding sites. The metabolic action of aluminum and vanadium ions released from the implant and their interference with normal biochemical functions of the human body are important intervention to consider. Today, developments are on the way to replace Ti6Al4V by low modulus (E < 50 GPa) β-type Zr-containing alloys such as Ti13Nb13Zr.

Moreover, the bond coat will reduce the thermal gradient at the substrate/coating interface induced by the rapid quenching of the molten particle splats. This fast quenching leads to deposition of amorphous calcium phosphate (ACP) with an associated decrease in resorption resistance [26] and hence, reduced *in vivo* performance, i.e. reduced longevity of the implants. Reduction of the interfacial thermal gradient will work towards prevention of formation of ACP and will impede thermal decomposition of HAp.

Furthermore, it may also cushion damage to the coating initiated by cyclic micromotions of the implant during movement of the patient in the initial phase of osseointegration [27].

Considering all these requirements, my research team embarked on a series of studies to design thermally sprayed HAp-bond coat systems that should have most, if not all the desired properties.

15.2.4.1.1 Calcium silicate bond coats

Studies by Lamy et al. [22] had confirmed that applying calcium silicate-based bond coats enhanced considerably the adhesion strength and resorption resistance of the bioceramic HAp top coat. In accord with expectations, these bond coats ought to be biocompatible since calcium silicate bioglasses and ceramics have been found to bond easily to living bone, and to form, by a biomimetic process, an apatite surface layer when exposed to simulated body fluid (SBF). The cyclic silicate trimer $(Si_3O_9)^{6-}$ group of pseudowollastonite is thought to act as an active site for heterogeneous, stereochemically promoted nucleation of HAp in simulated body fluid. In addition, earlier histological investigation had provided evidence that silicon may be allied to the initiation of mineralization of pre-osseous tissue in periosteal or endochondral ossification [28], presumably through functional Si-OH (silanol) groups that are known to induce apatite nucleation, and to provide bonding sites for cation-specific osteonectin attachment complexes on progenitor cells. We found that the advantageous combination of a dicalcium silicate bond coat with a HAp top coat yielded an as-sprayed adhesion strength of 32 MPa, which, however, on immersion in protein-free SBF at 37°C for 7 days decreased to 21 MPa [29], outside the limit of 35 MPa stipulated by Wintermantel and Ha [8]. This large decrease is presumably the outcome of a hydrolysis reaction of β-dicalcium silicate (larnite) formed during plasma spraying from the γ-dicalcium silicate with olivine structure (shannonite) of the initial spray powder. In addition, dicalcium silicate has an exceptionally large coefficient of thermal expansion (CTE) of 13×10^{-6} K^{-1}. Since the CTEs of titanium and HAp are both lower at 8.5×10^{-6} and 11×10^{-6} K^{-1}, respectively, there is a rather good match causing the residual coating stresses to be significantly minimized.

15.2.4.1.2 Titania bond coats

Previous research by other authors has revealed that even uncoated titanium alloy implants show reasonable bone integration, pointing to a strong bioactive nature of the otherwise bioinert titanium metal surface. This is thought to be caused by the only a few nanometers thick native titanium oxide layer that lend to titanium passivity and protection against environmental influences such as corrosion in biofluid. This oxide layer forms during partial hydrolysis of Ti-OH surface groups that can act as docking sites for bone growth-supporting proteins, similar to the function of Si-OH groups mentioned above.

These considerations provided for our research group a strong incentive to further study the microstructure of TiO$_2$/HAp coatings and their interface, respectively. A string of papers emerged, partly in cooperation with colleagues from the University of British Columbia, Vancouver, and University of Alberta, Edmonton [26, 30–34]. Among the results of investigation of cross sections of titania-HAp "duplex" coatings with 2D-SIMS imaging was the realization that plasma spraying induced diffusional processes at the coating interfaces as well as the variation of spatial distribution of minor and trace elements across the coating cross sections [33]. Although the SIMS images revealed very little chemical interaction of the titania bond coat with both the Ti6Al4V substrate and the calcium phosphate top coat, they showed that the bond coat still provided substantially increased adhesive bond strength.

Furthermore, the very thin layer of amorphous calcium phosphate (ACP), formed at the immediate interface to the solid substrate by rapid quenching of the outermost melt layer with heat transfer rates beyond 10^6 K/s, takes on a special significance, as its relatively high solubility in body fluid may be one of the leading causes of coating delamination [23].

Figure 15.1A shows a scanning transmission electron microscope (STEM) image of a cross section of a plasma-sprayed titania bond coat separating the Ti6Al4V substrate (right) from the innermost calcium phosphate layer consisting of ACP (left). Figure 15.1B provides an enlarged bright-field STEM image of the interface of the polycrystalline titania bond coat and the ACP layer. This image suggests a columnar deposit of elongated, partly twinned titania crystals that by selected area electron diffraction (SAED) imaging were found to consist of brookite, the metastable orthorhombic polymorph of titania. This brookite phase is entirely unexpected since conventional wisdom has it that substoichiometric Magnéli-type Ti$_n$O$_{2n-1}$ phases should form. However, brookite formation may be the result of a process akin to Oswald's rule of steps, whereby the steep temperature gradient attained during plasma spraying may cause the preferential formation

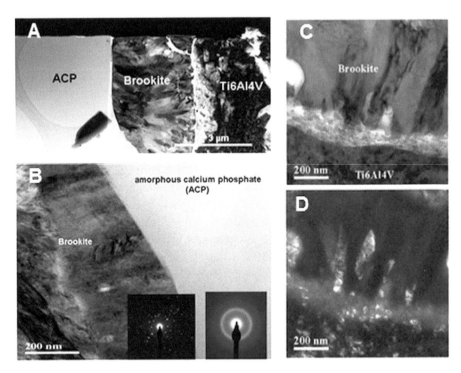

Fig. 15.1: A: STEM image of a cross section of a plasma-sprayed titania (brookite) bond coat-hydroxylapatite topcoat assembly. The sample was prepared by focused ion beam (FIB) cutting with Ga ions to conserve the true geometry of the coating interfaces. B: Bright field STEM image of the interface of a columnar polycrystalline titania (brookite) bond coat (left) and an amorphous calcium phosphate (ACP) top coat (right). The insets show the electron diffraction pattern of both phases [23]. C: Bright field STEM image of the interface bond coat/Ti6Al4V substrate. D: Dark field STEM image of C. © With permission from Elsevier.

of metastable phases. The columnar structure of the brookite layer (Figs. 15.1C, D) offers clues to the mechanism of bonding as the interdigitation of its finger-like columns with the nanocrystalline HAp may provide strengthening of the mechanical performance by increased resistance against shearing.

15.2.4.2 Thermal alteration of HAp during plasma spraying

The extremely high temperature in a plasma jet in excess of 15,000 K leads, even during the very short residence time of the HAp particles (hundreds of microseconds to few milliseconds, depending on particle density and size), to dehydroxylation and thermal decomposition by incongruent melting (Fig. 15.2A). This thermal decomposition of HAp in the hot plasma jet occurs in four consecutive steps (Tab. 15.1).

Based on this decomposition sequence, Graßmann and Heimann [35] developed a simple model of the in-flight evolution of individual calcium phosphate phases as shown schematically in Fig. 15.2B. During the short residence time of the particle in

Tab. 15.1: Thermal decomposition sequence of HAp.

Step 1:	$Ca_{10}(PO_4)_6(OH)_2$	→	$Ca_{10}(PO_4)_6(OH)_{2-x}O_x\square_x$	+	xH_2O
	HAp		OHAp		
Step 2:	$Ca_{10}(PO_4)_6(OH)_{2-x}O_x\square_x$	→	$Ca_{10}(PO_4)_6O_x\square_x$	+	$(1-x)H_2O$
	OHAp		OAp		
Step 3:	$Ca_{10}(PO_4)_6O_x\square_x$	→	$2Ca_3(PO_4)_2$	+	$Ca_4O(PO_4)_2$
	OAp		TCP		TTCP
Step 4a:		$Ca_3(PO_4)_2 \to 3\,CaO + P_2O_5$			
Step 4b:		$Ca_4O(PO_4)_2 \to 4\,CaO + P_2O_5$			

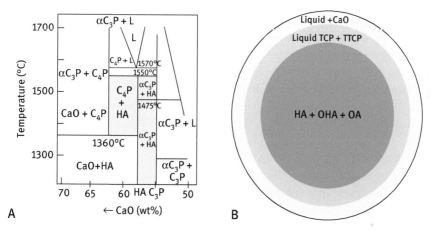

Fig. 15.2: A: Phase diagram of the quasi-ternary system CaO-P_2O_5-(H_2O) at a water partial pressure of 65.5 kPa. B: Schematic Graßmann-Heimann model [35] of the thermal decomposition of a spherical hydroxylapatite particle subjected to high temperature in a plasma jet [15].

the hot plasma jet, the innermost core is still at a temperature below 1550°C owing to the low thermal diffusivity of HAp. Hence, HAp and OHAp/OAp are the only stable phases (Tab. 15.1, steps 1 and 2).

The second shell, heated to a temperature above the incongruent melting point of HAp (1570°C), consists of a molten mixture of TCP and TTCP (Tab. 15.1, step 3). The outermost spherical shell of the particle comprises solid CaO + melt since evaporation of P_2O_5 shifts the composition along the liquidus (Fig. 15.2A) towards CaO-richer phases (Tab. 15.1, step 4). The temperature increases to well beyond 1730°C, and the only unmelted composition is CaO.

Among the thermally altered calcium phosphate phases observed during plasma spraying of HAp, the OAp phase takes on a special significance. OAp, $Ca_{10}O(PO_4)_6$, is the product of complete dehydroxylation of HAp but is known to be able to convert back to more or less stoichiometric HAp in the presence of water either during cooling of the as-sprayed coating in moist air or by reaction with biofluid *in vivo*. However, it should not be concealed that the existence of OAp as a thermodynamically stable compound has been subject to controversy, reaching as far back as the 1930s. To address this "OAp problem", work was performed that resulted in a recent paper in *American Mineralogist* [36]. Among other analytical techniques, in this contribution, I provided information on the phase content of plasma-sprayed HAp gained from nuclear magnetic resonance (NMR) spectroscopy. Figure 15.3 shows cross-polarized (CP) magic angle spinning NMR spectra of protons (Fig. 15.3A) and PO_4^{3-} tetrahedral groups (Fig. 15.3B) of typical plasma-sprayed HAp coatings. The assignment of individual bands is tentative, based on earlier work by myself [37] and one of my former PhD students [38]. The isotropically shifted L* band of the ^{1}H-NMR spectrum, shown in Fig. 15.3A, may be an, admittedly, weak signal of the existence of oxy- and/or OHAp-like conformations. More profitable is the ^{31}P-NMR spectrum shown in Fig. 15.3B.

Besides the A signal of well-ordered and highly crystalline HAp, apparently unchanged from its original composition, there are C and D signals that are thought to be related to PO_4^{3-} groups with strongly to very strongly distorted oxygen environments and few or no associated OH⁻ groups, typical for OHAp and OAp, respectively. The B signal is assigned to strongly distorted PO_4^{3-} tetrahedral groups without neighboring OH⁻ groups, indicating the presence of TCP and/or TTCP. Supporting CP-2D-^{1}H/^{31}P heteronuclear correlation NMR spectroscopy confirmed that the D band might indeed be a signal of OAp [39]. Since its chemical shift is identical to that of TTCP, we concluded that with increasing dehydroxylation of HAp the coating composition eventually approaches that of structurally closely related TTCP, in accord with the dehydroxylation sequence HAp → OHAp → OAp → TTCP shown in Tab. 15.1

15.2.4.3 Interaction of coatings with simulated body fluid and residual stresses

In addition to work mentioned above, my research group had studied in much detail the alteration of HAp coatings in contact with simulated body fluid (SBF) [16, 40–50].

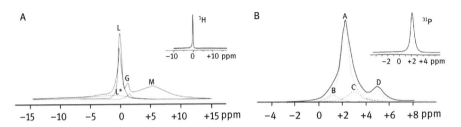

Fig. 15.3: CP-MAS-NMR spectra of an as-sprayed hydroxylapatite coating. A: ^{1}H-CP-MAS NMR spectrum (L: well ordered HAp; L*: OAp/OHAp; G: water). B: ^{31}P-CP-MAS NMR spectrum (A: well-ordered HAp; B: TCP/TTCP; C: OHAp; D: OAp) [36]. The insets show the spectra of phase-pure stoichiometric hydroxylapatite as a reference.

We also investigated the stress distribution in plasma-sprayed HAp coatings in cooperation with the Department of Physics, University of Cape Town, South Africa [51–56]. We also did some exploratory work on the *in vivo* osseoconductive performance of plasma-sprayed HAp coatings, without and with bond coats, in animal studies with a sheep model at the Universität Duisburg-Essen [57] and a dog model at Chulalongkorn University in Bangkok, Thailand [58, 59].

15.2.4.4 Other work

Although plasma-sprayed HAp coatings are still considered the gold standard of osseoconductive coatings, experiments were performed with plasma-sprayed transition metal-substituted calcium hexaorthophosphates of the type $Ca(Ti_nZr_{4-n})(PO_4)_6$, members of the NaSiCON (Na superionic conductor) structural family [60]. These ceramics show solubility in SBF at least one order of magnitude lower than that of other calcium orthophosphates including HAp and, in particular β-TCP and TTCP. Among their advantageous properties is weak solid-state ion conductivity that in the future may be utilized to design fourth-generation biomaterials (see Section 15.2.5).

Plasma-sprayed coatings of transition metal-substituted calcium hexaorthophosphates display sufficiently strong adhesion to Ti6Al4V implant substrates, with reasonable shear strength. Although considerable thermal decomposition due to incongruent melting occurred, there is some evidence that these bioinert decomposition products may lead to particle-mediated reinforcement of the coating microstructure thus improving their cohesive strength [61–63].

Cooperation with a major supplier of advanced ceramic medical products, CeramTec GmbH in Plochingen, has resulted in a study of the effect of sterilization with ionizing radiation on femoral heads of hip endoprostheses fashioned from Y-stabilized zirconia and alumina. It has been observed that irradiation with ionizing γ-radiation introduces color centers in Y-stabilized zirconia. Whereas this effect compromises neither the mechanical nor the biological performance of the femoral head, it has caused some concern among orthopedic surgeons. Indeed, irradiated and thus, bluish- or brownish-tinged zirconia femoral heads are known to have been rejected by orthopedic surgeons under suspicion of deleterious contamination. To get insight into the origin of the coloration and hence, to alleviate concerns of the surgeons, we irradiated alumina and zirconia femoral heads with both X-rays and γ-rays, and recorded and explained their color change using optical absorption spectroscopy, electron spin resonance spectroscopy, as well as thermoluminescence spectroscopy [64, 65].

15.2.5 Outlook on future developments and further research requirements

Progress in developing and applying novel or improved bioceramic coatings is uniquely connected to man's innate ability and need to learn from nature. However, whereas biology can inform technology at all procedural levels, i.e. materials, structures, mechanisms, machines, and control, there is still a gap between biology and its technological realization that remains to be bridged.

Looking back, biomaterials including bioceramics have conceptually evolved through three different generations. These are
- first-generation bioinert materials, which only have mechanical functions, such as alumina and zirconia, or highly crystalline HAp,
- second-generation bioactive and biodegradable materials, which have osseoconductive function, such as bone-like HAp or TCP, and
- third-generation biomaterials, which were designed to have osseoinductive function by stimulating specific cell responses at the molecular level such a HAp implanted with bone-growth stimulating agents [66].

Computational modeling is being applied for correlating gene expression profiling (genomics) with combinatorial material design strategies (materiomics) [67]. This effort also includes recognition that immune cells play a vital role in regulating bone dynamics. Hence, recently a paradigmatic shift was proposed [68] from traditional bone biomaterials to osteoimmunomodulatory materials. Along the same lines, incorporation into HAp coatings of bone-growth stimulating non-collagenous proteins such as recombinant human bone morphogenetic proteins (rhBMPs), osteocalcin, osteonectin, or silylated glycoproteins offer a wide field of novel research endeavor.

Recently, the idea of *fourth-generation biomaterials* was suggested [69], based on integrating electronic systems with the human body to provide powerful diagnostic as well as therapeutic tools for basic research and clinical use. The functionalities of such biomaterial systems include manipulating cellular bioelectric responses for tissue regeneration as well as monitoring cellular responses with the aim to communicate with host tissues via bioelectric signals. We anticipate that plasma-sprayed transition metal-substituted calcium orthophosphates with NaSiCON structure [60] will play a commanding role in the endeavor to develop and clinically test such novel fourth-generation bioceramics. Their elevated solid-state ionic conductivity may transmit signals adapted to this task. Consequently, future research should focus on development of a composite TiO_2–$CaTiZr_3(PO_4)_6$–HAp coating system for implants with the equivalent circuit of a "bio"-capacitor that by appropriate poling could store negative electrical charges close to the interface with the growing bone, thus enhancing bone apposition rate and bone density.

Notwithstanding the shortcomings of APS HAp coatings that include thermal decomposition, line-of-sight limitation, insufficient control of porosity, lack of load-carrying capability, and the inability to deposit coatings of less than about 20 μm thickness, plasma spraying is still the method of choice to apply coatings to the metallic parts of commercially supplied hip and knee endoprostheses as well as dental root implants. Currently, APS of HAp powder particles with diameters of tens to hundreds of micrometers is the most popular and the only Food and Drug Administration–approved method to coat implant surfaces for clinical use. However, novel promising deposition techniques include suspension plasma spraying [70, 71], solution precursor plasma spraying [72], and others [15].

However, even though deposition of HAp coatings by APS comes across as a mature and well research-supported technique, there is still need to address several problem areas. Consequently, during the past decades, many attempts have been made to optimize essential properties of osseoconductive bioceramic coatings. These properties include coating cohesion and adhesion, phase composition, homogeneous phase distribution, crystallinity, porosity and surface roughness, nano-structured surface morphology, residual coating stresses, and not in the least, coating thickness. Novel applications are about to emerge, including HAp-based targeted drug delivery vehicles, novel bone graft substitutes, and multifunctional nanoscopic bioceramic materials. However, despite these advances, the success story of calcium phosphates was somewhat downplayed in comparison with other achievements in the area of biomaterials. As pointed out clearly by Habraken et al. [73], this does not do justice to the enormous, yet still not fully exploited potential calcium phosphates possess in the area of biomineralization including coatings for biomedical implants.

In conclusion, despite much achievement in the field of bioceramics, there is still room for embarking on rewarding research designed to improving the deposition techniques as well as the mechanical, chemical, microstructural, and biological properties of bioceramic, most notably, HAp coatings.

References

[1] Worldwide hip and knee orthopedic surgical implant market shares, trend, growth, strategy, and forecast; 2016–2022. Available at: http://www.medgadget.com. Accessed March 13, 2016.
[2] De Jong WF. La substance minerale dans les os. Recl Trav Chim Pays-Bas Belg 1926;45:445–8.
[3] Posner AS, Perloff A, Diorio AF. Refinement of the hydroxyapatite structure. Acta Crystallogr 1958;11:308–9.
[4] Weng J, Liu X, Zhang X, Ji X. Thermal decomposition of hydroxyapatite structure induced by titanium and its dioxide. J Mater Sci Lett 1994;13:159–61.
[5] Pasteris JD, Wopenka B, Freeman JJ, Rogers K, Valsami-Jones E, van der Houwen JAM, Silva MJ. Lack of OH in nanocrystalline apatite as a function of degree of atomic order: implications for bone and biomaterials. Biomaterials 2004;25(2):229–38.
[6] Hench LL, Ethridge EC. Biomaterials. An interfacial approach. New York: Academic Press; 1982.
[7] Williams DF. Biocompatibility: an overview. In: Williams DF, editor. Concise encyclopedia of medical & dental materials. Oxford: Pergamon Press; 1990;52.
[8] Wintermantel E, Ha SW. Biokompatible Werkstoffe und Bauweisen. Implantate für Medizin und Umwelt. Berlin: Springer, 1996.
[9] LeGeros RZ, LeGeros JP. Phosphate minerals in human tissues. In: Nriagu JO, Moore PB, editors. Phosphate minerals. New York: Springer, 1984;351–85.
[10] Helmer JD, Driskell TD. Research on bioceramics. In: Symposium on use of ceramics as surgical implants. Clemson, SC: Clemson University, 1969.
[11] Garvie RC, Nicholson PS. Structure and thermodynamical properties of partially stabilized zirconia in the CaO-ZrO_2 system. J Am Ceram Soc 1972;55:152–7.
[12] Garvie RC, Hannink RHJ, Pascoe RT. Ceramic steel? Nature 1975;258:703–4.

[13] Cales B, Stefani Y. Yttria-stabilized zirconia for improved orthopedic prostheses. In: Encyclopedic handbook of biomaterials and bioengineering, part B: applications. Vol. 1. New York: Marcel Dekker, 1995;415–52.
[14] Heimann RB. Classic and advanced ceramics. From fundamentals to applications. Weinheim: Wiley-VCH, 2010.
[15] Heimann RB, Lehmann HD. Bioceramic coatings for medical implants. Weinheim: Wiley-VCH, 2015.
[16] Heimann RB. Plasma-sprayed hydroxylapatite-based coatings: chemical, mechanical, microstructural, and biomedical properties. J Thermal Spray Technol 2016;25(5):827–50.
[17] Heimann RB. Plasma spray coating. Principles and applications. 2nd ed. Weinheim, Wiley-VCH, 2008.
[18] Heimann RB. Thermal spraying of biomaterials. Surf Coat Technol 2006;201:2012–9.
[19] Heimann RB. The challenge and promise of low-temperature bioceramic coatings: an editorial. Surf Coat Technol 2016;301:1–5.
[20] Kokubo T, Yamaguchi S. Novel bioactive materials developed by surface-modified Ti metal and its alloys. Acta Biomater 2016;44:16–30.
[21] Lamy D. Development of plasma-sprayed hydroxyapatite coating systems for surgical implant devices. Unpublished Master thesis. University of Alberta, Edmonton, Alberta, Canada.
[22] Lamy D, Heimann RB, Pierre AC. Hydroxyapatite coatings with a bond coat on biomedical implants by plasma projection. J Mater Res 1996;11(3):680–6.
[23] Heimann RB, Wirth R. Formation and transformation of amorphous calcium phosphates on titanium alloy surfaces during atmospheric plasma spraying and their subsequent in vitro performance. Biomaterials 2006;27:823–31.
[24] Heimann RB, Lehmann HD. Recently patented work on thermally sprayed coatings for protection against wear and corrosion of engineered structures. Recent Patents on Mater Sci 2008;1(1):41–55.
[25] Lehmann HD, Heimann RB. Thermally sprayed thermal barrier coating (TBC) systems: a survey of recent patents. Recent Patents on Mater Sci, 2008;1(2):140–58.
[26] Heimann RB, Vu TA, Wayman ML. Bioceramic coatings: state-of-the-art and recent development trends. Eur J Mineral 1997;9:597–615.
[27] Søballe K. Hydroxylapatite ceramic coating for bone implant fixation. Mechanical and histological studies in dogs. Acta Orthop Scand 1993;64(Suppl. 244):58 pp.
[28] Carlisle EM. Silicon: a possible factor in bone calcification. Science 1970;167:279–80.
[29] Vu TA, Heimann RB. Effect of CaO on thermal decomposition during sintering of composite hydroxyapatite-zirconia mixtures for monolithic implants. J Mater Sci Lett 1997;16:437–9.
[30] Kurzweg H, Heimann RB, Troczynski T. Adhesion of thermally sprayed hydroxyapatite-bond coat systems measured by a novel peel test. J Mater Sci Mater Med 1998;9:9–16.
[31] Kurzweg H, Heimann RB, Troczynski T, Wayman ML. Development of plasma-sprayed bioceramic coatings with bond coats based on titania and zirconia. Biomaterials 1998;19:1507–11.
[32] Heimann RB. Design of novel plasma-sprayed hydroxyapatite-bond coat bioceramic systems. J Thermal Spray Technol 1999;8(4):597–604.
[33] Heimann RB, Kurzweg H, Ivey DG, Wayman ML. Microstructural and *in vitro* chemical investigations into plasma-sprayed bioceramic coatings. J Biomed Mater Res, Appl Biomater 1998;43:441–50.
[34] Heimann RB, Vu TA. Low-pressure plasma-sprayed (LPPS) bioceramic coatings with improved adhesion strength and resorption resistance. J Thermal Spray Technol 1997;6(2):145–9.
[35] Graßmann O, Heimann RB. Compositional and microstructural changes of engineered plasma-sprayed hydroxyapatite coatings on Ti6Al4V substrates during incubation in protein-free simulated body fluid. J Biomed Mater Res 2000;53(6):685–93.
[36] Heimann RB. Tracking the thermal decomposition of plasma-sprayed hydroxylapatite. Am Mineral 2015;100(11-2):2419–25.

[37] Heimann RB, Tran HV, Hartmann P. Laser-Raman and nuclear magnetic resonance (NMR) studies on plasma-sprayed hydroxyapatite coatings: influence of bioinert bond coats on phase composition and resorption kinetics in simulated body fluid. Materialwiss Werkstofftech 2003;34 (12):1163–9.
[38] Tran HV. Investigation into the thermal dehydroxylation and decomposition of hydroxyapatite during atmospheric plasma spraying: NMR and Raman spectroscopic study of as-sprayed coatings and coatings incubated in simulated body fluid. Unpublished PhD thesis. Department of Mineralogy, Technische Universität Bergakademie Freiberg, Germany; 2004.
[39] Hartmann P, Barth S, Vogel J, Jäger C. Investigations of structural changes in plasma-sprayed hydroxylapatite coatings. In: Rammlmair D, Mederer J, Oberthür Th, Heimann RB, Pentinghaus H, editors. Proc. 6th ICAM 2000. Appl Mineral Res Econ Technol Ecol Culture 2000;1:147–50. Rotterdam: A.A. Balkema.
[40] Götze J, Hildebrandt H, Heimann RB. Charakterisierung des in vitro-Resorptionsverhaltens von plasmagespritzten Hydroxylapatit-Schichten. BIOmaterialien 2001;2(1):54–60.
[41] Götze J, Heimann RB, Hildebrandt H, Gburek U. Microstructural investigations into calcium phosphate biomaterials by spatially resolved cathodoluminescence. Materialwiss Werkstofftechn 2001;32:130–6.
[42] Heimann RB, Graßmann O, Zumbrink T, Jennissen HP. Biomimetic processes during in vitro leaching of plasma-sprayed hydroxyapatite coatings for endoprosthetic application. Materialwiss Werkstofftechn 2001;32:913–21.
[43] Götze J, Hildebrandt H, Heimann RB. Microstructural changes in plasma-sprayed hydroxyapatite coatings during long-term incubation in simulated body fluid: evidence from cathodoluminescence. Abstracts CL2001, Cathodoluminescence in Geosciences: New insights from CL in combination with other techniques, Freiberg, September 6–8, 2001, 40–41.
[44] Heimann RB. *In vitro-* und *in vivo*-Verhalten von osteoconduktiven plasmagespritzen Ca-Ti-Zr-Phosphat-Beschichtungen auf Ti6Al4V-Substraten. BIOmaterialien 2006;7(1):29–37.
[45] Hesse C, Hengst M, Kleeberg R, Götze J. Influence of experimental parameters on spatial phase distribution in as-sprayed and incubated hydroxyapatite coatings. J Mater Sci Mater Med 2008;19(10):3235–41.
[46] Heimann RB. Characterization of as-plasma-sprayed and incubated hydroxyapatite coatings with high-resolution techniques. Materialwiss Werkstofftech 2009;40(1–2):23–30.
[49] Heimann RB. Laser-Raman, nuclear magnetic resonance (NMR) and electron energy loss (EEL) spectroscopic studies of plasma-sprayed hydroxyapatite coatings. In: Heimann RB, editor. Calcium phosphate. Structure, synthesis, properties, and applications. New York: Nova Science; 2012;215–30.
[50] Heimann RB. Structure, properties, and biomedical performance of osteoconductive bioceramic coatings. Surf Coat Technol 2013;233:27–38.
[51] Heimann RB, Graßmann O, Hempel M, Bucher R, Härting M. Phase content, resorption resistance and residual stresses of bioceramic coatings. In: Rammlmair D, Mederer J, Oberthür Th, Heimann RB, Pentinghaus H, eds. Appl Mineral Res Econ Technol Ecol Culture 2000;1:155–8. Proc. 6th ICAM 2000. Rotterdam, A.A. Balkema.
[52] Topic M, Ntsoane T, Hüttel T, Heimann RB. Microstructural characterisation and stress determination in as-plasma sprayed and incubated bioconductive hydroxyapatite coatings. Surf Coat Technol 2006;201(6):3633–41.
[53] Ntsoane TP, Topic M, Thovhogi T, Härting M, Robinson D, Heimann RB. Depth-resolved strain determination of hydroxyapatite coatings using high energy x-ray diffraction. Powder Diffract 2006;21(2):D083.
[54] Heimann RB, Ntsoane TP, Pineda-Vargas CA, Przybylowicz WJ, Topić M. Biomimetic formation of hydroxyapatite investigated by analytical techniques with high resolution. J Mater Sci Mater Med 2008;19:3295–302.

[55] Ntsoane TP, Topic M, Härting M, Heimann RB, Theron C. Spatial and depth-resolved studies of air plasma-sprayed hydroxyapatite coatings by means of diffraction techniques: part I. Surf Coat Technol 2016;294:153–63.

[56] Ntsoane TP, Theron C, Venter A, Topic M, Härting M, Heimann RB. In-vitro investigation of air plasma-sprayed hydroxyapatite coatings by diffraction techniques. Mater Res Proc 2016;2:485–490.

[57] Heimann RB, Schürmann N, Müller RT. *In vitro* and *in vivo* performance of Ti6Al4V implants with plasma-sprayed osteoconductive hydroxylapatite-bioinert titania bond coat "duplex" systems: an experimental study in sheep. J Mater Sci Mater Med 2004;15(9):1045–52.

[58] Heimann RB, Itiravivong P, Promasa A. *In vivo*-Untersuchungen zur Osteointegration von Hydroxylapatit-beschichteten Ti6Al4V-Implantaten mit und ohne bioinerter Titanoxid-Haftvermittlerschicht. BIOmaterialien 2004;5(1):38–43.

[59] Itiravivong P, Promasa A, Laiprasert T, Techapongworachai T, Kuptniratsaikul S, Thanakit V, Heimann RB. Comparison of tissue reaction and osteointegration of metal implants between hydroxyapatite/Ti alloy coat: an animal experimental study. J Med Assoc Thai 2003;86(2):S422–30.

[60] Heimann RB. Transition metal-substituted calcium orthophosphates with NaSiCON structure: a novel type of bioceramics. In: Heimann RB, editor. Calcium phosphate. Structure, synthesis, properties, and applications. New York: Nova Science; 2012;363–79.

[61] Schneider K, Heimann RB, Berger G. Untersuchungen im quaternären System $CaO-TiO_2-ZrO_2-P_2O_5$ im Hinblick auf die Verwendung als langzeitstabiler Knochenersatz am Beispiel des $CaTiZr_3(PO_4)_6$. Ber Dtsch Mineral Ges 1998;1:259.

[62] Schneider K. Entwicklung und Charakterisierung plasmagespritzter biokeramischer Schichten im quaternären System $CaO-TiO_2-ZrO_2-P_2O_5$. Unpublished master thesis. Technische Universität Bergakademie Freiberg, 2002.

[63] Schneider K, Heimann RB, Berger G. Plasma-sprayed coatings in the system $CaO-TiO_2-ZrO_2-P_2O_5$ for long-term stable endoprostheses. Materialwiss Werkstofftech 2001;32:166–71.

[64] Dietrich A, Heimann RB, Willmann G. The colour of medical-grade zirconia (Y-TZP). J Mater Sci Mater Med 1996;7:559–65.

[65] Heimann RB, Willmann G. Irradiation-induced colour changes of Y-TZP ceramics. Br Ceram Trans 1998;97:185–8.

[66] Navarro M, Michiardi A, Castaño, Planell JA. Biomaterials in orthopaedics. J R Soc Interface 2008;5(27):1137–58.

[67] Groen N, Guvendiren M, Rabitz H, Welsh WJ, Kohn J, de Boer J. Stepping into the Omics era: opportunities and challenges for biomaterials science. Acta Biomater 2016;34:133–42.

[68] Chen ZT, Klein T, Murray RZ, Crawford R, Chang J, Wu CT, Xiao Y. Osteoimmunomodulation for the development of advanced bone biomaterials. Mater Today 2016;19(6):304–21.

[69] Ning CY, Zhou L, Tan GX. Fourth-generation biomedical materials. Mater Today 2016;19(1):2–3.

[70] Bouyer E, Gitzhofer F, Boulos MI. The suspension plasma spraying of bioceramics by induction plasma spraying. J Mater 1997;49(2):58–62.

[71] Gross KA, Saber-Samandari S. Revealing mechanical properties of a suspension plasma sprayed coating with nanoindentation. Surf Coat Technol 2009;203:2995–9.

[72] Huang Y, Song L, Liu X, Xiao Y, Wu Y, Chen J, Wu F, Gu Z. Hydroxyapatite coatings deposited by liquid precursor plasma spraying: controlled dense and porous microstructures and osteoblastic cell responses. Biofabrication 2010;2(4):045003.

[73] Habraken W, Habibovic P, Epple M, Bohner M. Calcium phosphates in biomedical applications: materials for the future? Mater Today 2016;19(2):69–87.

Maurizio Romanelli, Fabio Capacci, Luca A. Pardi and Francesco Di Benedetto

16 A procedure to apply spectroscopic techniques in the investigation of silica-bearing industrial materials

16.1 Introduction

A substantial number of mineralogy and geochemistry studies have paid attention to the health issues related to crystalline silica (CS) [1–4]. Among them, silicosis, the most ancient professional disease ever recognized [5], and lung cancer, which has been assessed after the analysis of more than 30 years of clinical, epidemiological, and laboratory research by the International Agency for the Research on Cancer (IARC) [6] under the statement "Crystalline silica in the form of quartz or cristobalite dust is carcinogenic to humans", (page 396). This general interest about CS, a term including all the crystalline polymorphs of SiO_2, is due to the ubiquitous diffusion of SiO_2 in the earth's crust, in most of the industrial production processes, and in everyday environment [7].

In 2013, the world production of industrial sand and gravel was estimated to be 142 Mt, a massive amount compared, for instance, with the 77 Mt of Portland cement [8]. This large amount of CS is mainly used in the ceramics and glass productions (where CS is processed as a raw material) and in the metal foundry (where CS has a side use). Recently, a significant demand of CS has been coming also from hydrocarbon extraction by hydraulic fracturing (fracking) from low-porosity rocks and from sand blasting [8].

Workers operating in these industrial sectors are thus potentially exposed to CS, as well as those involved in mining and constructions. Estimating the number of potentially exposed workers is not straightforward. A study published in 2000 reveals that in EU, exposure to CS involves 3.2 million people (~2.3% of the total employed workers) [9]. In Europe, occupational pneumoconiosis causes almost 7200 deaths every year, while occupational lung disease prevalence is about 300/100,000 inhabitants. In China, state-of-the-art data on silicosis point to 440,000 workers with ascertained health consequences, a number probably underestimated, and to a trend of 10,000 new cases per year [10]. Nowadays, the problem of CS exposure concerns emerging countries, where high exposures, linked to the traditional ways of production, induce typical silicotigen effects with heavy or even deadly prognosis, as well as economically advanced countries. In the working sectors less affected by technology development, such as building, extraction, and manufacturing of stones, exposures are still higher than those set by the limiting values [11]. Moreover, clusters of silicosis related to less-known exposures can be revealed; one can cite for instance the cases of workers in casting, in jewellery, and, more recently, in sanding of jeans cloths, in manufacturing of "artificial stone" (a conglomerate made up by a fine intermixing of quartz microcrystals and polyester resins [12]), and in floor throwing, where quartz powders are used [13].

From review studies [1–4], the main findings obtained through a laboratory approach in the characterization of the effects of CS on human health are (i) the molecular scale of the interaction with the lung tissues, (ii) the surface nature of this interaction, (iii) the prevailing role played by radicals and by metal ions impurities in activating and modulating the toxicity (a schematic summary is presented in Tab. 16.1), and (iv) the grinding or mechanical activation immediately before the exposure (verified in "cell-free" tests [14], cell cultures [15–16], and test animals [17]). Thermal treatments can induce conflicting effects. On the one hand, they can favor the transition from quartz to cristobalite, which is very stable at high temperatures. On the other hand, they can progressively decrease the amount of surface radicals, thus favoring a decrease of their cytotoxic effects [18] and of their transforming potencies [19].

This variety of factors can conceivably be the reason for the large variability of CS risk [20]. In the last 15 years, some studies tried to overcome the lack of information on the physical, chemical, and toxicological properties of industrial CS-bearing materials [7, 21–35]. Here, one has to outline two important general results: (1) the prevailing role played by differences (even subtle ones) in the level of contaminants (e.g. aluminum, iron), in CS surface properties, grinding procedure, particle shape, and thermal treatment; apparently, these differences originate in the genetic history of each particular CS material considered; (b) the significant variability of the physicochemical features undergone by CS in each of the studied industrial processes are functions of its role (processed material, side material).

From the above considerations, the characterization of the physico-chemical features circumstantially related to the CS in the industrial processes is of paramount importance; indeed, the investigation of the actual dust breathed by the workers is crucial. The aim of the present study is to describe and discuss an investigation

Tab. 16.1: Species involved in toxicity: schematic summary.

Contaminant	Effect	Reference
Al-bearing compounds (e.g. clays)	Inhibition of severe effects	[52]
Carbon powders	Decreased induction of fibrotic damage (in rats)	[53–54]
	Decreased generation of free radicals and decreased cytotoxic effects in lung cells	[37]
Fe	In traces, increased induction of acute inflammation	[55]
	In traces, increased generation of free radicals	[56]
	If concentrated, increased in inhibition of cytotoxicity and cell transformation	[57]

procedure that can provide information on most of the cited circumstantial differences. Several examples showing the relevance of a detailed definition of the mineralogical properties of the industrial samples with their spectroscopic features are also discussed.

16.2 Activating factors in industrial CS-bearing materials: examples

The determination of the factors affecting the physico-chemical (and toxicological) properties of CS bearing materials is a not easy task. Indeed, they are numerous and often interrelated [4]. In the following examples, we want to bring into focus the CS-activating factors that can be studied independently of the particular production line. All the reported examples deal with powders coming from different industrial settings. The main aim of this section is to show that the use of several, mainly spectroscopic, techniques allows one to get information on the transformations undergone by CS-bearing materials in industrial processes. Comments on the importance of these transformations for the human health are also inserted.

16.2.1 Size distribution

Changes in size distribution and in the morphology of the crystals/particles may occur as a result of various processes, including grinding or thermal stress (Fig. 16.1). Several examples may be found in the literature that put in evidence the reduction of the particle size in CS-bearing materials during processing in association with thermal stress and/or mild mechanical activation [32, 36], different modes of physical aggregation among particles [21–37], and in the presence of natural or artificial coatings [28, 36]. All these features are generally not related to an experimentally evident change in the mineralogical composition nor in the total CS content.

The knowledge of the changes in size distribution and morphology of particles is important for four main reasons, i.e.:
1. There is an apparent correlation between the size distribution and the development of new surfaces. These, in turn, are related to an increased cytotoxicity under laboratory conditions [4, 27, 38]. The role of new surfaces, in fact, is considered one of the most significant in CS toxicity, as it has been pointed out in the IARC recommendation [6].
2. The larger the number of small-size particles, the higher the reactivity that they can finally experience. This fact is related to observed changes in both the chemical and radical speciation and the surface chemistry of CS particles.
3. It is largely demonstrated [39] that the particle size of bulk materials closely determines the amount of suspended dust in an indoor environment and thus the workers' exposure.

4. Studies report on some relationship among particle size, shape, crystallinity, and health effects, which could go beyond the simple dose considerations [40–41]. These relationships appear even more significant when nanosized particles are considered.

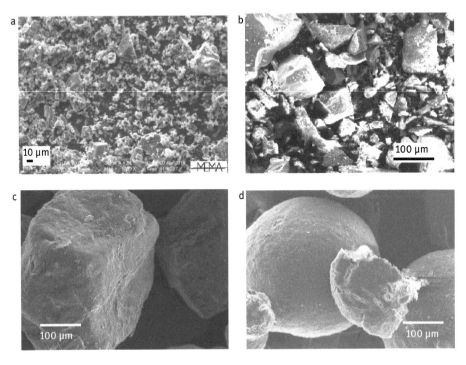

Fig. 16.1: Exemplar micrographs (magnification × 1000) exhibiting the variety of CS-bearing industrial materials: (a) quartz and cristobalite, mixed with gypsum for gold micro-casting process; (b) sands for blasting procedures; (c) sands for casting of non-ferrous alloys; (d) quartz mixed with kaolinite and feldspars in a raw ceramic mixture.

16.2.2 Speciation of inorganic radicals

The change in the population of long living inorganic radicals can be considered as one of the best markers of the reactivity occurring in CS-bearing materials. Numerous studies (e.g. [42] and references therein) pointed out, in fact, that almost all quartz-bearing geomaterials (i.e. those including most of the CS-bearing raw materials for industry) contain variable amounts of radiogenic radicals. These radicals are very stable (their half-life time being > 1 Ma; [42]) and recent studies [31] have pointed out that they can persist also after robust mechanical treatment, as e.g. crystal crushing and grinding. This radical speciation can be detected by conventional EPR experiments carried out at T < 70 K: above this temperature, delocalization effects broaden the signal [42]. The EPR signals of radiogenic radicals are observed in almost all the quartz-bearing materials immediately before their inlet in a production process [39], because the quartz used in the industrial production has a natural origin, i.e. is obtained by sands or gravels [8].

During the production process, the radical speciation of the material, in general, and of quartz in particular, can be modified or not, which depends on the type and strength of the intervening treatment. The lineage of the radical speciation related to the surface properties of CS could then be analytically followed by conventional EPR spectroscopy. Three typical situations can be envisaged (Fig. 16.2): (i) the radical species do not change (in raw materials after a robust milling under laboratory conditions [31]; in concrete production; in some stone manufacturing laboratories [39]); (ii) new radicals are added, but the radiogenic radicals persist (e.g. in ceramic production line [35]); (iii) the original radical speciation is depleted below the detection limit, and new radicals appear (e.g. in ceramic and cast iron productions [35–36]).

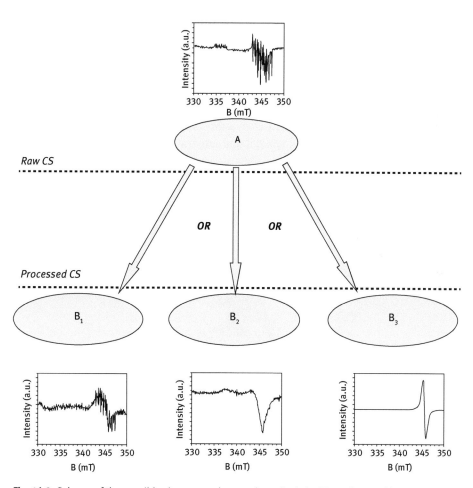

Fig. 16.2: Scheme of the possible changes undergone by radicals in CS, as detected by EPR spectra: (A) a raw industrial quartz (modified from [31]); (B1) a mechanically activated (under laboratory conditions) quartz powder (modified from [31]); (B2) a processed ceramic sample (modified from [35]); (B3) a quartz sand processed in the cast iron production line (modified from [36]). All spectra are registered at 35 K. Magnetic field values are expressed in millitesla.

The knowledge of the relationship between the radical and its structural neighborhood is necessary to detail the newly developed species. This information is in general not easily achievable by continuous-wave EPR. The inorganic radicals associated to quartz and CS-bearing materials often exhibit complex line patterns in their EPR spectra [43]. Conversely, once detected, the EPR spectrum of a specific radical, the investigation of the sample by means of pulsed EPR techniques (i.e. electron spin echo spectroscopy, ESE) allows one to get additional information on the nature, number, and geometrical arrangement of magnetic nuclei in the neighborhood of the radical. Although limited to a certain number of ESE-active nuclei such as H, C, Al, Na, Mn, and V, the information provided by this technique can give significant details concerning the type and stability of radicals. Successful results of this approach were obtained by Romanelli et al., who detailed the structure of the h_{Al} radicals (Fig. 16.3) [31] and who obtained the discrimination of the spectral features of h_{Al} and kaolinite radicals through the study of three-pulse ESE nuclear envelope modulation of the investigated samples [35].

16.2.3 Chemical and mineralogical contamination

In its more general consideration, the relationship between contaminants and CS activation can involve both the chemical and mineralogical species associated to CS [24–25, 27–29, 31, 34–35].

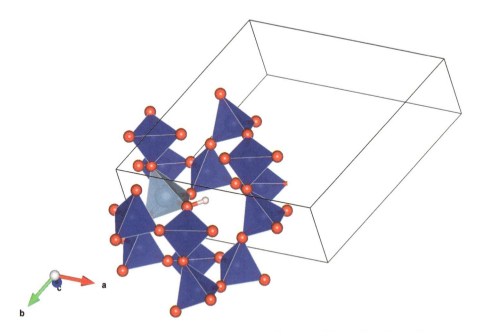

Fig. 16.3: Structural model of the h_{Al} radicals as described by Romanelli et al. [31]. Red, white, blue, and cyan show the O, H, Si and Al atoms, respectively.

Considering the chemical contaminants, straightforward methods to determine the change in the amount of minor and trace elements include spectrometric analyses [27, 31, 35], performed after the dissolution of the sample, and/or X-ray fluorescence spectroscopy [21, 25, 30]. This information, although valuable, lacks in the definition of the mineralogical phase in which the trace elements are occurring. Thus, an analytical approach associated to microscopic inspection should be preferred [28]. The determination through a single particle analysis, available in scanning electron microscopy (SEM) experimental set-up, appears particularly fruitful in this case. In Fig. 16.4a, the occurrence of residual detritus of Ir-bearing phases during the polishing of the artificial stone is shown. In this context, the contamination by Ir is certainly due to the use of reinforced steels, which often contain Ir as hardener. This example clearly shows how the identification of the mineralogical context of the contaminant allows one to relate it to its probable source in the processing line.

Spectroscopic investigations obtain information about the presence of contamination by redox active species. An example that can be considered is the identification of a Mn(II) contamination occurring when the material undergoes wet polishing during the processing of the artificial stone, as determined by cw EPR spectroscopy at 35 K (Fig. 16.4b) [34].

The complex relationship between CS and its chemical and mineralogical contaminants can be exemplified by the Fe speciation associated to raw industrial materials [33]. The Fe speciation in raw CS for industrial production was found to consist of at least three species (Fig. 16.5): besides crystalline Fe(III) oxide, which is the more obvious contaminant, metallic Fe and amorphous Fe(III) oxide were also detected. These two species suggest a variability of the redox properties of the CS materials (powders and sands), when properly activated in an industrial process. Elemental

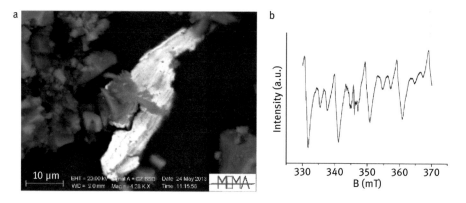

Fig. 16.4: (a) SEM back-scattered electron micrograph of the powders obtained by dry polishing of artificial stone: the fragment of an Ir-bearing phase is visible in light gray color; (b) cw EPR spectrum of the powders obtained by wet polishing of artificial stone: 5 of the 6 hyperfine lines of Mn(II) sextet as well as 8 of the 10 forbidden lines are clearly observable [58]; the narrow line at 346 mT is related to the inorganic radicals. Spectrum registered at 35 K, and magnetic field values are in millitesla.

Fe, in fact, is thermodynamically unstable in ordinary atmospheric condition and moisture >50% [44], whereas amorphous Fe(III) oxide could represent a metastable product prone to undergo to further reactions. Moreover, this species is supposed to activate silica, due to the fact that the coordination sphere of Fe is incomplete [45]. A further example of Fe reactivity in CS samples is provided by [46], who proved that tetrahedral Fe(III) replacing Si(IV) in quartz is almost completely removed after mechanical activation.

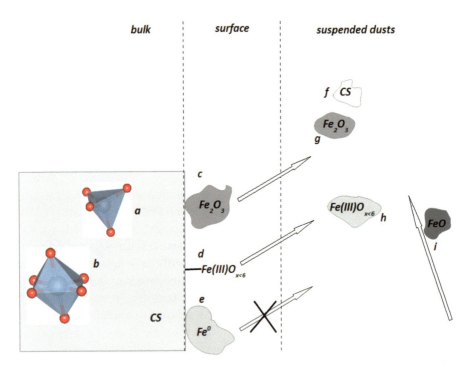

Fig. 16.5: Schematic representation of the main Fe species contributing to its speciation in relation to CS. In the bulk materials, Fe is internal to the CS structure, (a) tetrahedrally and (b) octahedrally coordinated, respectively [59]; at the CS surface, (c) hematite-like particles, (d) amorphous Fe(III) oxide with unsaturated first coordination shell, (e) elemental Fe; in the suspended dusts: (f), (g), and (h) suspended particles of CS, (c) and (d), respectively. In the suspended dusts, Fe occurs also as (i) wustite-like particles, which possibly originated from an external source, or by reaction of (e), which does not occur in this fraction [33].

16.2.4 Species selection in suspended dusts

Investigating the detailed nature of the CS-bearing materials involved in the industrial production does not satisfy the need to identify and characterize the dusts to which the workers are potentially exposed. As anticipated, most of the industrial samples have a particle size distribution with a reduced, or in few cases negligible, breathable fraction. Generally speaking, all the spectroscopic signatures determined in an industrial material are not necessarily preserved in its respirable fraction. Thus, it is worth to also investigate the suspended dusts. For this kind of samples, the proposal of a multianalytical procedure (see the next section) appears particularly suitable, as the total amount of dusts obtainable under conventional sampling procedures is very low (usually <1 mg of dusts over 8 sampling hours). Airborne dust samples exhibit specific concerns about representativeness and reproducibility; moreover, both environmental and personal samplings are not exclusively linked to a specific critical working place. Despite these drawbacks, the spectroscopic investigation of suspended dusts resulted to be an effective method in highlighting general changes in the Fe speciation with respect to the coarse dusts [33]. The XAS characterization performed by the authors of this article, in fact, evidenced the presence of at least two reactive species in the breathable dusts, i.e. Fe(II) oxide and the amorphous Fe(III) oxide, containing Fe with an unsaturated coordination sphere, that are not present in the bulk materials (Fig. 16.5). The latter species can be likely adsorbed at the quartz surface. Moreover, the same investigation allowed the exclusion of the presence of metallic Fe in the suspended dusts.

16.3 Procedure: sampling and investigation techniques

The study of industrial materials requires a proper design both in the sampling campaign and in the number and type of involved laboratory techniques. A paramount role is thus played by spectroscopic methods [47]. The basic assumption, in agreement with the IARC conclusions [6], is the fact that sampling the CS-bearing materials only at the beginning of the line cannot capture the whole variability of physical and chemical situations of the whole line. Thus, the identification of specific points along the production line where CS surface properties can be modified (hereafter labeled "critical locations") is mandatory. In order to capture the variability of workers' exposure to CS completely, a minimum of two different kinds of samples along a generic industrial line process should be taken (Fig. 16.6):
1. bulk materials directly processed/involved in the production and/or coarse dust, emerging from the production line and accumulated onto flat surfaces; these materials should be sampled before and after a critical location;
2. suspended dust, i.e. the fraction of dust that has the highest probability to be breathed by the workers.

The sampling campaign should take into account any available knowledge about the production process, including technical details (number, localization and nature of the thermal, chemical and mechanical processes involving CS) and hygiene surveillance details (ways and amount of the exposure to CS, distribution of the dust within the firm, results of the epidemiological surveillance). We want to emphasize that the

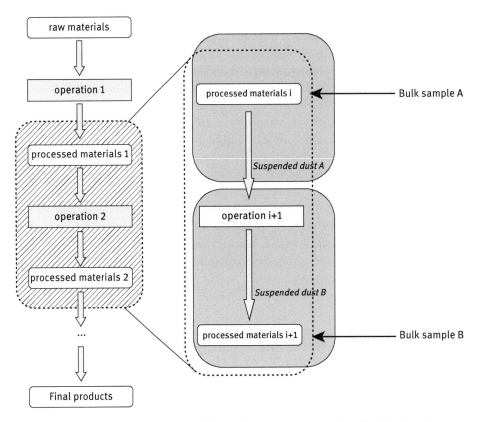

Fig. 16.6: Scheme of the possible design of a sampling campaign around a critical location (as defined in the text). Two stages of the industrial production line are identified, and corresponding bulk and suspended dusts are sampled.

choice of sampling bulk materials, i.e. materials having granulometric features that prevent their inhalation, is fundamental in providing information on the physical and chemical changes occurring to CS downstream along the production line. This information is then used for comparison with data obtained on suspended dust [33].

The details of the sampling of suspended dusts fraction are described elsewhere (see e.g. the NIOSH or the NIS guidelines provided by [48–49]). Here is useful to recall some general considerations:

i. The sampling procedures, which involve the use of personal pumps with cyclone impactors, should be realized as closest as possible to the conditions applied for surveillance operations, to get data comparable to the actual exposures measured in place.

ii. Only the breathable fraction (i.e. the fraction of dusts having a mean particle size <4 µm) should be sampled.

iii. Only polycarbonate filters should be used, because they are not interfering with the investigation techniques cited in the following. The drawbacks of the above investigation scheme must be also discussed. First, the amount of dusts effectively sampled onto the filter is very low (< 1 mg), thus generating problems with the detection limits of many techniques. Second, the sampling of a solid/gas solution is often non-ideal, and statistical representativeness is difficult to achieve. Accordingly, the choice of the sampler and its location in the working place are parameters to be optimized [50–51].

Numerous techniques can be applied in the physical and chemical characterization of CS-bearing materials to develop a deep knowledge of the fate of CS in the industrial process. Since inorganic radicals, hetero-ions (namely Fe, because of its recognized activity of modulation of the health effects of CS [6]) and/or associated minerals at the CS surface, and surface morphology can be considered the most critical agents from a toxicological point of view, a useful analytical procedure can be divided into three kinds of characterization (Tab. 16.2):

1. Minero-chemical characterization, carried out through X-ray powder diffraction (XRPD), SEM, and EDS microanalysis. This activity is aimed to provide information about the presence and amount of CS, the type and amount of mineralogical associated phases, the chemical composition of the material, the particle morphology, and the occurrence of fresh fracturing of the particles.
2. Characterization of the long-living radical species, carried out through EPR and ESE techniques [41]. This activity enables to detect markers of occurred reactivity before and after a selected critical location along the industrial line.
3. Characterization of the Fe speciation through EPR and XAS. This activity allows to trace the distribution of Fe among the observed mineralogical species and especially in relation to CS. Moreover, the total amount of Fe and the valence state of Fe species through the production line can be monitored.

Tab. 16.2: Characterization procedure: proposed techniques and expected results.

	Technique	Expected results
1	SEM/EDX (A)	Morphology and size of the particles, microchemical analysis, inclusions of microscale contaminants
	XRPD (B)	Phase composition, quantitative CS determination, average particle size
2	EPR/ESE (C)	Radical speciation, Fe(III) speciation, selected transition element determination
3	XAS (C)	Total Fe speciation, (possible) speciation of relevant ions.

A = destructive; B = partially destructive, the sample is modified by homogenization and, if needed, grinding; C = not destructive.

One of the advantages of choosing such a multianalytical procedure resides in the fact that an *a priori* knowledge of the changes occurred to the surface features of CS cannot be properly envisaged. Changes could be numerous and different in type and also concurrent [36]. Moreover, it appears appropriate to pursue a multianalytical approach able to couple the micromorphological characterization with the investigation of the sample at the "molecular level", at which the pathogenic effect is proposed to be operating [1–4].

A concluding methodological remark relates to the applicability of the present procedure. Apparently, the coupling of designed sampling strategy and multianalytical approach has a general application in the study of all kind of industrial CS-bearing material. However, the generalization of the results arising from different production lines (the comparison of results obtained in e.g. the ceramic and sand blasting processes) is basically hindered [39]. In our opinion, this multianalytical procedure could also be applied to characterize other toxic particulates, as asbestiform minerals and urban particulate matter.

16.4 Conclusions

The peculiar procedure proposed here points to the application of specific analytical techniques that can investigate chemical, physical, structural features, as well as radical speciation, to get information on the nature of the materials during the production process. More precisely, changes undergone by CS in the production processes can be fully characterized. The results can help in determining the critical location in the line productions where key factors are active. Namely, the comparison between data from bulk material and corresponding suspended dusts is important; indeed, it can contribute to clarify the reasons of the variability of manifestations of human diseases. We want to stress the necessity to address the future research on this subject as closest as possible to the dusts effectively breathed by the workers, sorting out the specificities of the most relevant production processes. Studies finalized to the investigation on the interaction between activated CS-bearing materials and biological materials need to be continued. The study of the molecular interactions between the inhaled material and biological fluids, cellular components, and tissues represent a further step for a comprehensive definition of the pathogenic mechanisms.

Acknowledgments

Many people contributed to this research, the results of which are summarized in this text. First of all, the authors want to thank Gabriele Fornaciai (Tuscan Environmental Protection Agency and University of Florence), to whom the manuscript is dedicated, for his constant contribution in this research subject. Specific and highly valuable contributions were also provided by G. Montegrossi (IGG-CNR) in the XAS experiments, E. Bafaro (Tuscan Environmental Protection Agency, Italy) in sampling and analyzing many materials, by A. Zoleo, A. Barbon, and M. Brustolon in the ESE measurements, by M. Innocenti (University of Florence) in the AFM measurements, and by D. Bartoli and C. Poli (Occupational Health Surveillance Unit of Empoli, Italy) in providing samples. All these people were also involved in deep and constant discussions of the ongoing results; the discussion also benefited from the advice of S. Silvestri (Tuscan Institute for the Study and Prevention of Cancer, Firenze, Italy), G. Sciarra and G. Scancarello (Public Health Laboratory, Siena, Italy), and B. Fubini and M. Tomatis (Scansetti Centre and University of Turin, Italy). Numerous other operators working in the different Occupational Health Surveillance Units of Tuscany were also involved, and they all are gratefully acknowledged. We acknowledge the Centro di Servizi di Microscopia Elettronica e Microanalisi (MEMA), University of Florence, for kindly granting the use of their scanning electron microscope, as well as Daniele Borrini and Maurizio Ulivi, for their assistance in the same investigations. The authors thank the Tuscany Administration for funding this research under the programs "Progetto di ricerca per l'individuazione delle cause di variazione della reattività superficiale della silice cristallina, nei principali comparti di lavoro toscani, in relazione alla sua potenziale patogenicità" and "Progetto di ricerca per il controllo del rischio d'esposizione a silice libera cristallina (SLC) nei comparti lavorativi toscani, lo studio delle proprietà chimico fisiche, morfologiche e biologiche delle polveri silicee in diversi comparti produttivi e lo studio dei meccanismi patogenetici e degli effetti dell'esposizione anche a basse dosi". The MIUR (Italy) is also acknowledged for the PRIN 2010/2011 project PROT. 2010MKHT9B as wells as the Italian CNR for support. F.D.B. and M.R. also benefited from departmental funding (EX 60%).

References

[1] Sahai N. Medical mineralogy and geochemistry: an interfacial science. Elements 2007;3:381–4.
[2] Schoonen MAA, Cohn CA, Roemer E, Laffers, R, Simon SR, O'Riordan T. Mineral-induced formation of reactive oxygen species. Rev Mineral Geochem 2006;64:179–221.
[3] Plumlee GS, Morman SA, Ziegler TL. The toxicological geochemistry of earth materials: an overview of processes and the interdisciplinary methods used to understand them. Rev Miner Geochem 2006;64:5–57.
[4] Fubini B, Fenoglio I. Toxic potential of mineral dusts. Elements 2007;3:407–14.
[5] Rosen G. A history of public health. JHU Press 1993;535 pp.
[6] WHO-IARC A review of human carcinogens: arsenic, metals, fibres, and dusts. IARC Monogr Eval Carc 2012;100 C:499 pp.
[7] Peluso MEM, Munnia A, Giese RW, Chellini E, Ceppi M, Capacci F. Oxidatively damaged DNA in the nasal epithelium of workers occupationally exposed to silica dust in Tuscany region, Italy. Mutagenesis 2015;30(4):519–25.

[8] USGS 2013 minerals yearbook. Silica [advance release]. 2015.
[9] Kauppinen T, Toikkanen J, Pedersen D, Young R, Ahrens W, Bovetta P, Hansen J, Kromhout H, Maqueda Blasco J, Mirabelli D, de la Orden-Rivera V, Pannett B, Plato N, Savela A, Vincent R, Kogevinas M. Occupational exposure to carcinogens in the European Union. Occup Environ Med 2000;57:10–8.
[10] CLB China Labour Bulletin; 2015. http://www.clb.org.hk/en/.
[11] ACGIH American Conference of Governmental Industrial Hygienists. Threshold limit values for chemical substances and biological exposure indices. Cincinnati (OH): ACGIH; 2006.
[12] Pavan C, Polimeni M, Tomatis M, Corazzari I, Turci F, Ghigo D, Fubini B. Abrasion of artificial stones as a new cause of an ancient disease. Physico-chemical features and cellular responses. Toxicol Sci 2016;153(1):4–17.
[13] Pérez-Alonso A, Córdoba-Doña JA, Millares-Lorenzo JL, Figueroa-Murillo E, García-Vadillo C, Romero-Morillos J. Outbreak of silicosis in Spanish quartz conglomerate workers. Int J Occup Environ Health 2014;20(1):26–32.
[14] Shi X, Dalal NS, Vallyathan V. ESR evidence for the hydroxyl radical formation in aqueous suspension of quartz particles and its possible significance to lipid peroxidation in silicosis. Toxicol Environ Health 1988;25(2):237–45.
[15] Dalal NS, Shi X, Vallyathan V. Role of free radicals in the mechanisms of hemolysis and lipid peroxidation by silica: comparative ESR and cytotoxicity studies. J Toxicol Environ Health 1990;29(3):307–16.
[16] Vallyathan V, Blake T, Leonard S, Greskevitch M, Jones W, Pack D, Schwegler-Berry D, Miller W, Castranova V. In vitro toxicity of silica substitutes used for abrasive blasting. Am J Ind Med 1999;36(S1):158–60.
[17] Vallyathan V, Castranova V, Pack D, Leonard S, Shumaker J, Hubbs AF, Shoemaker DA, Ramsey DM, Pretty JR, McLaurin JL. Freshly fractured quartz inhalation leads to enhanced lung injury and inflammation. Potential role of free radicals. Am J Respir Crit Care Med 1995;152(3):1003–9.
[18] Fubini B, Zanetti G, Altilia S, Tiozzo R, Lison D, Saffiotti U. Relationship between surface properties and cellular responses to crystalline silica: studies with heat-treated cristobalite. Chem Res Toxicol 1999;12:737–45.
[19] Elias Z, Poirot O, Danière MC, Terzetti F, Marande AM, Dzwigaj S, Pezerat H, Fenoglio I, Fubini B. Cytotoxic and transforming effects of silica particles with different surface properties in Syrian hamster embryo (SHE) cells. Toxicol In Vitro 2000;14(5):409–22.
[20] Cocco P, Dosemeci M, Rice C. Lung cancer among silica-exposed workers: the quest for truth between chance and necessity. Med Lav 2007;98(1):3–17.
[21] Clouter A, Brown D, Höhr D, Borm P, Donaldson K. Inflammatory effects of respirable quartz collected in workplaces versus standard DQ12 quartz: particle surface correlates. Toxicol Sci 2001;63(1):90–8.
[22] Tjoe Nij, E, Höhr, D, Borm, P, Burstyn I, Spierings J, Steffens F, Lumens M, Spee, T, Heederik, D. Variability in quartz exposure in the construction industry: implications for assessing exposure-response relations. J Occup Environ Hyg 2004; 1(3):191–8.
[23] Attik G, Brown R, Jackson P, Creutzenberg O, Aboukhamis I, Rihn BH. Internalization, cytotoxicity, apoptosis, and tumor necrosis factor-α expression in rat alveolar macrophages exposed to various dusts occurring in the ceramics industry. Inhal Toxicol 2008;20(12):1101–12.
[24] Creutzenberg O, Ziemann C, Hansen T, Ernst H, Jackson P, Cartlidge D, Brown R. In vivo study with quartz-containing ceramic dusts: inflammatory effects of two factory samples in lungs after intratracheal instillation in a 28-day study with rats. J Phys Conf Ser 2009;151:012031.
[25] Neghab M, Zadeh JH, Fakoorziba MR. Respiratory toxicity of raw materials used in ceramic production. Ind Health 2009;47(1):64–9.

[26] Ziemann C, Jackson P, Brown R, Attik G, Rihn BH, Creutzenberg O. Quartz-containing ceramic dusts: in vitro screening of the cytotoxic, genotoxic and pro-inflammatory potential of 5 factory samples. J Phys Conf Ser 2009;151:1–6.
[27] Øvrevik J, Myran T, Refsnes M, Låg M, Becher R, Hetland RB, Schwarze PE. Mineral particles of varying composition induce differential chemokine release from epithelial lung cells: importance of physico-chemical characteristics. Ann Occup Hyg 2005;49(3):219–31.
[28] Wendlandt RF, Harrison WJ, Vaughan DJ. Surface coatings on quartz grains in bentonites and their relevance to human health. Appl Geochem 2007;22:2290–306.
[29] Ghiazza M, Gazzano E, Bonelli B, Fenoglio I, Polimeni M, Ghigo D, Garrone E, Fubini B. Formation of a vitreous phase at the surface of some commercial diatomaceous earth prevents the onset of oxidative stress effects. Chem Res Toxicol 2009;22:136–45.
[30] Van Berlo D, Haberzettl P, Gerloff K, Li H, Scherbart AM, Albrecht C, Schins RPF. Investigation of the cytotoxic and proinflammatory effects of cement dusts in rat alveolar macrophages. Chem Res Toxicol:2009:22:1548–58.
[31] Romanelli M, Di Benedetto F, Bartali L, Innocenti M, Fornaciai G, Montegrossi G, Pardi LA, Zoleo A, Capacci F. ESEEM of industrial quartz powders: insights into crystal chemistry of Al defects. Phys Chem Miner 2012;39:479–90.
[32] Moroni B, Viti C, Cappelletti D. Exposure vs toxicity levels of airborne quartz, metal and carbon particles in cast iron foundries. J Expo Sci Env Epid 2014;24:42–50.
[33] Di Benedetto F, D'Acapito F, Capacci F, Fornaciai G, Innocenti M, Montegrossi G, Oberhauser W, Pardi LA, Romanelli M. Variability of the health effects of crystalline silica: Fe speciation in industrial quartz reagents and suspended dusts – insights from XAS spectroscopy. Phys Chem Minerals 2014;41:215–25.
[34] Di Benedetto F, Bartoli D, Banchi B, d'Acapito F, Farina G, Iaia T, Innocenti M, Montegrossi G, Poli C, Romanelli M, Scancarello G, Tarchi M, Zoleo A, Capacci F. An XAS and EPR study of industrial quartz-resin composites: preliminary results and health survey in Tuscany. In: International Mineralogical Association Meeting 2014, Johannesburg (South Africa), 1–6 Settembre 2014, Abstract Volume; 2014:163.
[35] Romanelli M, Di Benedetto F, Fornaciai G, Innocenti M, Montegrossi G, Pardi LA, Zoleo A, Capacci F. ESEEM of industrial silica-bearing powders: reactivity of defects during wet processing in the ceramics production. Phys Chem Minerals 2015;42:363–72.
[36] Di Benedetto F, Gazzano E, Tomatis M, Turci F, Pardi LA, Bronco S, Fornaciai G, Innocenti M, Montegrossi G, Muniz Miranda M, Zoleo A, Capacci F, Fubini B, Ghigo D, Romanelli M. Physico-chemical properties of quartz from industrial manufacturing and its cytotoxic effects on alveolar macrophages: the case of green sand mould casting for iron production. J Hazard Mater 2016;312:18–27.
[37] Ghiazza M, Tomatis M, Doublier S, Grendene F, Gazzano E, Ghigo D, Fubini B. Carbon in intimate contact with quartz reduces the biological activity of crystalline silica dusts. Chem Res Toxicol 2013;26:46–54.
[38] Castranova V, Vallyathan V, Ramsey DM, McLaurin JL, Pack D, Leonard S, Barger MW, Ma JY, Dalal NS, Teass A. Augmentation of pulmonary reactions to quartz inhalation by trace amounts of iron-containing particles. Environ Health Perspect 1997;105 (Suppl 5):1319–24.
[39] Capacci F, Carnevale F, Di Benedetto F. Silice libera cristallina nei luoghi di lavoro. Firenze: Giunti OS; 2010:382 pp. (in Italian).
[40] Warheit DB, Webb TR, Colvin VL, Reed KL, Sayes CM. Pulmonary bioassay studies with nanoscale and fine-quartz particles in rats: toxicity is not dependent upon particle size but on surface characteristics. Toxicol Sci 2007;95:270–80.
[41] Turci, F, Pavan, C, Leinardi, R, Tomatis, M, Pastero, L, Garry, D, Anguissola, S, Lison, D, Fubini, B. Revisiting the paradigm of silica pathogenicity with synthetic quartz crystals: the role of crystallinity and surface disorder. Particle Fibre Toxicol 2016;13(1):32.

[42] Ikeya M. New applications of electron paramagnetic resonance: ESR dating, dosimetry, and spectroscopy. World Scientific, Singapore; 1993:500 pp.

[43] Pan Y, Nilges MJ. Electron paramagnetic resonance spectroscopy: basic principles, experimental techniques and applications to earth and planetary sciences. Rev Miner Geochem 2014;78:655–90.

[44] Greenwood NN, Earnshaw, A. Chimica degli elementi. Vol. II. Padova (Italy): Piccin Nuova Libraria; 1992: 1582 pp.

[45] Ghiazza M, Scherbart AM, Fenoglio I, Grendene F, Turci F, Martra G, Albrecht C, Schins RPF, Fubini B. Surface iron inhibits quartz-induced cytotoxic and inflammatory responses in alveolar macrophages. Chem Res Toxicol 2011;24:99–110.

[46] Cortezão SU, Pontuschka WM, Da Rocha MSF, Blak AR. Depolarisation currents (TSDC) and paramagnetic resonance (EPR) of iron in amethyst. J Phys Chem Solids 2003;64:1151–5.

[47] Christidis GE, editor. European Mineralogical Union notes in mineralogy. Vol. 9: Advances in the characterization of industrial minerals. London: EMU; 2011:486 pp.

[48] NIOSH NIOSH hazard review – health effects of occupational exposure to respirable crystalline silica. Washington (DC): Department of Health and Human Services, Centers for Disease Control and Prevention, National Institute for Occupational Safety and Health; 2002.

[49] Network Italiano Silice. La valutazione dell'esposizione professionale a silice libera cristallina; 2015. Available at: https://www.inail.it/cs/internet/docs/ucm_220633.pdf; accession date September 11th, 2017.

[50] Sciarra G, Scancarello G, Vincentini M, Banchi B, Giomarelli A, Capacci F, Carnevale F. Silice libera cristallina: problematiche relative al campionamento e alla scelta del selettore. Giorn Igienisti Indust 2009;34(3):272–82.

[51] Sciarra G, Scancarello G, Banchi B, Vincentini M, Giomarelli A, Capacci F, Carnevale F. SLC: problematiche relativa al campionamento e alla scelta del selettore. In: Capacci F, Carnevale F, Di Benedetto F, editors. Silice libera cristallina nei luoghi di lavoro. Firenze: Giunti OS; 2010:122–35.

[52] Duffin R, Gilmour PS, Schins RP, Clouter A, Guy K, Brown DM, MacNee W, Borm PJ, Donaldson K, Stone V. Aluminium lactate treatment of DQ12 quartz inhibits its ability to cause inflammation, chemokine expression, and nuclear factor-kappaB activation. Toxicol Appl Pharmacol 2001;176:10–17.

[53] Le Bouffant L, Daniel H, Martin JC, Bruyere S. Effect of impurities and associated minerals on quartz toxicity. Ann Occup Hyg 1982;26:625–634.

[54] Martin JC, Daniel-Moussard H, Le Bouffant L, Policard A. The role of quartz in the development of coal workers' pneumoconiosis. Ann NY Acad Sci 1972;200:127–141.

[55] Ghio AJ, Kennedy TP, Whorton AR, Crumbliss AL, Hatch GE, Hoidal JR. Role of surface complexed iron in oxidant generation and lung inflammation induced by silicates. Am J Physiol - Lung C 1992;263(5), L511–8.

[56] Fubini B, Hubbard A. Reactive oxygen species (ROS) and reactive nitrogen species (RNS) generation by silica in inflammation and fibrosis. Free Radic Biol Med 2003;34:1507–16.

[57] Fubini B, Fenoglio I, Elias Z, Poirot O. Variability of biological responses to silicas: effect of origin, crystallinity, and state of surface on generation of reactive oxygen species and morphological transformation of mammalian cells. J Environ Pathol Toxicol Oncol 2001;20(Suppl 1): 95–108.

[58] Shepherd RA, Graham WRM. EPR of Mn^{2+} in polycrystalline dolomite. J Chem Phys 1984;81(12):6080–4.

[59] Rossman G. Coloured varieties of the silica minerals. In: Heaney PJ, Prewitt CT, Gibbs GV, editors. Rev Mineral 1994;29:433–67.

About the authors

Dr. Georg Amthauer is Professor Emeritus of Mineralogy at the Department of Chemistry and Physics of Materials of the University of Salzburg (AT). He studied mineralogy and crystallography at the University of Bonn (DE) and made his Dr. rer. nat. at the University of Saarbrücken (DE). He then worked as a postdoc at the University of Marburg (DE) and as a visiting associate at Caltech CA (USA). His main fields are physics of minerals and related compounds, crystal chemistry, and material science.

Dr. Behnam Akhavan is a research associate at the School of Physics of the University of Sydney (AU). He studied materials engineering from Shahid Beheshti University (IR). After his PhD at the University of South Australia, he worked as a visiting researcher at the Max Planck Institute for Polymer Research and Fraunhofer Institute of Microtechnology (DE). His main research focus is plasma surface engineering of particles for water purification.

Gerald Buck is doctorate student at Eberhard-Karls Universität Tübingen (DE) within the scope of the SFB TRR 141 program Biological Design and Integrative Structures with focus on the characterization and manufacturing of multifunctional highly porous ceramics. He studied geodynamics and applied mineralogy in Tübingen.

Fabio Capacci works at the Regional Health Agency of Tuscany in the Prevention and Safety Units (IT). He is specialized in work and occupational medicine. He authored more than 70 papers as well as edited monographs on *Crystalline Silica*, published by the Tuscany Regional Administration. Since 2012, he was appointed as National Coordinator of the Network Italiano Silice (NIS, Italian Network on Silica).

Dr. Antonio Checa is Professor and Chair of Paleontology at the Department of Stratigraphy and Paleontology, University of Granada (ES). He received his PhD in geology from the University of Granada. He is mainly interested in constructional morphology and biomineralization of the shells of molluscs and other invertebrates. The main goal is to determine the physical and biological controls acting on the fabrication of the shell microstructures. As a paleontologist, he is also interested in the evolution of the different types of exoskeletons since the emergence of molluscs, some 540 million years ago.

Dr. Francesco Di Benedetto is Professor of Mineralogy at the Department of Earth Sciences of the University of Florence (IT). His research interests include the crystal chemistry of minerals and synthetic materials, investigated through conventional physical methods and advanced spectroscopic methods as EPR, XAS, and SQUID magnetometry. His main topics are health effects of crystalline silica and synthesis and characterization of quaternary sulfides for solar energy conversion. In the last years, he also dedicated efforts in studying the technological answer to the limited availability of georesources.

Dr. Anton Eisenhauer is Professor of Marine Environmental Geology and Geochemistry at GEOMAR, Helmholtz Centre for Ocean Research, Kiel (DE). He received his PhD degree in environmental physics at Heidelberg University (DE), was a senior researcher at Caltech CA (USA) and assistant professor at the Institute of Geochemistry, Göttingen University (DE). In particular, he is interested in the application of isotope techniques as a tool to better constrain biogeochemical cycles for hydrology, marine and terrestrial biomineralization, and nuclear waste management.

Dr. Reto Gieré is Professor and Chair of the Department of Earth an Environmental Science at the University of Pennsylvania in Philadelphia. His research centers on environmental geochemistry, mineralogy, and petrology, with a focus on energy and waste as well as health impacts of atmospheric pollution.

Dr. Hermann Gies is Professor of Crystallography at the Department of Geology, Mineralogy and Geophysics of the Ruhr University Bochum (DE). He studied chemistry at the University of Kiel (DE), where he also got his PhD. During his career, he worked as a postdoc at the Universities of Aberdeen (UK) and British Columbia (CA) and as a visiting professor at the University of Minnesota (USA), the Institute of Material Science of Barcelona, and the Peking University. His main fields of research are crystal chemistry of microporous materials and the structure analysis of mineral surfaces.

Dr. Erika Griesshaber is a research associate at the Department of Earth and Environmental Sciences, LMU, Munich (DE). She received her PhD degree in geochemistry from the University of Cambridge (UK) and continued her career at MPI for Geochemistry in Mainz (DE) and at the Institute for Geology, Mineralogy and Geophysics, University of Bochum (DE). Her current research interests are processes in biomineralization, functionality principles, and material function relationships in carbonate biological hard tissues.

René Gunder is a research associate at Helmholtz-Zentrum Berlin für Materialien und Energie (HZB) in Berlin (DE) and co-responsible for the X-ray Corelab facility. He studied geomaterials science at Freie Universität Berlin and has a scholarship as a doctoral candidate at the MatSEC (Materials for Solar Energy Conversion) graduate school.

Dr. Robert B. Heimann is Professor Emeritus at TU Bergakademie Freiberg (DE) of Technical Mineralogy and Materials Science. He studied mineralogy at RWTH Aachen and Freie Universität Berlin, taught mineralogy at FU Berlin and Technical Universität Karlsruhe (DE), and was a researcher at McMaster University, Atomic Energy of Canada Limited, and Alberta Research Council (CA). During the last years, among his main research topics were bioceramic coatings for medical implants.

Dr. Soraya Heuss-Aßbichler is Professor of Mineralogy at the Department of Earth and Environmental Sciences, LM Universität München (DE). Her doctorate studies were focused on reaction mechanism in marbles. As postdoc fellow, she studied cation exchange experiments at high pressure and temperature. Her current research interests include environmental issues (incineration residues and recovery of heavy metal from aqueous solutions), processes in salt-rich systems, and classification of anthropogenic resources.

Antje Hirsch is a PhD student at the RWTH Aachen University (DE) at the Institute of Crystallography. After finishing her BSc in geosciences at the Christian-Albrechts-Universität zu Kiel (DE), she obtained her MSc in applied geosciences at RWTH Aachen University. Her scientific work is mainly focusing on monazite-type materials.

Dr. Markus Hoelzel is an instrument scientist for neutron powder diffraction at the Heinz Maier-Leibnitz Zentrum (MLZ), Technische Universität München, Garching near Munich (DE). He received his PhD in materials science at Darmstadt University of Technology, Darmstadt (DE). His research interests are *in situ* diffraction methods to study structure-property relationships in engineering alloys and technical relevant functional materials.

Dr. Karyn Jarvis is a research engineer at Swinburne University of Technology (AU). She completed her PhD at the University of South Australia and held postdoctoral appointments at the Australian Nuclear Science and Technology Organisation and the University of South Australia. Her research interests are surface modification, surface characterization, and thin film deposition. In her current work, she is investigating how plasma polymerization can be used to vary surface properties and therefore influence the interactions of proteins, bacteria and cells with surfaces.

Dr. Melanie John is currently a postdoc at the Department of Earth and Environmental Sciences, LMU Munich (DE). After studying geomaterials and geochemistry, she obtained her PhD in mineralogy at the LMU Munich. Her research is focused on precipitation and solubility processes as well as phase transformations in aqueous solutions. With her work, she bridges a gap between different research fields as nanoparticle synthesis and environmental topics as the recovery of heavy or noble metals from anthropogenic resources.

Dr. Christian M. Julien is presently emeritus member at the University Pierre et Marie Curie, Paris (UPMC). He received his engineering degree in physics from Conservatoire des Arts et Métiers, Paris. and obtained his PhD in materials science from UPMC. He has 35 years of research experience in the field of solid state ionics and materials for energy storage and conversion and has developed lithium-ion battery technology and thin-film lithium microbatteries.

Peter M. Kadletz finished his PhD research work at the Department of Earth and Environmental Sciences, LMU Munich (DE), on high-temperature shape memory alloys. He is currently a postdoc at the Nuclear Physics Institute (NPI) of the Czech Academy of Sciences in the group of Dr. Petr Lukáš, NPI Director, in Řež, and Dr. Petr Šittner in Prague (CZE). His interest is to improve methods of neutron radiography for the characterization of metallic materials, in particular shape memory alloys.

Katharina Klang is doctorate student at Eberhard-Karls Universität Tübingen (DE) within the scope of the SFB TRR 141 program Biological Design and Integrative Structures with focus on energy dissipation in load bearing systems. She studied geosciences (geology and paleontology) at Westphalian Wilhelm University Münster (DE).

Dr. Klemens Kelm is the team leader of the Central Analytical Research and Metallography facility, Aerospace Center, in Köln-Porz (DE). He received his PhD in inorganic chemistry at the University of Bonn (DE). He was research fellow at Caesar in Bonn and coordinator for TEM laboratories at the Center for Materials Analysis, Kiel University (DE). His research interests focus on microstructure and nanostructure investigation and mechanical properties analyses of hybrid materials (e.g. high-temperature ceramic matrix composites, functional materials such as biological and non-biological ceramics).

Christoph Lauer is a doctorate student at Eberhard-Karls Universität Tübingen (DE) within the scope of the SFB TRR 141 program Biological Design and Integrative Structures with focus on scaling problems in hierarchical structured biological materials. He studied Earth Sciences in Tübingen and Volcanology at University of Bristol (GB).

Leonhard Leppin is studying physics at Georg-August-Universität Göttingen and at Freie Universität Berlin (DE), including a one-year exchange scholarship at The Chinese University of Hong Kong. His bachelor's thesis was prepared at Helmholtz-Zentrum Berlin für Materialien und Energie (DE) and is the subject of a study presented in this book chapter.

Dr. Gregory R. Lumpkin is the program manager of Fuel and Resource Systems, Nuclear Fuel Cycle Research, at the Australian Nuclear Science and Technology Organisation (ANSTO) (AU). The focus of the program is to conduct basic research on minerals and materials for advanced energy applications, including nuclear fission and fusion systems and specialist materials for the high technology sector.

Dr. Bernd Marler is currently a senior researcher at the Department of Geology, Mineralogy and Geophysics of the Ruhr University Bochum (DE). After studying chemistry and mineralogy at the University of Kiel (DE), he obtained his PhD in crystallography at the same university. Afterwards, he spent one year as a postdoc at the University of Mulhouse (FR). His research is focused on the synthesis, modification, and structure analysis of zeolites and layered silicates.

Julien Marquardt is a doctoral student at the Helmholtz-Zentrum Berlin für Materialien und Energie (HZB) in collaboration with the graduate school MatSEC (Materials for Solar Energy Conversion) jointly by Dahlem Research School (DRS) of the Freie Universität Berlin (DE). He studied geo-materials science at the Freie Universität Berlin in the Department of Mineralogy and Petrology.

Dr. Alain Mauger attended the University of Paris where he graduated with a degree in solid-state physics and obtained his PhD degree in 1974 and Doctorat d'État in 1980. His work focused on the theory of impurities in semi-metals, electronic structure of solids, and magnetic semiconductors. Since 1992, he has been in university Paris 06, leading different groups on solid-state and complex matter physics, before joining IMPMC in 2007 to work on the materials science for Li-ion batteries.

Dr. Peter Majewski is Professor for Advanced Materials in the Future Industries Institute at University of South Australia since 2016. He studied geology and received his PhD in mineralogy at University of Hannover (DE). Afterwards, he worked at the Max-Planck-Institute for Metals Research (MPI-MF) Stuttgart (DE) and Ian Wark Research Institute (AU). From 2011 until 2016, he was professor and Head of School at the School of Engineering at UniSA (AU). His main research fields are nanomaterials synthesis and processing as well as nanomanufacturing.

Dr. Stefan Neumeier is a senior scientist at Forschungszentrum Juelich GmbH in the Institute of Energy and Climate Research. He studied chemistry and received his Dr. rer. nat. at the University of Duisburg-Essen (Germany) in inorganic chemistry with focus on nanomaterials. As a postdoc fellow, he dealt with the development of ceramic sensor materials at the RWTH Aachen University (Germany). Since 2008, his scientific work is dedicated to materials science for nuclear waste management with focus on ceramic materials.

Dr. Klaus G. Nickel is Professor of Applied Mineralogy at Eberhard-Karls Universität Tübingen (DE). After his studies in geology at Johannes Gutenberg University Mainz (DE) and University of Tasmania (AU), he was at Max Planck Institutes in Mainz and Stuttgart (DE). His field of work covers materials science of ceramics, glasses, binder materials, and biological materials. The subjects belong to application-oriented research topics like corrosion, relation between microstructure and mechanical properties or biomimetics.

Dr. Luca A. Pardi leads the High Field-High Frequency EPR lab (CNR-IFAM) in Pisa (IT). He graduated in chemistry at the University of Florence (IT). He started with characterization of molecular magnetic materials at the University of Florence and continued his studies at the Laboratoire Léon Brillouin in Saclay (FR) and at the High Magnetic Field Laboratory, Florida State University (USA). He is working on a variety of different systems including magnetic materials, geomaterials, and polymers.

Dr. Lars Peters is senior lecturer at the Institute of Crystallography of RWTH Aachen University (DE). He studied mineralogy (Dipl.-Min.) at Christian-Albrechts-University of Kiel (DE), where he also obtained his Dr. rer. nat. After postdoc positions at the Chemistry Department of Durham University (UK) and Crystallography at C.-A.-U. Kiel (DE), he accepted his current position in Aachen. His main research interests are solid solution effects in inorganic solids and their temperature and pressure changes, mainly studied by combining diffraction with spectroscopic methods.

Dr. Herbert Pöllmann is Professor of Mineralogy and Geochemistry at Martin-Luther-University Halle (DE). He studied mineralogy and chemistry at Universität Erlangen (DE) and received there his PhD degree. His main research projects are focused on the LDH's (manganese and zinc) and their intercalation properties, use of natural and artificial pozzolans as additive to cements, special cements, and reservoir minerals and their formation from industrial wastes.

Dr. Daniel Rettenwander is currently a postdoctoral associate at the Department of Material Science and Engineering at the Massachusetts Institute of Technology (USA). After studying chemistry at the University of Graz (AT), he obtained his PhD in materials science at the University of Salzburg (AT). His main research is conducted to structure-property relationships in energy-related materials.

Dr. Georg Roth is Professor of Crystallography at RWTH Aachen University (DE). He studied mineralogy and crystallography and received his doctoral degree at the Westfälische Wilhelms Universität Münster (DE). As a postdoc fellow, he was at the Department of Materials Science, Stanford University (USA), Philips-Universität Marburg (DE), and Research Center Karlsruhe (DE). His main research areas are structure-property relations in materials, among them solid electrolytes, superconductors, Fullerenes, low-dimensional magnets, multiferroics, functional polymers, and materials for nuclear waste disposal.

Dr. Maurizio Romanelli graduated with a degree in chemistry (PhD physical chemistry) and is Professor Emeritus of Physical Chemistry at the Department of Earth Sciences, University of Florence. His main research skills are pulsed and c.w. electron paramagnetic resonance. He authored more than 100 scientific publications. His research main topics are changes in the physical and chemical properties of silica powders in production processes of different firms and structural and magnetic properties of sulfur minerals, both natural and synthetized.

Dr. Marthe Rousseau attended the National Museum of Natural History in Paris where she graduated with a degree in biology and obtained her PhD degree in 2004. Her work focused on the structure and formation of nacre. Since 2008, she has been as CNRS researcher working at the Faculty of Medicine in Nancy on the biological properties of nacre on bone. Recently she moved to the faculty of medicine in Saint Etienne to work on nacre and osteoporosis at SAINBIOSE.

Dr. Tsutomu Sato is Professor of Environmental Geology at the Hokkaido University (JP). He studied engineering and started as researcher at Japan Atomic Energy Research Institute before he took a position as associate professor at Kanazawa University (JP). Much of his current research focuses on Environmental Mineralogy and Geochemistry, with particular interest in radioactivity and nuclear waste engineering.

Korbinian Schiebel is currently a doctoral student at the Department of Earth and Environmental Sciences, LMU Munich (DE). He obtained his MSc degree in earth sciences at LMU Munich, with research work on symmetry analysis of NiTi shape memory alloys. His is now working on neutron radiography of water distribution in clay-mineral–bound molding sands in relation to their mechanical stability.

PD Dr. Hartmut Schlenz is the group leader of the Structure Research Group at the Institute for Energy and Climate Research IEK-6 at Forschungszentrum Jülich (Ge). He studied mineralogy at the Universities of Münster (DE) and Toledo, Ohio (USA) and received his Dr. rer. nat. at the University of Münster. He finished his habilitation at the University of Bonn (Ge) in 2003 using synchrotron radiation for advanced structural studies. Currently, his scientific focus is on structure analysis of crystalline solid solutions and amorphous solids and the conditioning of nuclear waste using ceramics and geopolymers.

Dr. Wolfgang W. Schmahl is Professor and Chair of Crystallography/Material Science at the Department of Earth and Environmental Sciences, LMU Munich (DE). He received his PhD degree in crystallography at CAU Kiel (DE) and was a postdoc at the University of Cambridge (UK) and TU Darmstadt (DE). Subsequently he was professor at the Universities of Tübingen (DE) and Bochum (DE). His main research interests are structural characterization of biogenic geomaterials and structure-property relationships of technologically relevant smart materials such as shape memory alloys.

Dr. Katherine L. Smith is the Counsellor Nuclear at the Australian Embassy and Permanent Mission to the United Nations in Vienna. This position is funded by ANSTO. Her research interests include radiation damage effects in ceramics.

Dr. Susan Schorr is Professor of Geo-Materials Science at the Freie Universität Berlin (DE) and head of Department Structure and Dynamics of Energy Materials at Helmholtz-Zentrum Berlin für Materialien und Energie (HZB). She studied crystallography at Humboldt University Berlin and received a PhD in physics from the Technical University Berlin. She was a postdoc at Hahn-Meitner-Institute Berlin, University Leipzig (DE) and guest scientist at the Los Alamos National Laboratory (USA). Her research focuses on correlations among off-stoichiometry, point defects, and physical properties of compound semiconductors.

Dr. Stefan Stöber is the head of the Laboratory "Phase Analysis" at the Institute of Geosciences Mineralogy/Geochemistry at Martin-Luther-University Halle (DE). He studied mineralogy/geochemistry at Universität Erlangen (DE) and at the Martin-Luther-Universität Halle, where he received his PhD degree in the field of cement chemistry and finished his habilitation theses in 2011. His research is focused on the structure properties of perovskites at non ambient temperatures and of layered double hydroxides (LDHs).

Dr. Reinhard Wagner is currently a postdoctoral researcher at the Department of Chemistry and Physics of Materials at the University of Salzburg (AT). After studying geology and applied mineralogy in Salzburg, he obtained a PhD in materials science. In his PhD thesis, he investigated crystal-chemical and structural features of Li-oxide garnets.

Xiaofei Yin is a doctoral student at the Department of Earth and Environmental Sciences, LMU in Munich (DE). She studied materials science at Fudan University, Shanghai (CN), and received her MSc in the framework of the Advanced Materials Science Program from the LMU, TU, and Augsburg Universities (DE) by research on shape memory alloys with neutron diffraction. Her research interests focus on biological mineralization, material characterization of biological, and biomimetic composite materials.

Dr. Andreas Ziegler is a senior researcher at the Central Facility for Electron Microscopy, University of Ulm (DE). After his PhD study in biology at University of Regensburg (DE), he was a postdoc research fellow at the Department of Molecular Physiology and Biological Physics, University of Virginia (USA). His main research interests are transepithelial calcium transport in biological mineralization, the mechanisms in biological mineralization, and the interrelation among the structure, composition, and function of crustacean skeletal elements.

Index

3-aminopropyl trimethoxysilane 100
1,7-octadiene 107

γ-radiation 311

ab initio data 235
absorber 65, 73, 76, 77, 84, 85, 88, 91
absorption coefficient 73
ACC 262
actinide 171, 174, 187, 212, 226
adsorbents 159, 160
adsorption 110, 154, 165
air/oxygen plasma 107
allylamine 100
alpha-decay dose 236
ALPS 155
alumina 280, 312
aluminophosphates 52
alunite 179
Am 171
$AmAlO_3$ 214
amine-functionalized silica 100
amorphization 230, 235
amorphous 230, 237
amorphous calcium carbonate 246, 274
amorphous calcium phosphate 262, 305, 307
anisotropic 264, 281
antiferromagnetic effect 187
antisite defects 29, 30
apatite 173, 175
applied mineralogy 153
aqueous durability 174, 227, 229
aragonite 250, 252, 285, 286, 289
aseptic loosening 305
atmospheric plasma spraying 304
atomic force microscopy 250, 251, 287
atomic position 76
atomistic modeling 187, 189, 233, 234, 237
attenuation length 77
austenite 113, 117, 120, 121, 122, 123, 125, 127, 131

backfill 225
backfill material 224
band gap 73, 75, 92
band gap energy 76
$BaPrO_3$ 212
$BaPuO_3$ 212
battery 8, 14, 23, 27
Bearpaw 207
bioactive 285, 303, 307
bioceramics 301, 303, 304, 312
biocompatibility 303
bioinert 303, 307
bio-inspiration 285, 289
biological calcite 250
biological carbonate 246, 248, 264, 266, 287
biomaterial 258, 285, 261, 266, 287, 294, 298, 303, 312
biomaterial functionalization 260
biomedical implants 313
biomimetic 266, 273, 275, 280, 281
biomineral 285, 286
biomineralization 285, 313
biopolymer matrix 245, 261
biotolerant 303
bivalve 245, 246, 248, 252, 255, 258, 285, 286
bond coat 305, 307
bond distance 187
bone 287, 288, 289, 302, 306
bone graft substitutes 313
borosilicate glass 172, 202, 223, 232, 233
brachiopod 245, 246, 248, 255
brannerite 225, 226, 228, 232
Brillouin zone 122, 123
britholite 173, 176
brittle foam 277
brookite 307, 308
brownmillerite 197, 204
building material 273, 280, 282

calcium carbonate 285, 289
calcium phosphate 301, 306, 313

calcium silicate 306
Calvet-type twin calorimeter 186
capacity 33
carbonate crystal 254, 260
carbon coating 27, 31
cathode 33, 37
cathode for Li-ion batteries 26, 28
cathode material 23, 25
$CaTiO_3$ 203
cell constant 187
cement 161, 197
ceramic 171, 273, 280, 281
ceramic host phases 172
ceramic nuclear waste form 233
ceramic waste form 223
chabazite 41, 47, 50, 51
chalcopyrite 74
charge 33
chemical durability 187
chemical stability 14
clay mineral 164, 165
clinoptilolite 41, 47, 50, 155
Cm 171
Cmcm 209
collagen 302
columnar nacre 252
compound semiconductor 73, 74
compressive strength 277
computer simulations 187, 189
conchoidal fracturing 274, 275
concrete 50, 273, 280, 281
condensation 59, 60, 61
conditioning 172, 189
conductivity 8, 34, 73
connectivity 279
copper 107, 137, 138, 139, 143, 144, 146
copper ferrite 140, 143, 145
co-precipitation 141, 143, 145, 159
coral 255, 265
crandallite 179
critical dose 230
crushing zone material 277
crystalline silica 317
crystalline-to-amorphous transformation 229
crystallographic defects 78
crystallographic sites 226
crystal structure of garnet 4
Crystal structure of LiMPO4 olivine 26
Cs 160, 199
Cs ball 165, 166
cuprite 139, 143

daughter product 226
decontamination 50, 157
deep geologic repositories 237
deep geologic repository 224, 232
defect migration 237
defect 27, 56, 65, 76, 231, 233, 235, 236
dehydroxylation 302, 303, 308
delafossite 140, 142, 144, 146
density functional theory 187, 234
density functionals 234
derivative phase 120, 121, 123
DFT calculation 187, 235
differential scanning calorimeter 186
displacive 113, 124
disposal 153, 155, 158, 161, 167, 171, 172, 174, 202
dissipation 278
dissolution 232, 233
domain size 77, 78, 81, 83, 85
drop solution calorimeter 186
durability 99, 171, 173, 223, 225, 232
durable 226

EBSD 95, 250, 252, 253, 254, 261, 266
elastic moduli 187
electrochemical stability 8, 9, 14
electrolyte 8, 28, 29, 36
endoprostheses 301, 305
energy dissipation 279, 280
energy storage 3, 52, 62, 99
enthalpies of formation 186, 187
enthalpy of mixing 186, 187
environmental mineralogy 153
EPR 320, 322, 323, 327
eulytine 179
extracellular matrix 245, 247
extra-framework 42, 47, 58

fabrication 92, 99, 225
fabrics 245, 261, 264
failure 264, 277, 281
femoral heads 311
ferrihydrite 138, 141, 144, 146
ferrite 143, 145, 146
ferrocyanide 155, 157, 159, 160
Fe speciation 323, 325, 327

ferroelastic 113, 116, 125, 126, 128, 132
ferroic 115, 116
first-generation bioinert materials 312
fission product 171, 174, 190, 199, 202, 203, 209, 226, 233, 234, 236
florencite 173, 174, 179
fourth-generation biomaterial 311, 312
fractal 259
fractal (dendritic) microstructure 259
freeze-dried 280
Frenkel defect 235, 237
fuel debris 154
Fukushima 50, 153, 157, 165, 167
functionality 84, 105, 312
functionally graded concrete 281

garnet 3, 4, 5, 173
garnet structure 4, 11, 16
garnet supergroup 4, 6
gastropod 245, 248, 255, 285, 287
geological apatite 301
geologic formation 225
geopolymer 161
glass 47, 161, 166, 223
goethite 138, 140, 141, 143, 146, 267
gold 137, 138, 140, 144
gorceixite 179
goyazite 179
GR 141, 142, 143, 144
graceful failure 280, 281
gray phase 213
grazing incidence X-ray diffraction 77
green rust 141
group-subgroup 122, 123, 197, 208

half-lives 172
hardness 7, 23, 24, 203, 225, 275
health 137, 153, 157, 317, 319, 327
heat capacity 186, 187
heat insulation 282
heat storage 52, 62, 66
heavy metal 143, 145, 146, 306
He bubbles 231
hematite 138, 140, 141, 267
heterosite 24, 25
high-level waste 197, 198, 223, 224
high-symmetry 115, 116
high-temperature calorimetry 186
HLW sludge 201

^{1}H-NMR spectrum 310
hollandite 199, 200, 202, 214, 226, 227, 228, 232, 233
host phase 227, 234
Hubbard U parameter 188
hydrophobic 65, 99, 107
hydroxylapatite 301

illite 163
Imma 209, 210
immobilization 161, 171, 172, 176, 197, 198, 202, 223
implant 275, 288, 304, 307
implantation 287, 288, 289
infrared 181, 183
inorganic radicals 320, 327
interim storage 158, 159, 224
intermediate-level waste 171, 200, 201
ionic conductivity 12, 13, 15, 30, 179
ionotropic gelation 280, 281
iron-(hydro)oxide 141
irradiation 188, 198, 231, 235, 237
IR spectrum 184
irreducible representation 120, 124, 125
isoelectric point 101
Isolueshite 210
isopod 245, 262, 265

Kaiserstuhl 211
kesterite 73, 74, 84, 91
Khibina 207, 209, 210
Kola Peninsula 205, 207, 209
kosnarite 173, 174, 177
Krivovichev 210

Landau theory 115
lanthanide 174, 186, 187, 189, 200, 236
lanthanides' contraction 183
larnite 307
La Trappe monastery 211
latrappite 211
leaching 232
Li cation 10, 13, 57, 58, 62
$LiCoPO_4$ and LiNiPO4 olivine 35
$LiFePO_4$ olivine 23, 27
Li-ion batteries 8, 23, 25, 29, 33, 37
Li-ion diffusion 15, 16
Li$^+$-ion diffusion pathway 10
Li-ion dynamics 15

Li(Mn,Fe)PO4 olivine 34
Li-oxide garnet 3, 8, 16
Li-stuffed oxide garnet 8, 9, 11, 12, 13, 16
lithiophilite 24, 34
loparite 200, 204, 207, 210
loparite K 207, 208, 209
Lovozero 205, 207
low-level waste 171
low-symmetry 117
lueshite 203, 204, 205, 207

magnetite 138, 141, 142, 143, 201
martensite 113, 114, 117, 120, 121, 122, 123, 124, 125, 126, 127, 131, 132
martensitic 113, 114, 117, 118, 120, 131, 132
meltdown 154
melt solution calorimetry 186
membrane 52, 245, 246
mesocrystal 274, 287
metal canister 225
metal (M)-oxide 141
metamictization 231
(M)-ferrite 141
Mg-calcite 274, 275
microcracking 229, 231
microparticle 166
microstrain 77, 81, 83, 85, 93, 96
microstructural analysis 78
microstructure 14, 77, 84, 88, 91, 113, 252, 255, 264
microstructure analysis 73, 76, 84, 95
mineral analogue 232
mineral organization 245, 264
minor actinides 171, 187, 213
Mitchell 197, 204
molecular dynamics 187, 233
monazite 171, 173, 174, 181, 183, 187, 188, 189, 233
Mont Saint-Hilaire 49, 205
mordenite 41, 43, 47, 50, 155
MOX 213
multibarrier HLW repository 225
murataite 173, 226

nacre 250, 285, 286, 287, 288, 289, 290
nacre crystals 250
nacre tablet 250, 252, 286
$NaFePO_4$ olivine 35, 36, 37
Na-ion batteries 35, 36, 37

$NaNbO_3$ 204, 205, 206, 207, 210
nanoparticles 137, 140, 142, 245, 250, 266
NASICON 8, 28, 179
NaSiCON 311, 312
native copper 139, 143
native metals 146
natural environment 146, 227, 237
natural garnets 3, 6
natural REE-perovskites 200
neutron absorbers 227
niobian perovskite 211
NiTi 113, 114, 120, 128, 130, 132
(NMR) spectroscopy 310
noise dampening 282
Np 171, 172, 180, 187, 202, 212, 213
nuclear waste form 171, 185, 187, 188, 189, 225, 227, 232, 233, 234, 236
nuclear waste management 171, 174, 189

OAp phase 309
oil removal 107, 109
oil removal efficiency 109, 110
Oklo natural reactor 175
oleophilic particles 107
olivine 23, 30, 32, 37, 307
olivine-based Na-ion cell 35
order parameter 115, 116, 118, 119, 120, 125, 126, 127, 128, 129, 130, 132
ordering 13, 36, 75
organic 8, 43, 46, 49, 53, 54, 56, 57, 59, 100, 164, 281, 285, 286
orthophosphate 24, 174, 176, 178, 188
osseoconductive bioceramic coatings 313
osseoconductivity 304
osseoinductivity 304
ossification 304
osteoconductivity 303, 304
osteogenesis 288
oxide zone 138
oxygen plasmas 104

pair potentials 233, 235
paraelastic 113, 116
paraphase 116, 118
parent phase 115, 117, 120, 121, 123, 127
penetration experiment 279
perovskite 8, 173, 226, 197, 198, 199, 201, 202, 203, 204, 207, 212, 213, 214, 225, 226, 227, 232, 233, 234

perovskite space 204
perovskite-type structure 198, 208, 209, 212, 213, 215
Perowski 203
phase transformation 15, 113
phase transition 12, 85, 113, 114, 115, 116, 117, 120, 121, 122, 124, 126, 127, 128, 132, 186, 203, 206, 207
phosphates 23, 140, 171, 173, 174, 179, 186, 187, 189
phospho-olivine 24, 26
photovoltaic 73, 74, 77, 91, 197
plasma polymerization 99, 100, 102, 103, 105, 107, 110
plasma spraying 303, 304, 308, 312
plumbogummite 179
^{31}P-NMR spectrum 310
$P\bar{m}3m$ 127
point defects 77, 78, 235, 236
polyhedral framework 5
polyhedra volume ratio 208
porosity 45, 48, 248, 276, 280, 282
porosity distribution 280
position 75
precipitation 139, 140, 142, 144, 145, 285
preparation methods 16
pseudoelasticity 113
pseudowollastonite 306
Pu 171, 172, 174, 180, 187, 199, 210, 212, 213
pyralspite series 7
pyrochlore 173, 200, 202, 214, 215, 225, 226, 227, 230, 232, 233, 236
pyrochlore ceramic 229
pyrophosphate 181

radiation 161, 164, 173, 224
radiation damage 174, 177, 189, 200, 225, 227, 230, 231, 233, 234
radiation effects 173, 234, 235
radiation resistance 171, 174, 187, 188, 189, 229
radioactive materials 155, 189, 223
radioactive waste 171, 172, 198, 210
radionuclides 157, 159, 164, 171, 172, 176, 190, 198, 201, 223, 234
radiotoxicity 172
Raman 31, 95, 181, 183
Raman modes 185
Raman spectrum 31, 184
recoil nucleus 224

recovery 137, 145, 146, 153, 154, 167
regular solid solutions 186
relaxation 237
removal 100, 107, 110, 154, 155, 159
repository 171, 172, 186, 223, 224, 232
resource 137, 144, 146
rotary plasma generator 100
Ruddlesdon-Popper 197

sea urchin 254, 255, 261, 264, 273
secondary resource 137
second-generation bioactive and biodegradable materials 312
shannonite 307
shape memory 113, 114, 116, 123, 125
sheet nacre 252
sicklerite 24, 25
silicate garnet 5, 6, 7
silicates 23, 173
silicotitanate 155
silver 137, 138, 139, 142, 145
simulated body fluid 304, 310
simulations 187, 189, 233, 234, 235, 236, 237
single crystals 73, 203, 254, 264
single-phase ceramics 173
site preference 14
SO_2 107
SO_3 107
SO_4 107
SO_3H 107
SO_4H 107
solar energy conversion 73
solid electrolyte 3
solidification 161
solid Li-ion conductor 8, 12
solubility 139, 232, 234
SOxH 104
space group $I\bar{4}3d$ 14
space group $I4_1/acd$ 6, 12
space group $Ia\bar{3}d$ 4, 8, 9, 12, 14, 16
specific capacity 29, 34, 36
spent nuclear fuel 172, 202, 223, 233
Sr 12, 42, 160, 179, 202, 203, 204, 212, 213, 285, 302
$SrPuO_3$ 212
$SrZrO_3$ 209
stabilization mechanism 13
stilbite 41, 57
strengite 25

structural features of Li-oxide garnets 9
structure 75, 201, 235
structure-property relationships 15
substitution 7, 12, 13, 14, 75, 176, 179, 180, 199, 200, 201, 202, 204, 208, 209, 210, 234, 235, 303
sulfate-reducing 138
sulfonate-functionalized silica 103
sulfoxide 107
supercalcine ceramic 198
supercomputer 189
superelasticity 113, 114
supergene 139, 199
supergene metal deposits 144, 146
supergene ore deposits 138, 139
supergene silver deposits 139
surface energy 279
swelling 60, 227, 231
symmetry analysis 113, 114, 123, 126
symmetrical stretching 185
symmetry 74, 114, 116, 117, 120, 121, 122, 123, 124, 125, 126, 127, 128, 132
synchrotron 62, 181
Synroc A 199
Synroc B 199
Synroc C 200, 201, 202, 227, 232, 233
Synroc D 201, 202, 227, 232, 233
Synroc E 202
Synroc F 202, 227
synthesis methods 137
synthetic garnets 7
systematic of Li-oxide garnets 11

tausonite 204, 205, 214
tenorite 139, 143
textures 245, 252, 255, 264
thermal conductivities 187
thermal decomposition 304, 306, 308
thermal gradient 306
thermal treatment 84, 276, 318
thermodynamic potential 115, 116, 117
thin film batteries 16
thiol 107
thiophene 103, 106, 107
third-generation biomaterials 312
thorianite 173
thorite 173

Th-phosphate-diphosphate 173, 180
Th-rich loparite 209
tilt system 197, 203, 206, 207, 212
titanite 173, 198, 200
titanium alloy 307
titanium oxide 157, 307
ToF-SIMS 101, 106
topologies 44, 53, 59, 60
topsoil 158
transition metals 145
translational symmetry 124, 128
transmutation 172, 213, 234
triphylite 24, 25, 34
type-structure 209

U 171
ugrandite series 7
unit cell 183, 212
unit-cell parameter 5, 6, 12, 14
uraninite 173, 201, 227

vacancies 13, 235, 302, 303
vermiculite 163
voltage-capacity curve 30
volume expansion 230

waste loading 173, 198, 225, 227
waste streams 171, 172
wastewater 49, 137, 144, 145, 146, 154
water molecules 42, 46, 47, 48, 55, 57, 62, 63, 65
water purification 107
weathered biotite 163, 164, 165, 166

XAS 325, 327
xenotime 173, 176, 179, 187, 189
XPS 100, 103
X-ray diffraction 28, 29, 36, 79, 181, 183, 203

Young's moduli 187
Y-stabilized zirconia 311

zeolite 41, 155, 160, 161
Zeta potential 102
zircon 173, 176, 233
zirconia 173, 201, 233, 312
zirconolite 173, 199, 200, 201, 202, 214, 225, 226, 227, 228, 230, 232, 233, 235